Lecture Notes in Mathematics

Edited by A. Dold and B. Eck

Subseries: USSR

Adviser: L. D. Faddeev, Leningrac

1334

Yu. G. Borisovich Yu. E. Gliklikh (Eds.)

Global Analysis – Studies and Applications III

Springer-Verlag

Berlin Heidelberg New York London Paris Tokyo

Editors

Yuriĭ G. Borisovich
Yuriĭ E. Gliklikh
Department of Mathematics, Voronezh State University
394693, Voronezh, USSR

Consulting Editor

A. M. Vershik
Department of Mathematics and Mechanics, Leningrad State University
198904, Petrodvorets, Leningrad, USSR

The articles in this volume are translations of articles that appeared originally in Russian in the books "Global'nyĭ analiz i matematicheskaya fizika" and "Geometriya i teoriya osobennosteĭ v nelineĭnykh uravneniyakh" published by Voronezh University Press in 1987.

Mathematics Subject Classification (1980): 58-02, 58C, 58D, 58E, 58F, 58G

ISBN 3-540-50019-7 Springer-Verlag Berlin Heidelberg New York
ISBN 0-387-50019-7 Springer-Verlag New York Berlin Heidelberg

Printing and binding: Druckhaus Beltz, Hemsbach/Bergstr.
2146/3140-543210

PREFACE

This Lecture Notes volume, a sequel to volumes 1108 and 1214, presents to English speaking readers further two issues of the Voronezh University Press series "Novoe v global'nom analize" (New Developments in Global Analysis). These issues, published in Russian in March, 1987 and in November, 1987, are entitled "Global Analysis and Mathematical Physics" and "Geometry and Theory of Singularities in Nonlinear Equations", respectively. For this translation volume we have selected the articles of survey (expository) character and papers giving detailed accounts of research results. Short communications and the survey article of Yu.P.Solov'ev (March, 1987) are omitted here (the material of the latter will be published in English elsewhere).

The articles are ordered chronologically. The date of publication in Russian is indicated in the Contents.

Assistance of all the members of the series editorial board, especially A.T.Fomenko, A.S.Mishchenko, S.P.Novikov, M.M.Postnikov, A.M.Vershik, is highly appreciated.

Yu.G.Borisovich
Yu.E.Gliklikh

December, 1987

CONTENTS

MARCH, 1987

NOVEMBER, 1987

PLATEAU OPERATOR AND BIFURCATIONS OF TWO-DIMENSIONAL MINIMAL SURFACES

A.Yu.Borisovich

Voronezh Institute
of Civil Engineers
394680, Voronezh, USSR

The classical Plateau problem of finding a two-dimensional surface of a given topological type for a minimal surface (MS) passing through any rectifiable curve has been solved in the works of J. Douglas, T. Rado, and R. Courant (see review [1] and references therein). The study of the multidimensional case has necessitated a modification of the classical problem statement and a generalization of such concepts as surface and its boundary. The first existence theorems for this case were proved later by C.B. Morrey Jr., H. Federer, W.H. Fleming, E.R. Reifenberg, E. De Giorgi, and F.J. Almgren Jr.; and analytical properties at interior points of an MS in $\mathbb{R}^n$, n = 4,5,6,7, were proved by F.J. Almgren Jr. and J. Simons. For n = 3, it is a classical result, and for n = 8 the interior regularity theorem is not valid (see review [1] and references therein). A.T. Fomenko [1] has proposed a new approach to the formulation of the multidimensional Plateau problem, which has made it possible to prove the existence theorems and study the regularity of MS in Riemannian manifolds under very general assumptions. With this approach, the Plateau problem is considered in the class of all surfaces which are continuous images of manifolds with boundary homeomorphic to a fixed manifold A. In this case the Hausdorff measure attains a minimum in all dimensions, and an MS is shown to be regular almost everywhere. The methods of global analysis, as applied to the Plateau problem, have been developed in [3].

The bifurcation of minimal surfaces is a far less studied problem. It is well known that an MS in $\mathbb{R}^3$, which spans a given contour, may experience qualitative changes as the contour is deformed. These changes are quite diverse; many of them have been observed and studied in experiments with soap films (first experiments of this kind were performed as early as in the 19th century by Plateau; numerous soap-film experiments have been described by A.T. Fomenko [2], see also the work of R. Courant [4]).

In particular, the experiments have revealed the bifurcation of MS families formed under contour deformation. T. Poston [5] studied experimentally the bifurcation of MS families which arises under continuous deformation of a special contour (the so-called Douglas contour) and found that the bifurcation can be described in terms of the "Whitney cusp".

A.T. Fomenko and A. Tuzhilin [2] proposed another approach to the description of MS bifurcations; their idea lies in studying branching points of the graph of the multivalued area functional defined on the space of contours. They analysed the Poston example and performed physical experiments to demonstrate that the corresponding bifurcation diagram is described by the "swallow-tail" scheme. This approach can also be applied to study the stability of bifurcating MS families (see [2], p. 122).

As to a general analytic method of studying MS bifurcations, it has not been developed until very recently, although M.J. Beeson and A.J. Trombar [21] have calculated a particular example of the bifurcation of MS families from the Enneper (minimal) surface - the "cusp catastrophe". It should be noted that the Morse-Palaise-Smale-Simons multidimensional theory of variations [6-9] offers, as was pointed out by A.T. Fomenko, an apparatus ("extremal index" theory) which enables one to infer on the existence of so-called "conjugate contours". However, the "conjugacy" property introduced for a smooth contour is only a necessary condition for the existence of bifurcations on this contour.

On the other hand, functional analysis offers effective methods of studying the branching and bifurcation of solutions to non-linear operator equations in Banach spaces, which go back to Lyapunov and Schmidt (see, for example, [11-13]). In the case of Fredholm operators, the methods of singularity theory are also quite successful [14,15]. The possibility of applying the operator methods of bifurcation theory to the Plateau problem was noticed as early as in 1982 at the seminar on global analysis held at Voronezh State University (VSU) (V.G. Zvyagin and Yu.I. Sapronov).

No bifurcation can exist over a convex domain in the non-parameter two-dimensional problem in $\mathbb{R}^3$, which follows from the existence and uniqueness theorems proved by S.N. Bernshtein [16] and his followers. Furthermore, it was found [17] that the MS operator in this problem is a diffeomorphism of Hölder spaces.

In the parametric two-dimensional problem the MS operator is of much more complicated form, since the MS with the radius vector $\vec{r}(u,v)$ is defined by a complicated system $F[\vec{r}] = \vec{0}$ of three Euler-Lagrange non-linear differential equations for the area functional $S[\vec{r}] = \iint_{\Omega} \sqrt{EG - F^2}\, dudv$, which makes it difficult to study not only bifurcations but also the existence problems. Nevertheless, in the parametric problem for an MS in $\mathbb{R}^3$ bifurcations can be studied using a general operator approach. This is just the topic of the present paper, which extends the methods and results reported in [18,19]. The approach can also be applied to the case of a minimal hypersurface of an arbitrary dimension.

The results of the work have been reported at the seminar on global analysis at VSU; the author is greatly indebted to all the participants of the Seminar for fruitful discussions, and also to A.T. Fomenko for interest in the work (conversation in June, 1984 and kind offer of the part of monograph [2] containing the description of soap-film experiments).

1. Formulation of the bifurcation problem. The Plateau operator

The problem on the bifurcation of a family of minimal surfaces is formulated, in a general case, as follows. Let us consider in $\mathbb{R}^3$ a two-dimensional regular MS, $\vec{r}(u,v)$, defined in conformal coordinates. Boundary conditions, dependent on the parameter λ, are assumed to be specified for surfaces close to $\vec{r}$. As the parameter λ varies continuously, we shall study bifurcations from $\vec{r}$ of continuous families of other MS which also satisfy the boundary conditions.

The surfaces adjacent to $\vec{r}$ will be considered as sections of a bundle normal to $\vec{r}$. This approach leads to a single scalar equation for a close minimal surface:

$$F(w) = 0 \qquad (1.1)$$

from which the normal coordinate w is determined as a function $w = w(u,v)$ of the two curvilinear coordinates on the surface $\vec{r}$. The surface $\vec{r}$ itself is defined by the section $w_0(u,v) = 0$.

The area functional for those sections, $S(w)$, and its Euler operator $F(w)$ are defined by the relations

$$S(w) = S[\vec{r} + w\vec{n}] \quad , \quad F(w) = \left(F[\vec{r} + w\vec{n}], \vec{n} \right)_{\mathbb{R}^3} \quad , \tag{1.2}$$

where $\vec{n}(u,v) = |\vec{r}_u \times \vec{r}_v|^{-1} (\vec{r}_u \times \vec{r}_v)$ is the Gaussian mapping (unit normal to the surface $\vec{r}$).

Theorem 1. The analytic structure of $S(w)$ and $F(w)$ is as follows:

$$S(w) = \iint_{\Omega} \mathcal{D}_4^{1/2} \, du\,dv \quad , \tag{1.3}$$

$$F(w) = -\frac{1}{4} \mathcal{D}_4^{-3/2} \cdot \left(\mathcal{A}_6 w_{uu} - 2\mathcal{B}_6 w_{uv} + C_6 w_{vv} + \mathcal{E}_7 \right) . \tag{1.4}$$

The expression in brackets is a quasilinear elliptic second-order differential operator; $\mathcal{D}_4$, $\mathcal{A}_6$, $\mathcal{B}_6$, C_6, and $\mathcal{E}_7$ are polynomials (with variable coefficients) in w, w_u, and w_v of orders 4, 6, 6, 6, and 7, respectively. These polynomials include: $k(u,v) = +\vec{n}_u^{\,2}$, the Jacobian of n ; $E(u,v) = \vec{r}_u^{\,2}$, the coefficient of first quadratic form; and $\ell(u,v) = (\vec{n}, \vec{r}_{uu})$ and $m(u,v) = (\vec{n}, \vec{r}_{uv})$ the coefficients of second quadratic form of the surface $\vec{r}$. For example, the polynomial $\mathcal{D}_4$ is of the form

$$\mathcal{D}_4 = E^2 + E \cdot (w_u^2 + w_v^2 - 2kw^2) + 2w(\ell w_u^2 + 2m w_u w_v - \ell w_v^2) + \\ + k w^2 \cdot (w_u^2 + w_v^2 + k w^2) \tag{1.5}$$

Thus, the problem on the bifurcation, from $\vec{r}$, of families of other MS is reduced to a problem on the branching, from $w_0 = 0$, of families of small solutions of Eq. (1.1) for given boundary conditions with the parameter λ . In this paper we consider two types of boundary conditions, that is two problems on the bifurcation of MS families.

Let the parameter $\lambda = (\alpha, \beta) \in \mathbb{R}^2_+$ (i.e. $\alpha > 0$, $\beta > 0$) and the domain $\Omega_\lambda \subset \mathbb{R}^2$ be defined by the conditions (see Fig. 1)

$$-\alpha < u < \alpha \quad , \quad -\beta < v < \beta \quad . \tag{1.6}$$

We consider in the domain Ω_λ the following boundary-value problems.

Problem I. $$\begin{cases} F(w) = 0 \quad , & (u,v) \in \Omega_\lambda \ , \\ w = 0 \quad , & (u,v) \in \partial\Omega_\lambda \ , \end{cases}$$

where the solution $w(u,v)$ is sought in the Sobolev functional space $W_2^k(\Omega_\lambda)$; k is an integer $\geqslant 2$.

Problem II. $$\begin{aligned} &F(w) = 0 \quad , && (u,v) \in \Omega_\lambda \ , \\ &w(-\alpha, v) = w(\alpha, v) = 0 \quad , && -\beta \leqslant v \leqslant \beta \ , \end{aligned}$$

where the solution is sought in the space $\tilde{W}_2^k(\Omega_\lambda)$ of functions $w(u,v) \in W_2^k(\Omega_\lambda)$ (k is an integer $\geqslant 2$) satisfying the boundary conditions

$$\frac{\partial^n}{\partial v^n} w(u,-\beta) = \frac{\partial^n}{\partial v^n} w(u,\beta) \quad , -\alpha \leqslant u \leqslant \alpha \ , \ n = 0,1,\ldots,k-1 \ . \tag{1.7}$$

Obviously, the space $\tilde{W}_2^k(\Omega_\lambda)$ is isomorphic to the Sobolev space of k-smooth functions defined in the band $-\alpha \leqslant u \leqslant \alpha$ and periodic in v with a period $T = 2\beta$ ($w(u,v) = w(u, v+T)$).

The boundary conditions in Problems I and II have the following geometrical meaning.

In Problem I we consider, on a minimal surface $\vec{r}$, a two-parameter family of curvilinear rectangulars Q_λ with boundary Γ_λ (the rectangular domain Ω_λ is the projection of Q_λ onto the plane (u,v)) and study the bifurcation from $\vec{r}$ ($w_0 \equiv 0$) of families of other MS (i.e. $w_\lambda(u,v)$) passing through the contour Γ_λ ($w_\lambda = 0$ on $\partial\Omega$; see Fig. 1).

Problem II is of greatest interest for those β values for which the functions k, E, l, and m appearing in the coefficients of the operator F are periodic in v with a period of $T = 2\beta$. In this case any solution $w(u,v)$ of Problem II can uniquely be continued (with the same period T) in the band $-\alpha \leq u \leq \alpha$, and is in this band a solution of Problem II, since $F(w)$ is also T-periodic in v. For each β we consider on $\vec{r}$ two curves, Γ_α^+ and Γ_α^- ($\tilde{\Gamma}_\alpha = \Gamma_\alpha^+ \cup \Gamma_\alpha^-$), defined in the coordinates (u,v) by the conditions $u = +\alpha$ and $u = -\alpha$, respectively. We shall also study the bifurcation, from $\vec{r}$ ($w_0 = 0$), of families of other MS (i.e. - $w_\lambda(u,v)$) passing through the curves Γ_α^- ($w_\lambda(-1,v) = 0$), Γ_α^+ ($w_\lambda(1,v) = 0$) and T-periodic in v ($T = 2\beta$). This boundary-value problem arises in the study of bifurcations of families of MS with a complicated topological genus (Fig. 2a) or of families of non-compact MS (Fig. 2b).

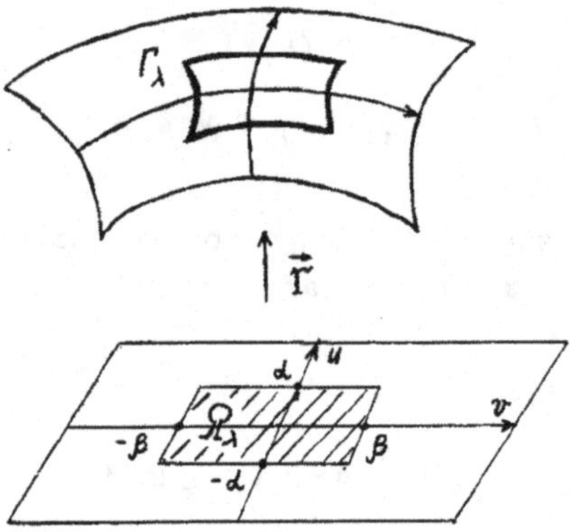

Fig. 1

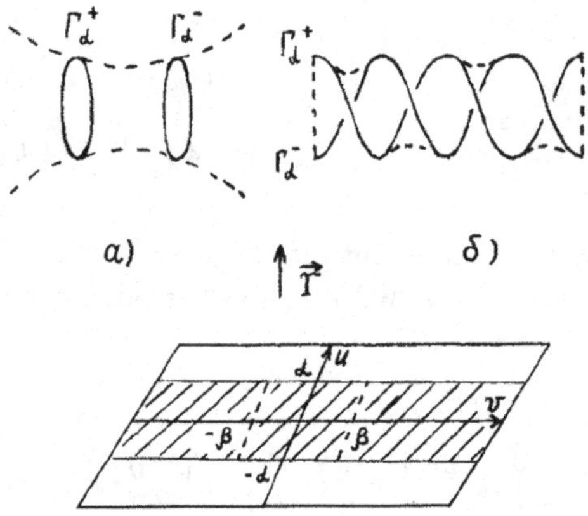

Fig. 2

By virtue of the change ($u \to u/\alpha$, $v \to v/\beta$) Problem I is reduced to a problem in the standard domain $\Omega = \Omega_{(1,1)}$, and the parameter $\lambda = (\alpha, \beta)$ naturally turns into the operator F. At the next stage we obtain the operator equation

$$P(w, \lambda) = [0, 0] \tag{1.8}$$

where the non-linear operator P (with the parameter λ) defined according to the rule $P(w,\lambda) = [F(w,\lambda), w|_{\partial\Omega}]$ is called the Plateau operator.

In Problem II, the change $(u \to u/\alpha\ ,\ v \to v/\beta)$ also leads to the operator equation

$$\tilde{P}(w,\lambda) = [0,0,0] \quad , \tag{1.9}$$

where the Plateau operator is now of the form $\tilde{P}(w,\lambda) = [F(w,\lambda), w(-1,v), w(1,v)]$.

In paper [19] we constructed the operators P and $\tilde{P}$, proved their continuous differentiability, Fredholm character, and other properties, provided the surface $\vec{r}$ is a catenoid. The results presented below extend the results of Ref. [19] to the case of an arbitrary regular MS.

<u>Theorem II</u>. For integer $k \geq 2$, the Plateau operators P and $\tilde{P}$ act in the following Sobolev spaces:

$$P: W_2^k(\Omega)_\varepsilon \times \mathbb{R}_+^2 \longrightarrow W_2^{k-2}(\Omega) \times B_2^{k-1/2}(\partial\Omega) \quad , \tag{1.10}$$

$$\tilde{P}: \tilde{W}_2^k(\Omega)_\varepsilon \times \mathbb{R}_+^2 \longrightarrow W_2^{k-2}(\Omega) \times \left(\tilde{B}_2^{k-1/2}[-1,1]\right)^2 . \tag{1.11}$$

1) P is a C^α-smooth operator in the variables (w, λ) and its Frechet derivative with respect to the first variable, $P_w(w,\lambda)$ is a linear Fredholm operator of zero index ($P \in \Phi_0^{(w)}C^\alpha$).

2) $\tilde{P} \in C^\alpha$, but $\tilde{P}_w(w,\lambda)$ is not necessarily a Fredholm operator. In particular, if $\lambda = (\alpha, \beta)$, and the functions k, E, ℓ, and m appearing in the coefficients of the operator F are 2 -periodic in v, then

$$\tilde{P}(\cdot,\lambda): \tilde{W}_2^k(\Omega)_\varepsilon \longrightarrow \tilde{W}_2^{k-2}(\Omega) \times \left(\tilde{B}_2^{k-1/2}[-1,1]\right)^2 \quad , \tag{1.12}$$

and the linear operator $\tilde{P}_w(w,\lambda) \in \Phi_0$ for all $w \in W_2^k(\Omega)_\varepsilon$.

3) Here $\tilde{B}_2^{k-1/2}([-1,1])$ is the Besov-Sobolevsky space of functions which are periodic on $\mathbb{R}^1$ with a period of $T = 2$.

4) The operator P ($\tilde{P}$) is not defined on the entire space $W_2^k(\Omega) \times \mathbb{R}_+^2$ ($\tilde{W}_2^k(\Omega) \times \mathbb{R}_+^2$), but only in a certain open neighbourhood of the

set $\{(0, \lambda)/ \lambda \in \mathbb{R}^2_+\}$. This condition is quite sufficient for studying bifurcations.

In what follows we shall need the following statement which is a consequence of the variational nature of Problems I and II.

Lemma 1. The linear differential operator F_w is self-adjoint on subspaces $\overset{0}{W}{}^k_2(\Omega) = \{w \in W^k_2(\Omega) : w|_{\partial\Omega} = 0\}$ and $\overset{0}{\widetilde{W}}{}^k_2(\Omega) = \{ w \in \widetilde{W}^k_2(\Omega) : w(-1,v) = w(1,v) = 0\}$, $k \geq 2$, with respect to the scalar product in $L_2(\Omega)$. In other words,

$$\langle F_w(w,\lambda)h, g\rangle = \langle h, F_w(w,\lambda)g\rangle, \tag{1.13}$$

for any $h,g \in \overset{0}{W}{}^k_2(\Omega)$ and any $(w,\lambda) \in W^k_2(\Omega)_\varepsilon \times \mathbb{R}^2_+$; for any $h,g \in \overset{0}{\widetilde{W}}{}^k_2(\Omega)$ and any $(w,\lambda) \in \widetilde{W}^k_2(\Omega)_\varepsilon \times \mathbb{R}^2_+$; $\langle f,g\rangle = \iint_\Omega f\, g\, du dv$.

2. The necessary condition for bifurcation

For each $\lambda \in \mathbb{R}^2_+$ we consider in $W^k_2(\Omega)$ a finite-dimensional subspace $N(\lambda) = \operatorname{Ker} P_w(0,\lambda)$, i.e. the set of solutions of the operator equation $P_w(0,\lambda)h \equiv [F_w(0,\lambda)h, h|_{\partial\Omega}] = [0,0]$; $P_w(0,\lambda)$ is the Frechet derivative with respect to the first variable calculated at the point $(0,\lambda)$.

Lemma 2. The subspace $N(\lambda)$ is a set of solutions (to within the change $u \to u/\alpha$, $v \to v/\beta$) for the following boundary-value problem:

$$\begin{cases} h_{uu} + h_{vv} + 2K(u,v)h = 0, & (u,v) \in \Omega_\lambda, \quad (2.1)\\ h = 0, & (u,v) \in \partial\Omega_\lambda. \quad (2.2)\end{cases}$$

To prove the lemma, suffice it to verify that $F_w(0,\lambda)h = -\alpha^{-2}h_{uu} - \beta^{-2}h_{vv} - 2k(\alpha u, \beta v)h$.

For each $\lambda \in \mathbb{R}^2_+$ we consider in $\widetilde{W}^k_2(\Omega)$ a finite-dimensional subspace $\widetilde{N}(\lambda) = \operatorname{Ker} \widetilde{P}_w(0,\lambda)$ - the set of solutions to the operator equation

$\tilde{P}_w(0,\lambda)h \equiv [\, F_w(0,\lambda)h,\ h(-1,v),\ h(1,v)] = [0,0,0]$.

Lemma 3. The subspace $N(\lambda)$ is (to within the change $u \to u/\alpha$, $v \to v/\beta$) a set of solutions for the following boundary-value problem:

$$\begin{cases} h_{uu} + h_{vv} + 2K(u,v)\,h = 0\ , & (u,v) \in \Omega_\lambda\ , \\ h(-\alpha,v) = h(\alpha,v) = 0 & , \ -\beta \le v \le \beta\ . \end{cases} \qquad \begin{matrix}(2.3)\\(2.4)\end{matrix}$$

Theorem III. (The necessary condition for bifurcation of MS families)
1) Let $\lambda = \lambda_0$. The necessary condition for the bifurcation of families of small solutions to Eq. (1.8) (Problem I) is

$$\dim N(\lambda_0) \neq 0 \qquad (2.5)$$

2) Let $\lambda = \lambda_0$. The necessary condition for the bifurcation of families of small solutions to Eq. (1.9) (Problem II) is that $\operatorname{Im} \tilde{P}_w(0,\lambda)$ should be a closed subspace in $\tilde{W}_2^{k-2}(\Omega)$ and that

$$\dim \tilde{N}(\lambda_0) \neq 0 \qquad (2.6)\ ,$$

where $\operatorname{Im} \tilde{P}_w(0,\lambda_0)$ is the image of the space $\tilde{W}_2^k(\Omega)$ under the action of the linear operator $P_w(0,\lambda_0)$; in particular, this image is closed if $k(u,v)$ is periodic in v with a period of $T = 2\beta_0$ ($\lambda_0 = (\alpha_0, \beta_0)$).

Proof. If, for example, in Problem I $\dim N(\lambda_0) = 0$, the operator $P_w(0,\lambda_0)$ is a linear isomorphism and, according to the implicit mapping theorem, operator equation (1.6) has in the neighbourhood of the point $(0,\lambda_0)$ only one branch of solutions, $M_0 = \{(0,\lambda)/\ \lambda \in \mathbb{R}^2_+\}$.

Lemma 4. (The absence of bifurcation of MS families under the variation of the contour along the normal to the initial minimal surface.)
1)Suppose that for $\lambda = \lambda_0$ condition (2.5) (Problem I) is not satisfied. Then, there will be no bifurcation of MS families not only under the variation of the contour Γ_{λ_0} along the surface $\vec{r}$ by the parameter λ, but also under the variation of this contour along the normal to $\vec{r}$ by a small function $\varphi \in B_2^{k-1/2}(\partial\Omega)$.
2) Suppose that for $\lambda_0 = (\alpha_0, \beta_0) \in \mathbb{R}^2_+$ condition (2.6) (Problem II) is not satisfied, and the subspace $\operatorname{Im} \tilde{P}_w(0,\lambda_0)$ is closed. Then,

there will be no bifurcation of MS families under the variation of the curves Γ_α^+ and Γ_α^- along the normal to the surface $\vec{r}$ by small functions ψ_1, ψ_2 $\in$ $\widetilde{B}_2^{k-1/2}[-1,1]$.

Proof. In Problem I, for example, it follows from the inverse operator theorem that $P_W(0, \lambda)$ is a local diffeomorphism in the neighbourhood of $W_0 \equiv 0$, and the equation $P(W, \lambda_0) = [0, \varphi]$ has only one solution, which is small in the space $W_2^k(\Omega)$ for small $\varphi \in B_2^{k-1/2}(\partial\Omega)$.

In the Morse-Palaise-Smale-Simons variational problem (see,e.g. [8,9]) one can introduce the concept of an adjoint contour Γ, provided Γ^{p-1} is a smooth boundary of a compact minimal p-dimensional submanifold M^p in an n-dimensional Riemannian manifold ($p < n$). The condition that Γ should be smooth is caused by the fact that submanifolds with boundary Γ, which are close to M^p (sections of the bundle normal to M^p) are considered in the class of C^∞-smooth functions.

Let us consider, on a two-dimensional minimal surface $\vec{r}(u,v)$ defined in $\mathbb{R}^3$ by conformal coordinates (u,v), a piecewise smooth contour Γ which has finitely many corner points and bounds a domain Q on the surface $\vec{r}$ (Q is projected on the plane (u,v) into a domain Ξ), so that the corresponding theorems for the Sobolev spaces $W_2^k(\Xi)$ are valid.

Definition 1. A contour Γ is called adjoint if problem (2.1)-(2.2) for the domain Ξ has non-zero solutions in $W_2^k(\Xi)$; the dimension of the solution subspace is called the 0-index of the adjoint contour and is denoted by $\mathrm{ind}_0 \Gamma$.

The theorems on the regularity of solutions to boundary-value problems imply that Definition 1 does not depend on the choice of the integer $k \geq 2$.

Definition 2. The sum of dimensions of eigensubspaces in $W_2^k(\Xi)$ corresponding to negative eigenvalues of the problem

$$\begin{cases} h_{uu} + h_{vv} + 2K(u,v)h = -\lambda h \ , & (u,v) \in \overset{\nabla}{\Delta} \ , \qquad (2.6) \\ h = 0 \ , & (u,v) \in \partial \overset{\nabla}{\Delta} \ . \qquad (2.7) \end{cases}$$

is called the total index of a non-adjoint contour Γ ; the total index of Γ is denoted by ind Γ .

Definitions 1 and 2 can be applied to contours Γ_λ considered in Problem I. If Γ_λ is an adjoint contour (dim $N(\lambda) \neq 0$), its 0-index is equal to dim $N(\lambda)$.

The most important result in adjoint contour theory is as follows: under the homotopy of a non-adjoint contour Γ over a surface $\vec{r}$ into a point, in the class of smooth contours (the area of the bounded domain monotonically decreases, and each subsequent domain lies inside the preceding one), the sum of 0-indices of the arising adjoint contours does not depend on the way the homotopy is carried out, and is equal to the total index.

By analogy, we now introduce the concept of adjoint contour for Problem II.

Definition 3. The contour $\tilde{\Gamma}_\alpha = \Gamma_\alpha^+ \cup \Gamma_\alpha^-$ is called adjoint in Problem II (with period $T = 2\beta$), if dim $\tilde{N}(\lambda) \neq 0$ ($\lambda = (\alpha, \beta)$). Correspondingly, $\mathrm{ind}_0\, \tilde{\Gamma}_\alpha (T) = \dim \tilde{N}(\lambda)$. If $\tilde{\Gamma}_\alpha$ is a non-adjoint contour, its total index ind $\tilde{\Gamma}_\alpha (T)$ is said to be the sum of dimensions of eigensubspaces in $\tilde{W}_2^k(\Omega_\lambda)$ corresponding to negative eigenvalues of problem (2.6),(2.4).

It follows from Theorem III that the condition that the contour is adjoint is only the necessary condition for bifurcation of MS families on this contour.

3. Calculation of adjoint contours on catenoid, helicoid, and Scherk and Enneper surfaces

We now apply the general approach developed in Items 1,2 to study bifurcations of MS families from catenoid, helicoid, and Scherk and

Enneper surfaces in Problems I and II.

Among surfaces of rotation, only catenoid is a minimal surface; it is obtained when the curve $y = a\cosh(x/a)$, $a > 0$, is rotated about the OX-axis. Helicoid is the only MS among line surfaces; its generator (straight line normal to the OZ-axis) moves at a constant velocity V along the OZ-axis and simultaneously rotates about this axis at a constant angular speed ω. The Scherk surface is the only MS among displacement surfaces; it can be globally defined as the graph of the function $z = \ln\cos x - \ln\cos y$. The Enneper surface is an algebraic MS, $\vec{r} = (-uv^2 + 1/3u^3 - u, -1/3v^3 + u^2v + v, v^2-u^2)$, which has several self-intersection lines.

Let us introduce the following conformal coordinates on catenoid, helicoid, and Scherk and Enneper surfaces, respectively:

$$\vec{r} = (u, \cosh u \cos v, \cosh u \sin v), \quad a = 1; \tag{3.1}$$

$$\vec{r} = (\sinh u \cos v, \sinh u \sin v, v), \quad \omega = V = 1; \tag{3.2}$$

$$\vec{r} = (v, \operatorname{arctg}(\sinh u/\cos v), \tfrac{1}{2}\ln(\sinh^2 u + \cos^2 v)); \tag{3.3}$$

$$\vec{r} = (e^u\cos v - \tfrac{1}{3}e^{3u}\cos 3v, -e^u\sin v - \tfrac{1}{3}e^{3u}\sin 3v, e^{2u}\cos 2v); \tag{3.4}$$

	k	E	ℓ	m
(3.1)	$\cosh^{-2} u$	$\cosh^2 u$	-1	0
(3.2)	$\cosh^{-2} u$	$\cosh^2 u$	0	-1
(3.3)	$\cosh^{-2} u$	$\frac{\cosh^2 u}{\sinh^2 u + \cos^2 v}$	$\frac{-\cosh u \cos v}{\sinh^2 u + \cos^2 v}$	$\frac{-\sinh u \sin v}{\sinh^2 u + \cos^2 v}$
(3.4)	$\cosh^{-2} u$	$4e^{4u}\cosh^2 u$	$2e^{2u}\cos 2v$	$2e^{2u}\sin 2v$

Note that adjoint contours in Problems I and II arise on all the four MS for the same critical values of the parameter λ, since for all the surfaces $k = 1/\cosh^2 u$.

To determine these critical λ values, one has to solve boundary-value problems (2.1)-(2.2) and (2.3)-(2.4) for $k = 1/\cosh^2 u$, and find the subspaces $N(\lambda)$ and $\tilde{N}(\lambda)$, respectively.

In $\mathbb{R}^2_+$ we consider the set $\Lambda = \bigcup \Lambda_n$ (n = 1,2, ...), where the curves

$$\Lambda_n = \left\{ \lambda : \lambda = (\alpha, \beta),\ \alpha = -\tfrac{1}{2} \ln g(t),\ \beta = \frac{\pi n}{2t},\ 0 < t < 1 \right\} \tag{3.5}$$

are shown in Fig. 3, and the function $g:(0,1) \longrightarrow (0,1)$ is given implicitly by the equation

$$(1-g)(1-g^t) - t\,(1+g)(1+g^t) = 0 . \tag{3.6}$$

The function $g(t)$ (its graph) is calculated approximately by a computer.

Theorem IV. If $\lambda \bar{\in} \Lambda$, then $\dim N(\lambda) = 0$. If $\lambda \in \Lambda$, then $\dim N(\lambda) = 1$, and the subspace $N(\lambda)$ is generated by the element $e(u,v)$, where

$$e(u,v) = \begin{cases} \cos(\pi n v/2) \cdot z(u,\lambda), & n = 1,3,5,\ldots, \\ \sin(\pi n v/2) \cdot z(u,\lambda), & n = 2,4,6,\ldots, \end{cases} \tag{3.7}$$

$$z(u,\lambda) = \tanh(\alpha u) \cosh(t \alpha u) - t \sinh(t \alpha u) . \tag{3.8}$$

In $\mathbb{R}^2_+$ we consider the set $\tilde{\Lambda} = \Lambda^* \cup \Lambda_{2n}$ (n = 1,2, ...) shown in Fig. 4, where $\Lambda^* = \{ \lambda \,/\, \lambda = (\alpha_*, \beta),\ \beta > 0 \}$ and α_* is the only positive root of the equation

$$\alpha \cdot \tanh \alpha = 1 \tag{3.9}$$

computed approximately by the method of false position ($\alpha = 1.195$).

Theorem V. If $\lambda \bar{\in} \tilde{\Lambda}$, then $\dim \tilde{N}(\lambda) = 0$. If $\lambda \in \Lambda^*$, then $\dim \tilde{N}(\lambda) = 1$, and the subspace $\tilde{N}(\lambda)$ is generated by the element

$\tilde{e}(u,v)$,

$$\tilde{e}(u,v) = \alpha_* u \sinh(\alpha_* u) - \cosh(\alpha_* u) \quad . \tag{3.10}$$

If $\lambda \in \cup \Lambda_{2n}$ (n = 1,2, ...), then $\dim \tilde{N}(\lambda) = 2$, and the subspace $\tilde{N}(\lambda)$ is generated by elements $\tilde{e}_1(u,v)$ and $\tilde{e}_2(u,v)$, where

$$\begin{aligned} \tilde{e}_1(u,v) &= \cos(2\pi n v)\, z(u,\lambda) \quad , \\ \tilde{e}_2(u,v) &= \sin(2\pi n v)\, z(u,v) \quad , \end{aligned} \tag{3.11}$$

and the function $z(u, \lambda)$ is defined by relation (3.8).

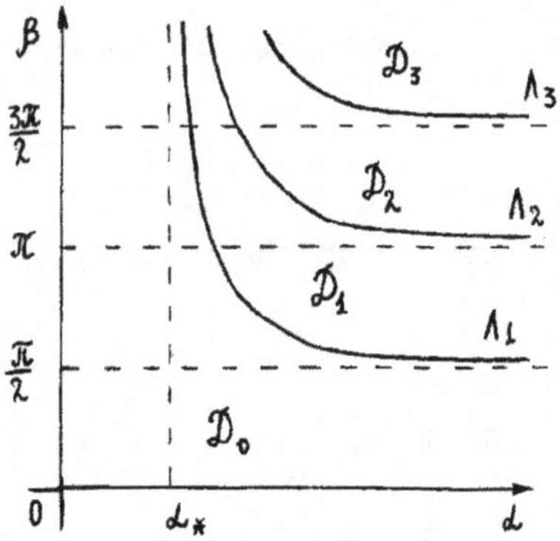

Fig. 3

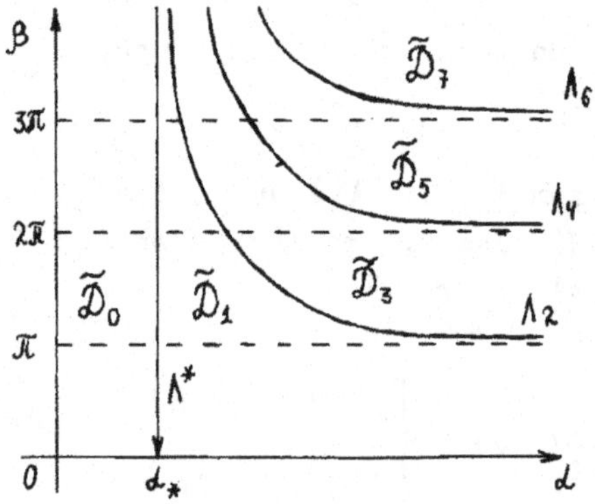

Fig. 4

<u>Proof of Theorems IV and V</u>. To determine the kernels of $N(\lambda)$ and $\tilde{N}(\lambda)$, we solve, using the Fourier method, boundary-value problems (2.1)-(2.2) and (2.3)-(2.4) for $k = 1/\cosh^2 u$. After separation of variables, the main problem is to integrate the differential equation

$$\mathcal{U}''(u) + \left(\frac{2}{\cosh^2 u} - \lambda\right) \mathcal{U}(u) = 0 \quad , \quad \lambda \geq 0 \quad , \tag{3.12}$$

and to find its general solution:

$$\lambda = 0 , \qquad \mathcal{U}(u) = C_1 \tanh u + C_2(u \tanh u - 1) , \qquad (3.13)$$

$$\lambda > 0 , \lambda \neq 1 , \qquad \mathcal{U}(u) = C_1(\tanh u - \sqrt{\lambda})e^{\sqrt{\lambda}u} + C_2(\tanh u + \sqrt{\lambda})e^{-\sqrt{\lambda}u} , \qquad (3.14)$$

$$\lambda = 1 , \qquad \mathcal{U}(u) = C_1 \frac{1}{\cosh u} + C_2 (\sinh u + \frac{u}{\cosh u}) . \qquad (3.15)$$

Substituting the solution into boundary conditions (2.2) and (2.4), we obtain $N(\lambda)$ and $\tilde{N}(\lambda)$.

It follows from Theorems III, IV, and V that adjoint contours in Problems I and II (the necessary condition of bifurcation) arise on catenoid, helicoid, and on Scherk and Enneper surfaces only for parameter values $\lambda \in \Lambda$ (Problem I) and $\lambda \in \tilde{\Lambda}$ (Problem II). We now calculate the 0-indices of these adjoint contours.

Lemma 5. If in Problem I $\lambda \in \Lambda$, then $\mathrm{ind}_0\, \Gamma_\lambda = 1$. If $\lambda \in D_n$, $n = 0,1,2, \ldots$ (see Fig. 3), then $\mathrm{ind}\, \Gamma = n$.

For example, if $\lambda \in D_0$, then ind $\Gamma_\lambda = 0$, and the contour Γ_λ does not contain homotopic piecewise-smooth adjoint contours (a finite number of corner points; the absence of bifurcation of MS families).

Lemma 6. If in Problem II $\lambda = (\alpha_*, \beta) \in \Lambda^*$ (i.e. $\alpha = \alpha_*$), then $\mathrm{ind}_0\, \tilde{\Gamma}_{\alpha_*}(T) = 1$ for all $T > 0$. If $\lambda \in \Lambda_{2n}$, $n = 1,2,3, \ldots$ (see Fig. 4), then $\mathrm{ind}_0\, \tilde{\Gamma}_\alpha(T) = 2$. If $\lambda \in D_n$, $n = 0,1,3,5,7,$ then the total index $\mathrm{ind}\, \tilde{\Gamma}_\alpha(T) = n$.

4. Sufficient conditions of bifurcation for adjoint contours on catenoid, helicoid, and on Scherk and Enneper surfaces

Topological and analytic methods have been developed for studying the existence of solutions and bifurcations in the case of Fredholm operators (see [10-13]). We now apply these methods to the operators P and $\tilde{P}$ to establish whether there exists bifurcation of MS families on adjoint contours considered in Item 3 for the four classical MS.

In Problem I as the sufficient condition of bifurcation at the point

$\lambda_0 \in \Lambda$ we take Theorem 3.2.2 in L. Nirenberg's monograph [13] (3.2.2 theorem). Conditions (i) and (ii) of the theorem are satisfied because P is a smooth Fredholm operator (Theorem II); condition (iii) (unidimensionality of the subspace $N(\lambda_0) = \ker P_W(0, \lambda_0)$) is satisfied by virtue of Theorem IV. The condition

$$P_{W\beta}(0,\lambda_0)e \;\bar{\in}\; R\left(P_W(0,\lambda_0)\right) \tag{4.1}$$

is verified by straightforward calculation using relations (3.7), (3.8), and Lemma 1 (for details, see Ref. [19]). As a result we obtain

Theorem VI. (On bifurcation of MS families in Problem I)
If $\lambda_0 = (\alpha_0, \beta_0) \in \Lambda$, then λ_0 is the bifurcation value of the parameter, and the set of solutions to the operator equation $P(W, \alpha_0, \beta) = [0,0]$ in a neighbourhood of the point $(0, \beta_0) \in W_2^k(\Omega) \times \mathbb{R}_+^1$ consists of two C^1-smooth curves $M_0 = \{(0,\beta)/\beta \in \mathbb{R}^1\}$ and $M_1 = \{(W(s), \beta(s))/ -\varepsilon < s < \varepsilon\}$, where

$$W(s) = es + q(s), \quad q(0) = q_s'(0) = 0, \quad \beta(0) = \beta_0, \tag{4.2}$$

the curves intersecting only at point $(0, \beta_0)$.

Theorem 3.2.2 of Ref. [13] can also be applied to Problem II to give the following statement.

Theorem VII. (On bifurcation of MS families in Problem II)
Let $\lambda_0 = (\alpha_*, \beta_0)$, where $\beta_0 > 0$ (i.e. $\lambda_0 \in \Lambda^*$) for catenoid and helicoid, $\beta_0 = \pi p$ (p = 1,2, ...) for the Scherk surface, and $\beta_0 = \pi p/2$ (p = 1,2, ...) for the Enneper surface. In this case λ_0 is the bifurcation value of the parameter λ, and the set of solutions to the operator equation $\tilde{P}(W, \alpha, \beta_0) = [0,0,0]$ in a neighbourhood of the point $(0, \alpha_*) \in \tilde{W}_2^k(\Omega) \times \mathbb{R}_+^1$ consists of two smooth curves $\tilde{M}_0 = \{(0,\alpha)/\alpha \in \mathbb{R}^1\}$ and $\tilde{M}_1 = \{(\tilde{W}(s), \alpha(s))/ -\varepsilon' < s < \varepsilon\}$, where

$$\tilde{W}(s) = \tilde{e}s + \tilde{q}(s), \quad \tilde{q}(0) = \tilde{q}_s'(0) = 0, \quad \alpha(0) = \alpha_*, \tag{4.3}$$

the curves intersecting only at the point $(0, \alpha_*)$.

If in Problem II $\lambda_0 \in \tilde{\Lambda} / \Lambda^*$, a further analysis is required.

Note that in Problem I bifurcation of MS families arises due to the variation of the adjoint contour Γ_{λ_0} ($\lambda_0 = (\alpha_0, \beta_0)$) by the parameter β and in Problem II, due to the variation of the adjoint contour $\tilde{\Gamma}_{\alpha_*}$ by the parameter α .

Now a few words about the location of the curves Γ_λ and $\tilde{\Gamma}_\alpha$ on the four classical minimal surfaces.

On catenoid, the contour Γ_λ does not have self-intersections for $\beta < \pi$ (see Fig. 5a); unfortunately, bifurcation on this contour for $\lambda \in \Lambda$ is not reported in Refs. [18,19] . The curves Γ_α^+ and Γ_α^- are two coaxial circles of equal radii (Fig. 2a); bifurcation for $\lambda_0 = (\alpha_*, 2\pi)$ (Problem II) was known earlier in the class of surfaces of rotation, the branch M_1 being formed by adjoint catenoids (see, e.g., [2] , p. 116). It is known from physical experiments that the initial catenoid becomes unstable for $\alpha > \alpha_*$, and cannot be obtained as a soap film.

On helicoid, the contour Γ_λ does not have self-intersections either. The curves Γ_α^+ and Γ_α^- are two coaxial infinite helixes of equal radii (see Fig. 2b). An increase in the parameter α can be interpreted, geometrically, as an increase in the ratio $\omega / V \sim \alpha$. Bifurcation for $\alpha = \alpha_*$ was studied earlier in physical experiments (see [2] , p. 114) which demonstrated that in this case two ribbon (not line) minimal surfaces are bifurcated from helicoid.

If the Scherk surface is defined in coordinates (3.3), there exist on it two infinitely distant points for $v = \pi/2 + \pi n$ ($n \in \mathbb{Z}$).

On the Enneper surface, the curves Γ_λ do not have self-intersections for $\alpha < \ln\sqrt{3}$ and $\beta < \pi$. The curves Γ_α^+ and Γ_α^- do not exhibit self-intersections for $\alpha < \ln\sqrt{3}$; if we pass from coordinates (3.4) to the coordinates $(-uv^2 + \frac{1}{3}u^3 - u,\ -\frac{1}{3}v^3 + u^2v + v,\ v^2 - u^2)$, then Γ_α^+ and Γ_α^- turn into two circles with common centre at (0,0) and with radii $R_1 = e^{\alpha}$ and $R_2 = e^{-\alpha}$ (see Fig. 5b).

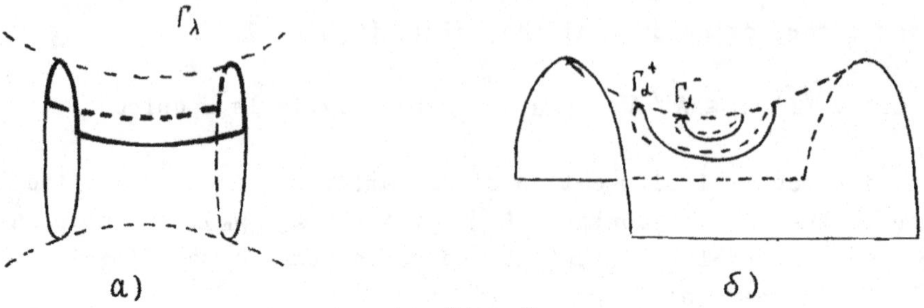

Fig. 5

At the seminar on global analysis (VSU, 1982) Yu.I. Sapronov put forward the following hypothesis: if a closed contour Γ in $\mathbb{R}^3$ is orthogonally projected, without self-intersections, onto a certain plane, and if this property is stable relative to small deformations of Γ, no bifurcation of MS families arise on the contour Γ. The bifurcation contours Γ_λ found in the present paper satisfy the condition: the set $\vec{n}(\Omega_\lambda)$ on a unit sphere $S_1^2 \subset \mathbb{R}^3$ is not entirely contained in any closed hemisphere on S_1^2 (Ω_λ is the projection of the domain $Q_\lambda \subset \vec{r}$ bounded by the curve Γ_λ onto the plane (u,v)). This confirms Sapronov's conclusion made on basis of soap-film experiments.

In Problems I and II, in the neighbourhood of the bifurcation point $(0, \lambda_0)$, one can pass to the Lyapunov-Schmidt branching equation $\varphi(\xi, \lambda) = 0$, $\varphi(\cdot, \lambda): \mathbb{R}^n \to \mathbb{R}^n$, $n = \dim N(\lambda)$ ($\tilde{N}(\lambda)$ for Problem II), and study the type of singularity for the branching of small solutions. The families of small solutions $\xi(\lambda)$ to the branching equation define families of MS ($W_\lambda(u,v)$) which are close to the initial MS, $\vec{r}$. Yu.I. Sapronov and Yu.N. Zavarovsky (see, e.g. [23]) have proposed a modification of this method for variational problems - a so-called method of the key function $\Phi(\xi, \lambda)$, $\operatorname{grad}_\xi \Phi = \varphi$ - which considerably reduces computation procedure. In publications to follow we shall describe the algorithm of calculating singularity types for the critical points $\xi_0 = 0$ of the function $\Phi(\xi, \lambda_0)$ for the bifurcation values of the parameter β (α) determined in Problems I and II. For example, in the case of bifurcation from catenoid and helicoid in Problem II ($\beta = \pi$) calculations carried out in accordance with this algorithm give singularities of type A_2 ("fold") and A_3 ("cusp"), respectively.

The author is grateful to A.A. Talashev, translation editor.

REFERENCES

1. Fomenko A.T., Variational methods in topology.- Moscow, Nauka, 1982 (in Russian).
2. Fomenko A.T., Topological variational problems.- Moscow State University, 1984 (in Russian).
3. Trombar A.J., On the number of simply connected minimal surfaces spanning a curve.- Memoirs of the Amer. Math. Soc., v. 12, No. 194 (1977).
4. Courant R., Dirichlet's principle, conformal mapping, and minimal surfaces.- New York, 1950.
5. Poston T., "The Problem of Plateau, an introduction to the whole Mathematics".- Mimeographical Notes - Summer Conference, Trieste, 1971.
6. Smale S., Morse theory and a non-linear generalization of the Dirichlet problem.- Annals of Math., v. 80 (1964), 382-396.
7. Palaise R. and Smale S., A generalized Morse theory.- Bull. Amer. Math. Soc., v. 70, No. 1 (1964), 165-172.
8. Smale S., On the Morse index theorem.- J. Math. Mech., v. 14, No. 6 (1965), 1049-1055.
9. Simons J., Minimal varieties in Riemannian manifold.- Annals of Math., v. 88, No. 1 (1968), 62-105.
10. Borisovich Yu.G., Zvyagin V.G., and Sapronov Yu.I., Nonlinear Fredholm Mappings and the Lerey-Schauder Theory.- Usp. Mat. Nauk, v. 32, No. 4 (1977), 3-54 (in Russian).
11. Wainberg M.M. and Trenogin V.A., Branching Theory of Solutions of Nonolinear Equations.- Moscow, Nauka, 1969 (in Russian).
12. Krasnosel'sky M.A. Vainikko G.M., Zabreiko P.P., Rutitsky Ya.B., and Stetsenko V.Ya., Approximate Solution of Operator Equation.- Moscow, Nauka, 1969 (in Russian).
13. Nirenberg L., Topics in Nonlinear Functional Analysis.- Courant Inst. of Math. Sci., New York, 1974.
14. Zachepa V.R. and Sapronov Yu.I., On Local Analysis of Nonlinear Fredholm Equations.- Transactions of Steklov Math. Inst., v. 154: Topology (1983), 113-117 (in Russian).
15. Khimshiashvili G.N., On Small Solutions of Nonlinear Fredholm Equations.- Vestn. Moscow State University, Ser. Math. Mech., 1977, No. 2, 27-31 (in Russian).
16. Bernshtein S.N., Partial differential equations.- Moscow, 1960 (in Russian).
17. Borisovich A.Yu. and Zvyagin V.G., On global reversibility of nonlinear operators generated by boundary-value problems.- Approximate methods of studying differential equations and their applications - Kuibyshev, 1983, 27-33 (in Russian).
18. Borisovich A.Yu., A certain geometrical application of the theorem on simple bifurcation point.- Application of topology in modern analysis .- Voronezh, 1985, 172-174 (in Russian).

19. Borisovich A.Yu., On the operator-functional method of studying bifurcations of a minimal surface. 1985, deposited at VINITI, No. 4442-85 (in Russian).

20. Besov O.V., Il'in V.P., and Nikolsky S.M., Integral representations of functions and embedding theorems.- Moscow, Nauka, 1975 (in Russian).

21. Beeson M.J. and Trombar A.J., The cusp catastrophe of Thom in the bifurcation on minimal surfaces.- Manuscr. math., v. 46, No.1-3 (1984), 273-308.

22. Borisovich A.Yu., Reduction of the bifurcation problem for minimal surfaces to operator equations and determination of bifurcations from catenoid, helicoid, and Scherk and Enneper surfaces.- Usp. Matem. Nauk, 1986, v. 41, No. 5(251), 165-166 (in Russian).

23. Zavarovsky Yu.N., Standard form of the key function for the Kirchhoff generalized equation.- Usp. Matem. Nauk, 1983, v. 38, No. 3, 177-178 (in Russian).

HOMOLOGICAL METHODS IN THE THEORY OF PERIODIC AND EQUIVARIANT MAPS

Yu.G.Borisovich
Department of Mathematics
Voronezh State University
Universitetskaya pl., 1,
394693, Voronezh, USSR

T.N.Fomenko
Department of Mathematics
Moscow Institute of
Steel and Alloys
Leninsky prospekt, 4,
117936, Moscow, USSR

Introduction.

During some recent years, quite a number of works were made both in the Soviet Union and abroad dealing with the investigation of various classes of groups actions and equivariant maps of topological spaces. Among the main methods being used in this field we should mention, first of all, the (co)homological methods widely exploiting the algebraic apparatus. Besides, some important results in this direction were obtained for the first time by geometrical methods for the metric spaces. The employment of the extraordinary cohomology theories in this subject (the Conner-Floyd theory and its development) involves a peculiar synthesis of the mentioned trends.

Up to now, a large number of results have been collected in the field under discussion which were obtained by the use of homological methods. The aim of the present paper is to summarize them (without claiming the full review of these methods) and to compare these results with geometrical ones concerning the same problems.

The paper consists of three sections. The first section summarizes the main ideas of the P.A.Smith theory [1] on the group Z_k actions and some new results obtained by Ya.A.Izrailevich, E.M.Mukhamadiev and T.N.Fomenko (Shchelokova). The second section deals with the development of the Smith theory by the use of the Borel spectral sequence method [2] . It also contains some results obtained in this direction by Yu.G.Borisovich, Ya.A.Izrailevich and T.N.Fomenko. Theorems 6 and 8 are original. At the same time we list some geometrical results concerning the same problems which are, to our mind, the most striking ones (the works by M.A.Krasnosel'skiĭ, P.P.Zabreĭko, Z.I.Balanov and S.D.Brodskiĭ). The third section deals with the investigations of actions of an arbitrary finite group or a compact not quite unconnected topological group on cohomological spheres and equivariant maps. All results considered in this section are original and belong to T.N.Fomenko.

In this paper the authors fixed their attention on the questions which are close to their own research fields, that is why brief reviews in sections 1 and 2 do not claim to be complete.

The authors hope, however, that the information given here may become a starting point for further investigations in this interesting and extensive field.

I. Smith theory.

Let X be a paracompact Hausdorff (or locally compact of finite combinatorial dimension) space with finite generated groups of Alexandrov-Čech integer cohomologies. Let $T : X \longrightarrow X$ be a homeomorphic action of the group Z_K, $K \geq 2$ on X. It will be recalled that the action T is called free if $Tx^q \neq x$, $x \in X$, $q = 1,2,\dots, k-1$. Action T is called a semi-free one if it is free outside the set $F = \{x \in X \mid Tx = x\}$ of fixed points of T. Below the cohomological groups are considered with the coefficients in Z_k.

The following long exact cohomological sequences are the main in Smith theory:

$$\begin{array}{l} \dots \longrightarrow H^i_\sigma \longrightarrow H^i(X) \xrightarrow{j} H^i(F) + H^i_\delta \longrightarrow H^{i+1}_\sigma \longrightarrow \dots \\ \dots \longrightarrow H^i_\delta \longrightarrow H^i(X) \xrightarrow{j} H^i(F) + H^i_\sigma \longrightarrow H^{i+1}_\delta \longrightarrow \dots \end{array} \qquad (I)$$

where H^*_σ, H^*_δ are cohomological groups of the so-called Smith special complexes associated with the action of cochain operators $\delta = I - T$ and $\sigma = I + T + \dots + T^{k-1}$. P.A.Smith proved the exactness of sequences (I) at simple k [1,3]. However, the whole theory extends to the situation when $k \geq 2$ is arbitrary under the condition that the action T of group Z_k is semi-free [4]. Alternating the connecting homomorphisms of the exact sequences (I) one can obtain the following compositions of mappings:

$$\begin{array}{lll} S_T : & H^0_\sigma + H^*(F) \longrightarrow H^n_\rho \\ S_T : & H^0_\delta + H^*(F) \longrightarrow H^n_\rho \end{array} \qquad \rho = \sigma \quad \text{or} \quad \rho = \delta \quad (2)$$

which we shall call full Smith homomorphisms. With the help of these homomorphisms all known invariants of action T in Smith theory are determined. We shall present some of them.

Let X be a cohomological sphere over the ring of the coefficients of Z_k of the formal dimension n (below Z_k is a cohomological n-sphere) and the action T preserves its orientation, i.e. deg T = +I. If T is free ($F = \emptyset$) then it is easy to show that the homomorphisms δ^o, σ^o, j^n_σ, j^n_δ are the automorphisms of the group Z_k and S_T is naturally constructed upto the homomorphism:

$$J^{\emptyset}_T = (j^n)^{-1} \cdot S_T \cdot (\sigma^o)^{-1} : H^o(X) \longrightarrow H^n(X) ,$$

$S_T : H^o_\sigma \longrightarrow H^n_\rho$ also being an automorphism of the group Z_k.

<u>Definition I</u>. Smith index of a free action T of the group Z_K on a Z_K-cohomological n-sphere X ($n > 0$) is a generator $ind^{\emptyset}_T \in Z_K$ defined by the equality $J^{\emptyset}_T(U^o) = ind^{\emptyset}_T \cdot U^n$ where $U^n \in H^n(X)$ is a Z_K-orientation X, U^o is a ring unit in $H^*(X)$.

In the case when $F \neq \emptyset$ from the sequences (I) it is easy to show that $H^o_\sigma = H^o_\delta = 0$ and j^n_σ, j^n_δ are still the automorphisms of Z_K. In this case a full Smith homomorphism S_T has the form:

$$J^F_T = (j^n)^{-1}\ S^F_T : H^o(F)/Jmj^o_\sigma \oplus H^1(F) \oplus \ldots \oplus H^m(F) \longrightarrow H^n(X)$$

There are two variants with respect to the cohomologies of the set F:

1) $H^o(F)/Jmj^o = 0$, then $H^o(F) = Z_K$ and there exists a unique dimension m, $0 < m \leq n$, such that $H^m(F) = Z_K$, i.e. F is a Z_K-cohomological sphere of a formal dimension m, $m > 0$.

2) if $H^o(F)/Jmj^o \neq 0$, then $H^o(F) = Z_K \oplus Z_K$, $H^i(F) = 0$, $i > 0$, i.e. F is a Z_k- cohomological 0-sphere, in particular, $H^o(F)/Jmj^o = Z_K$.

Variants 1), 2) are in fact Smith theorem which states that a set of fixed points of the action of the group Z_k (K is simple or the action is semi-free) on Z_K- cohomological n-sphere X is itself a Z_K-cohomological m-sphere and $-I \leq m \leq n$ ($\emptyset$ is understood as a sphere of dimensionality -I).

Thus, at $F \neq \emptyset$ the homomorphism $J^F_T : H^m(F)/Jmj^m_\rho \longrightarrow H^n(X)$ is always defined, where m is a formal dimension of F, $Jmj^m_\rho = 0$ at $0 \leq m < n$ and $J^F_T = (j^n)^{-1}$, if $m = n$.

<u>Definition 2</u>. Smith index of a semi-free action T ($F \neq \emptyset$) of the group Z_K on a Z_K-cohomological n-sphere X is a generator $ind^F_T \in Z_K$ defined by the equality $J^F_T(U^m) = ind^F_T \cdot U^n$ where U^n, U^m are Z_K-orientations of X and F respectively. At $m = 0$ U^m is replaced by the corresponding generator of the group $H^o(F)/Jmj^o_\sigma = Z_K$.

The definitions of Smith indices 1,2 are given in [1,10,4] where they are formulated somewhat differently. Our variant seems to be more natural.

Likewise using Smith homomorphism $\widetilde{S}_T$ one can define indices $\widetilde{\mathrm{ind}}_T^{\emptyset}$ and $\widetilde{\mathrm{ind}}_T^F$. However in reality the homomorphisms S_T and $\widetilde{S}_T$ coincide (see, e.g. [3,5]).

Smith indices as elements of the group Z_K can also be defined in more general situations than the case of a cohomological sphere.

For example, if $H^0(X) = H^n(X) = Z_K$, $H^i(X) = 0$ at $i > n$ and the homomorphism of the change of coefficients $H^n(X; Z) \longrightarrow H^n(X; Z_K)$ is an epimorphism, then under the free action T the homomorphism $\mathcal{I}_T^{\emptyset}$ described above for cohomological spheres is defined. However, in this case it is not clear when the corresponding Smith index will be a generator in Z_K. A space with such conditions is considered in [6] where it is called an oriented n-space.

Similarly Smith indices are defined for the action of the group Z_K on Z_K-cohomological connected orientable n-manifold M (the definition and properties of cohomological manifolds see, e.g. in [7]). Then, in particular, there are isomorphisms:

$$H^n(M \times M,\ M \times M \setminus \Delta) \xrightarrow{\lambda} H^n(M, M \setminus x) \xrightarrow{jx} H^n(M) \qquad (x \in M)$$

at which the orientation $U \in H^n(M)$, the fundamental class $W \in H^n(M \times M, M \times M \setminus \Delta)$ and the element $W_x \in H^n(M, M \setminus x)$ of the orienting family correspond to each other. It is known that at prime K (or any K and semi-free action) the set F of fixed points of the action of the group Z_K is either empty or each its component $F_\alpha \subset F$ is a Z_K-cohomological m_α -manifold, $0 \le m \le n$. For every such component as above for the spheres Smith index $\mathrm{ind}_T^F \in Z_K$ is defined which is a generator in Z_K.

Using Smith homomorphisms and indices and also the representations of Lefshetz cohomological class in [8] is proved the following

<u>Theorem I</u>. Let M_1, M_2 be Z_K-cohomological connected oriented n-manifolds ($n > 0$); $T_i : M_i \longrightarrow M_i$ be actions of the group Z_K (K is prime) with the set F_i of fixed points ($i = 1,2$). Let $f, g : M_1 \longrightarrow M_2$ be equivariant mappings and $F_{i\alpha} \subset F_i$ be such components of the sets F_i ($i = 1,2$) that $f(F_{1\alpha}) \subset F_{2\alpha}$, $g(F_{1\alpha}) \subset F_{2\alpha}$. Let also the dimensions of the components $F_{1\alpha}$ and $F_{2\alpha}$ coicide and be equal to m, $0 \le m \le n$. Then for Lefshetz numbers Λ_{fg} and $\Lambda_{fg}^{F_{1\alpha}} = \Lambda_{f|F_{1\alpha}, g|F_{1\alpha}}$ for the pairs of mappings(f,g) and their restrictions on $F_{1\alpha}$ the following comparison takes place:

$$\Lambda_{f,g} = \mathrm{ind}_{T_1}^{F_{1\alpha}} \cdot \Lambda_{f,g}^{F_{1\alpha}} \cdot \left[\mathrm{ind}_{T_2}^{F_{2\alpha}}\right]_K^{-1} \pmod{K} \qquad (3)$$

Formula (3) generalizes (in the category of cohomological manifolds) the result [6] on Lefshetz number of one map and allows us to consider two different actions of the group Z_K and two nonidentical equivariant mappings. This formula under the condition that M_1 and M_2 are cohomological spheres also provides the known result [9] which states that at $K>2$ the difference of dimensions $n - m$ should be even. From the same formula follows also Floyd's result on the equality of Euler characteristics $\chi(F) = \chi(F_\alpha)$ [26].

In many problems it is required to compare Smith indices of actions of different groups Z_{K_1}, Z_{K_2}, in particular, action T of the group Z_{gl} and action T^q of its subgroup Z_1. The analysis of Smith sequences (I) gives the following result [10]:

Theorem 2. Let action T of the group $Z_K (K>2)$, $K = q\cdot l$, $1<q<K$, be given on Z_K-cohomological n-sphere. Then for Smith indices we have:

$$\mathrm{ind}_T^{\emptyset} = \left(\frac{q}{(q,K)}\right)^{\frac{n+1}{2}} \cdot \mathrm{ind}_{T^q}^{\emptyset} \qquad \left(\mathrm{mod}\,\frac{K}{(q,K)}\right) \qquad (4)$$

where (q,K) is the greatest common divisor of numbers q and K.

This result is generalized in [4] for a semi-free action.

Smith indices allow to obtain useful information on the degree of an equivariant mapping, since this mapping evidently commutes with Smith homomorphisms.

Theorem 3. Let X_1, X_2 be Z_K-cohomological n-spheres (or Z_K-cohomological connected oriented n-manifolds). Let $T_i : T_i \to X_i (i = 1,2)$ be semi-free actions of the group Z_K; let $f : X_1 \to X_2$ be equivariant mapping. Let the sets of fixed points F_1, F_2 (or their components, chosen so that $f(F_1) \subset F_2$) have equal formal dimensions. Then in Z_K the following equalities take place:

1) if $F_1 = F_2 = \emptyset$, i.e. the actions are free, then

$$\mathrm{ind}_{T_1}^{\emptyset} = \deg f \cdot \mathrm{ind}_{T_2}^{\emptyset} \pmod{K} \qquad (5)$$

2) if the sets F_1, F_2 are not empty, then

$$\mathrm{ind}_{T_1}^{F_1} \cdot \deg(f|_{F_1}) = \deg f \cdot \mathrm{ind}_{T_2}^{F_2} \pmod{K} \qquad (6),$$

where deg f, $\deg(f/F_1)$ are the degrees of mapping f and its restrictions on the set F_1 calculated over the ring of coefficients Z_K.

Formulas (5) and (6) are given in [1,4,10] .

The problems connected with equivariant mappings and nonsemi-free actions of a cyclic group Z_K($K>2$) have no place in Smith theory and require additional algebraic technique as it will be shown below. As an example, we shall give one of these problems successfully solved by M.A.Krasnosel'skiĭ in 1955 by geometrical methods:

<u>Theorem 4</u> [11] . Let actions T_1, T_2 of the groups Z_K and Z_q respectively be given on the euclidean shere S^n, the action Z_K being free and q being the divisor of K. Then the degrees of two arbitrary continuous equivariant mappings coincide modulo K.

The original proof of M.A.Krasnosel'skiĭ uses geometrical methods of mappings extension . Below we shall consider cohomological methods connected with a spectral sequence of the so-called Borel fibration for the given action of the group Z_K ($K>2$) which, in particular, allow to obtain the statement of theorems 3-4 for cohomological spheres.

2. Borel spectral sequence method.

Let the space X satisfy the same conditions as at the beginning of section I and let $T : X \longrightarrow X$ be the action of the group $Z_K(K>2)$. Denote by E_K and B_K the space and base of the universal principal fiber bundle for Z_K. It is known that $E_K = S^\infty$ is an infinitely dimensional sphere with a canonical free action Z_K and B_K is an infinitely dimensional lense L_K (see [2,12]). Consider a diagonal action Z_K on $X \times E_K$ and the projection $\pi_K : X \times E_K \longrightarrow E_K$ which is evidently equivariant. Its factor-mapping $\pi : X_k = \frac{X \times E_k}{Z_k} \longrightarrow B_k = L_k$ is a fibration with a fibre X. This is Borel fibration of the given action Z_k on X. Calculations connected with its spectral sequence of cohomologies lead to stronger results in the theory of equivariant mappings, as it is shown below.

The member $E_2(X) = E_2^{p,q}(X)$ of this spectral sequence has the form ([2,5,13]):

$$E_2^{p,q}(X) = \begin{cases} \dfrac{\operatorname{Ker}\delta(q)}{\operatorname{Im}\sigma(q)}, & p \text{ is even} \\[2ex] \dfrac{\operatorname{Ker}\sigma(q)}{\operatorname{Im}\delta(q)}, & p \text{ is odd} \end{cases}$$

where $\delta^{(q)} = I - T^{(q)}$; $H^q(X) \to H^q(X)$, $\sigma^{(q)} = I + T^{(q)} + \ldots + (T^{(q)})^{k-1}$: $H^q(X) \to H^q(X)$ are homomorphisms induced in the groups of cohomologies by mappings δ, σ : $X \to X$.

Consider a more general case. Let actions $T_1 : X \to X$, $T_2 : Y \to Y$ of the group Z_k be given on spaces X,Y satisfying the conditions enumerated at the beginning of section I. Let $f : X \to Y$ be an equivariant mapping. Extending naturally (identically on segment I) actions T_1 and T_2 up to the action T of the group Z_k on the cylinder C_f of the mapping f, we can consider instead of f the inclusion $i : X \to C_f$, where the image i(X) is identical to $X \times \{0\}$ in the cylinder C_f. Since C_f is homotopically equivalent to Y and i induces in cohomologies the same homomorphisms as f, it is natural to call an exact sequence of the pair (C_f, X) an exact sequence of the mapping f and to denote by $H^q(Y,X)$ the groups $H^q(C_f,X)$. Thus, we shall juxtapose any equivariant mapping $f : X \to Y$ and the pair (Y,X) whose cohomologies will be calculated from an exact sequence of the mapping f.

Consider the spectral sequence of Borel relative fibration with the fibre (Y,X) (meaning, as above, (C_f,X)). The idea of the introduction of the equivariant mapping index can be formulated as follows. Let X,Y be Z_k-cohomological spheres of dimensionalities m and n respectively. Then at different correlations of dimensions m and n the following chains of mappings are considered:

I) at $m < n$

$$\varphi_{n-m}(Y,X) = \varphi_{n-m} : H^n(Y) \xleftarrow{j} H^n(Y,X) \xleftarrow{\gamma} E_2^{0,n} \xrightarrow[\tau]{} E_{n-m}^{0,n} \xrightarrow{d_{n-m}} E_{n-m}^{n-m,m+1} \longrightarrow$$
$$\longrightarrow E_2^{n-m,m+1} \xrightarrow[\gamma]{} H^{m+1}(Y,X) \xleftarrow{\alpha} H^m(X)$$

2) at $m > n$ (or $m = n$, and $\deg f = 0$)

$$\varphi_{m-n+2}(Y,X) = \varphi_{m-n+2} : H^m(X) \xrightarrow{\alpha} H^{m+1}(Y,X) \xleftarrow{\gamma} E_2^{0,m+1} \xleftarrow[\tau]{} E_{m-n+2}^{0,m+1} \xrightarrow{d_{m-n+2}}$$
$$\longrightarrow E_{m-n+2}^{m-n+2,n} \xrightarrow[\tau]{} E_2^{m-n+2,n} \xrightarrow[\gamma]{} H^n(Y,X) \xrightarrow{j} H^n(Y)$$

3) at $m = n$ and $\deg f \neq 0$ (in Z_k)

$$\varphi_0(Y,X) = f^* : H^n(Y) \longrightarrow H^n(X)$$

where α, $j_{p,q}$ are homomorphisms of an exact sequence of the pair (Y,X); $\gamma : E_2 = H^p(B_k ; H^q(Y,X)) = H^q(Y,X)$ is a canonical isomorphism to the group of coefficients, τ is a choice of a representative of the subfactorgroup E_q^{**} in E_2^{**}, d_* is a differential of the spectral sequence.

Definition 3 ([15, 16]). The equivariance index of the mapping $f : X \to Y$ of Z_k-cohomological spheres X and Y of the dimensions m and n respectively is an element I(f) of the group Z_k determined from the equality:

a) $\varphi_{n-m}(Y,X)(U^n) = I(f) \cdot U^m$ if $m < n$ or $m = n$ and $\deg f \neq 0$,

b) $\varphi_{m-n+2}(Y,X)(U^m) = I(f) \cdot U^n$ if $m > n$ or $m = n$ and $\deg f = 0$

The homomorphisms $\varphi_{(Y,X)}$ were determined above, U^n, U^m are orientations of the spheres Y, X respectively.

In the case when $f = i$ is an inclusion and $m \leq n$, the equivariance index is in fact determined and investigated in [13,14,18], where it is called a relative spectral index. There also was considered an absolute spectral index, an analogue of the relative one for the pair $(Y, \emptyset)$. Let us recall this notion.

Definition 4. [13,14] . Let Y be Z_k-cohomological sphere with the action T of the group Z_k. Considering the spectral sequence of the Borel fibration with the fibre Y one can construct the homomorphism:

$$\varphi_n(Y, \emptyset) = \varphi_n : H^n(Y) \xleftarrow{\gamma} E_2^{0,n} \xleftarrow{\tau} E_n^{0,n} \xrightarrow{d} E_n^{n+1,0} \xrightarrow{\tau} E_2^{n+1,0} \xrightarrow{\gamma} H^0(Y)$$

where γ, τ are determined as above, d is a differential. The spectral index $\mathrm{Ind}_T^{\emptyset} \in Z_k$ is determined from the equality: $\varphi_{(Y,\emptyset)}(U^n) = \mathrm{Ind}_T^{\emptyset} \cdot U^0$, where U^n is an orientation of Y, U^0 is a ring unit in $H^*(Y)$.

The homomorphism $\varphi_{(Y,\emptyset)}$ obviously coincides with the Borel fibration transgression (see e.g. [12]). If the action T has a fixed point on Y then there exists a section of the Borel fibration determined by this point. Then the transgression is trivial since it is the obstruction to the existence of a section. That's why in this case $\mathrm{Ind}_T^{\emptyset} = \mathrm{Ind}_{(Y,\emptyset)} = 0$ (in Z_k). Similarly one can show that if the action T has a fixed point in Y outside the image f(X), then $I_{(Y,X)} = 0$. And vice versa, if the action T is free on Y (or on $Y \setminus f(X)$ respectively), then $I_{(Y,\emptyset)}$ (or $I_{(Y,X)}$) will be a generator in Z_k.

It is easy to see that if (X_1, A_1) and (X_2, A_2) are pairs of Z_k-cohomological spheres with the actions T_1, T_2 of the group Z_k and $f : (X_1, A_1) \to (X_2, A_2)$ is an equivariant mapping, such that $\dim X_1 = \dim X_2 = n$, $\dim A_2 = \dim A_1 = m$, then at $m \leq n$ (and $\deg j_1 \neq 0$, $\deg j_2 \neq 0$ if $m=n$) one of the following diagrams is commutative:

$$\begin{array}{ccc} H^n(X_2) & \xrightarrow{f^*} & H^n(X_1) \\ \downarrow{\scriptstyle \varphi(X_2,A_2)} & & \downarrow{\scriptstyle \varphi(X_1,A_1)} \\ H^m(A_2) & \xrightarrow{(f|A_1)^*} & H^m(A_1) \end{array} \quad (7) \qquad \begin{array}{ccc} H^n(X_2) & \xrightarrow{f^*} & H^n(X_1) \\ \downarrow{\scriptstyle \varphi(X_2,\emptyset)} & & \downarrow{\scriptstyle \varphi(X_1\emptyset)} \\ H^o(X_2) & \xrightarrow{f^o=id} & H^o(X_1) \end{array} \quad (8)$$

where the homomorphisms $\varphi_{(X_2,\emptyset)}$, $\varphi_{(X_2,\emptyset)}$ in (8) are obvious modifications of the homomorphisms $\varphi_{(X_i,A_i)}$ at $A_1 = A_2 = \emptyset$. From diagrams (7),(8) immediately follow the formulas (see [13,14,18]):

$I_{(X_1,A_1)} \cdot \deg f = \deg(f|A_1) \cdot I_{(X_2,A_2)}$ if A_1, $A_2 \neq \emptyset$, i.e. $m>-I$(9)

$I_{(X_1,\emptyset)} \cdot \deg f = I_{(X_2,\emptyset)}$ if $A_1 = A_2 = \emptyset$, i.e. $m = -I$ (10)

Since formulas (9), (10) are full analogues of formulas (5),(6) from section I which comprise Smith indices, the question naturally arises on the relation between Smith indices and the equivariance indices when both are defined. The comparison of the cochain structures of the corresponding homomorphisms S_T^F, φ_{n-m}, S_T, φ_n shows that takes place

Theorem 5. ([17]). Let X be Z_k-cohomological sphere with a semi-free action T of the group Z_k and the set F of fixed points be a Z_k-cohomological sphere of dimension m, $-1 \leq m \leq n$, $F \subset X$. Then Smith homomorphisms S_T^F, $\widetilde{S}_T^F$ and a spectral homomorphism φ_{n-m} are defined and reversible in Z_k and takes place the equality:

$$\varphi_{n-m} = \begin{cases} (\widetilde{S_T^F})^{-1} & n-m = 2 \pmod 4 \\ -(S_T^F)^{-1} & n-m = 0 \pmod 4 \\ (\widetilde{S_T^F})^{-1} & n-m = 1 \pmod 4 \\ -(S_T^F)^{-1} & n-m = 3 \pmod 4 \end{cases} \qquad (11)$$

The next naturally arising question is the question on the connection between three equivariance indices $I_{(Y,X)}$, $I_{X,A)}$, $I_{(Y,A)}$ determined for the triplet of spaces (Y, X, A) with the actions T_1, T_2, T_3 of the group Z_k connected by the equivariant mappings $A \xrightarrow{f} X \xrightarrow{g} Y$. Let, as above, the spaces A, X, Y be Z_k-cohomological spheres, dim A= = r, dim X = m, dim Y = n, and the exact cohomological sequence (over Z_k) have the form:

$$\ldots \longrightarrow H^q(Y,X) \xrightarrow{\widehat{\alpha}} H^q(Y,A) \xrightarrow{\widehat{\gamma}} H^q(X,A) \xrightarrow{\widehat{\gamma}} H^{q+1}(Y,X) \longrightarrow \ldots \qquad (12)$$

Then arises a diagram of homomorphisms:

$$\begin{array}{ccccc} & & H^m(X) & & \\ & \varphi_*(X,A) \diagup & & \diagdown \varphi_*(Y,X) & \\ H^r(A) & & \overline{\quad\varphi^*_*(Y,A)\quad} & & H^n(Y) \end{array} \qquad (13)$$

where the directions of the arrows are not indicated and the lower indices of the spectral homomorphisms are replaced by asterisks, since all this (by virtue of the definition of equivarience indices) essentially depends on the relation of the dimensions r, m, n.

It was shown in [15] that under the relation $r<m<n$ diagram (13) commutes and it was announced there that this is true for all other relations. As it will be shown below the validity of this statement in some cases requires an additional condition. Besides we'd like to find out if there are such variants of relations between r, m, n when any of the three equivariance indices must be zero. We shall formulate the answer in the form of the following theorem.

<u>Theorem 6</u>. (T.N.Fomenko). Let for the triplet of spaces (Y,X,A) the conditions described above be fulfilled. Then

1) if $r<m<n$, then, generally speaking, the indices $I_{(X,A)}$, $I_{(Y,A)}$, $I_{(Y,X)}$ should not equal zero and for the differentials d_{m-r}, d_{n-r}, and d_{n-m} of spectral sequences of Borel fibrations of the pairs (X,A), (Y,A) and (Y,X) respectively the following exact sequence takes place:

$$0 \to \ker d_{n-m} \xrightarrow{\alpha} \ker d_{n-r} \xrightarrow{\beta} \ker d_{m-r} \to H^{m+1}(Y,X)/\operatorname{Im} d_{n-m} \xrightarrow{\tau} H^{r+1}(Y,A)/\operatorname{Im} d_{n-r} \xrightarrow{\gamma} H^{r+1}(X,A)/\operatorname{Im} d_{m-r} \to 0 \quad (14)$$

where the homomorphisms α, ν, γ are induced by the corresponding homomorphisms $\hat{\alpha}$, $\hat{\nu}$, $\hat{\gamma}$ of the sequence (12); if $a \in \ker d_{n-r}$ then $\beta(a) = \hat{\gamma}^{-1} \cdot d_{n-m} \cdot \hat{\alpha}^{-1}(a) \in \ker d_{m-r}$; for the representative $\bar{c} \in H^{m+1}(Y,X)$ of the class $c \in H^{m+1}(Y,X)/\operatorname{Im} d_{n-m}$ the image $\tau(c) = [\hat{\gamma}^{-1} \cdot d_{m-r} \cdot \hat{\nu}^{-1}(\bar{c})]$, where the square brackets denote the transition to the factor-class in $H^{r+1}(X,A)/\operatorname{Im} d_{m-r}$.

2) if $r<n<m$, then it is necessary that $d_{m-n+2} = 0$ and $d_{n-r} = 0$. Then the standard exact sequence of the differential d_{m-r}:

$$0 \to \ker d_{m-r} \to H^{m+1}(X,A) \xrightarrow{d_{m-r}} H^{r+1}(X,A) \to H^{r+1}(X,A)/\operatorname{Im} d_{m-r} \to 0 \quad (15)$$

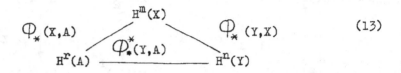

is the analogue of the exact sequence (14).

3) if $m<r<n$, then it is necessary that $d_{n-r}=0$ and $d_{r-m+2}=0$ and the exact sequence of the differential d_{n-m}:

$$0 \to \mathrm{Ker}\, d_{n-m} \to H^n(Y,X) \to H^{m+1}(Y,X) \to \frac{H^{m+1}(Y,X)}{\mathrm{Im}\, d_{n-m}} \to 0 \qquad (16)$$

is the analogue of (15).

4) if $m<n<r$, then the differentials (and the indices $I_{(X,A)}$, $I_{(Y,A)}$, $I_{(Y,X)}$) should not be, generally speaking, equal to zero and the analogue of (14) is the exact sequence:

$$0 \to \mathrm{Ker}\, d_{r-n+2} \to \mathrm{Ker} d_{r-m+2} \to \mathrm{Ker} d_{n-m} \to \frac{H^n(Y,A)}{\mathrm{Im} d_{r-n+2}} \to \frac{H^m(X,A)}{\mathrm{Im} d_{r-m+2}} \to$$

$$\to \frac{H^{m+1}(Y,X)}{\mathrm{Im} d_{n-m}} \to 0, \qquad (17)$$

the homomorphisms of which are analogous to the corresponding homomorphisms of (14).

5) if $n<r<m$, then the differentials (and the indices) should not equal zero and the sequence:

$$0 \to \mathrm{Ker} d_{m-r} \to \mathrm{Ker} d_{m-n+2} \to \mathrm{Ker} d_{r-n+2} \to \frac{H^{r+1}(X,A)}{\mathrm{Im} d_{m-r}} \to \frac{H^n(Y,X)}{\mathrm{Im} d_{m-n+2}} \to$$

$$\to \frac{H^n(Y,A)}{\mathrm{Im} d_{r-n+2}} \to 0 \qquad (18)$$

is exact and analogous to (17) and (14).

6) if, finally, $n<m<r$, then it is necessary that $d_{r-m+2}=0$ and $d_{m-n+2}=0$ and the exact sequence of the differential d_{r-n+2}:

$$0 \to \mathrm{Ker} d_{r-n+2} \to H^{r+1}(Y,A) \to H^n(Y,A) \to \frac{H^n(Y,A)}{\mathrm{Im}\, d_{r-n+2}} \to 0$$

is the analogue of (16). Diagram (13) is commutative in cases 1,4,5. In cases 2,3,6 it is trivially commutative only if all the three differentials (and the spectral indices) equal zero.

Note that in the formulation of theorem 6 the cases of even (1,4, 5) and odd(2,3, 6) rearrangements of dimensions r,m,n are pointed out in respect to the initial relation $r<m<n$.

The proof of theorem 6 follows from the analysis of spectral sequences of Borel fibrations for the pairs (X,A), (Y,A), (Y,X), their convergence to the cohomologies of fibration spaces, the exact sequence of the triplet of the total spaces of the fibrations and the the differentials cochain structure considered in [17] .

Note also that in the inequalities giving the relations between the dimensions r,m,n one can also admit the signs of equality with the following necessary condition which is the consequence of the definition of the equivariance indices: if any of the dimensions r,m,n coincide, e.g. $m = n$, then under the non-zero degree of the corresponding mapping this case should be included in the situation when $m < n$. If the degree of mapping of the corresponding spaces is zero then this case is included in the situation when $m > n$. The same is true for all other pairs of dimensions. Now let us give one of the topological consequences of theorem 6 (see, e.g. [18] where it is derived from Smith theory):

<u>Lemma 1</u>. Let $f: X_1 \longrightarrow X_2$ be an equivariant mapping of the Z_k-cohomological spheres of the dimensions m and n respectively. Let the actions T_1, T_2 of the group Z_k be semi-free on X_1, X_2. The dimensions of the sets F_1, F_2 of the fixed points are the same and equal to r and the degree $\deg(f|F_1) \neq 0$. Then it is necessary that $m \leq n$.

The proof of this lemma easily follows from Smith theory. Here we shall show how it follows from theorem 6. Indeed, let us assume that $m > n$. Then we have the following diagram:

$$\begin{array}{ccccc} & H^m(X_1) & \xrightarrow{\ \varphi_{m-n+2}\ } & H^n(X_2) & \\ \varphi_{m-r} & \downarrow & \overset{\gamma}{\swarrow} & \downarrow & \varphi_{n-r} \\ & H^r(F_1) & \xleftarrow{\ (f/F_1)^*\ } & H^r(F_2) & \end{array}$$

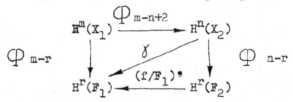

where the homomorphisms φ_{m-r}, φ_{n-r} define the equivariance indices $I(j_1)$, $I(j_2)$ of the inclusions $j_1: F_1 \longrightarrow X_1$, $j_2 : F_2 \longrightarrow X_2$ respectively. The homomorphisms φ_{m-n+2} and $(f|F_1)^*$ define the equivariance indices $I(f)$, $I(f|F_1) = \deg(f/F_1)$ respectively. The homomorphism γ define the equivariance index $I(f \cdot j_1)$ of the superposition $f \cdot j_1 = j_2 \cdot (f|F_1)$. Since the actions T_1, T_2 are semi-free, the indices $I(j_1)$ and $I(j_2)$ are the generators in Zk. Besides, $I(f|F_1) \neq 0$ according to the condition. The relation between the dimensions in the right-hand lower triangle of diagram (19) satisfies condition 1) of theorem 6 and consequently according to the statement of the theorem $I(f \cdot j_1) \neq 0$ in Z_k. But on the other hand, the relation between the dimensions of the spaces in the left-hand upper triangle of diagram (19) satisfies condition 2) of theorem 6 and consequently $I(f) = I(f \cdot j_1) = 0$ in Z_k. The obtained contradiction proves the lemma.

If in the conditions of lemma 1 $\deg(f|F_1) = 0$, then it is known ([19]), the case $m>n$ is quite possible. Indeed, in this case the right-hand triangle of diagram (19) satisfies, according to the relations between the dimensions , condition 3) of theorem 6 and consequently $I(f \cdot j_1) = I(f|F_1) = 0$. The contradiction disappears. Concrete examples of such equivariant mappings in the case of the action of the group Z_k can be found in the same paper [19].

Let us consider now a more general situation when the group Z_k acts non-semifreely on Z_k-cohomological spheres X_1, X_2. Let S_1, S_2 be sets of stationary points (i.e. the points with non-trivial isotropy groups) of the actions T_1, T_2 respectively. Let $f : X_1 \to X_2$ be an equivariant mapping. For the study of the equivariance index in this case we shall need

<u>Theorem 7</u> [20]. Let $k = k_1 \cdot k_2 \cdot \ldots \cdot K_l$, where $k_j = p_j^{\alpha_j}$ is the degree of a prime number p_j, be the canonical primary decomposition of the order k of the group Z_k. Then $S_i = \bigcup_{j=1}^{\ell} S_{ij}$ $(i = 1,2)$ where S_{ij} is a Z_{k_j}-cohomological m_j-sphere $(-1 \leq m_j \leq n)$ which is a set of stationary points of the action T^{k_j} of the subgroup Z_{k_j} of the group Z_k on X_i and $\deg(T_i|S_{ij}) = \deg T_i \pmod{k_j}$.

Using theorem 7 one can calculate the equivariance index $I(f)$ through the indices $I(f|S_{ij})$. Consider the commutative diagram of the equivariant mappings:

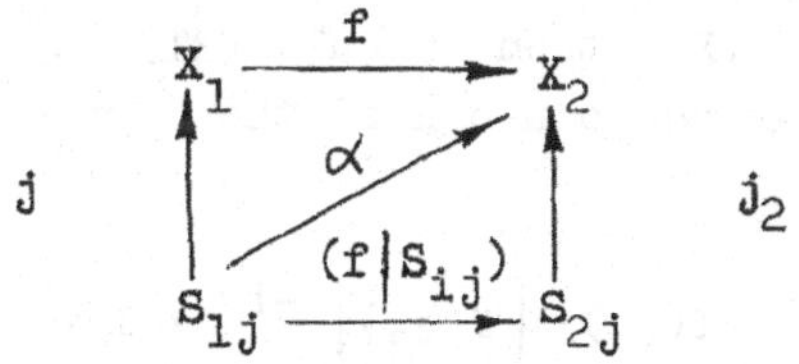

where $\alpha = f \cdot j_1 = j_2 \cdot (f|S_{1j})$; j_1, j_2 are inclusions, $\dim X_1 = m$, $\dim X_2 = n$, $\dim S_{1j} = r$, $\dim S_{2j} = q$ are the dimensions of cohomological spheres. Then from theorem 6 follows <u>lemma 2</u> (see also [15]). In the described conditions under all admissible relations of the dimensions r, q, m, n the following formula (in Z_{k_j}) takes place

$$I(f) = \left(I(j_1) \cdot \left[I(j_2)\right]^{-1} \right)^{\lambda} \cdot I(f|S_{ij}), \quad \lambda = \begin{cases} 1, & m<n \\ -1, & m>n \end{cases} \qquad (20)$$

with the indices $I(f)$ and $I(f|S_{ij})$ being simultaneously equal to zero in (20), if the dimensions r, q, m, n are in one of the following three relations: 1) $r<q<n<m$, 2) $q<r<m<n$, 3) $q<r<n<m$.

Thus, according to Lemma 2, the equivariance index I(f) can be non-trivial in (20) under the following relations of dimensions $r<q<m<n$, $r<m<q<n$, $q<n<r<m$.

Note that the case $S_{1j} = S_{2j} = \emptyset$ fits into the proposed scheme if an empty set is considered (as it is usually assumed) to be a cohomological (-1)-sphere and $I(f|S_{1j}) = 1$

By way of one more topological consequence from Theorem 6 and Lemma 2 let us formulate the following statement generalizing Lemma 1.

Theorem 8 (T.N.Fomenko). There is no equivariant mapping $f: X_1 \to X_2$ of Z_k-cohomological spheres ($\dim X_1 = m$, $\dim X_2 = n$) with the actions of the group Z_{p^α} such that the sets of fixed points F_1, F_2 are empty, the sets of stationary points are Z_{p^α}-cohomological spheres with $\dim S_1 = r$, $\dim S_2 = q$ and one of the following conditions is satisfied: a) $r<q<n<m$, b) $q<r<m<n$, c) $q<r<n<m$. The condition $F_1 = F_2 = \emptyset$ can be replaced by the conditions: $\dim F_1 = \dim F_2$ and $\deg(f|F_1) \neq \emptyset$ in Z_{p^α}.

Let us return to the actions of the group Z_k. With the use of comparisons (20) (mod k_j) for each $k_j = p_j^{\alpha_j}$ from the canonical primary decomposition of the order k of the group Z_k in [15] the following formula is obtained

$$I(f) = \sum_{j=1}^{l} \frac{k}{k_j} \cdot \left[\frac{K}{k_j}\right]^{-1}_{k_j} \cdot \left(I(j_1) \cdot \left[I(j_2)\right]^{-1}_{k_j} \right)^{\lambda} \cdot I(f|S_{1j}), \quad \lambda = \begin{cases} 1, & m<n \\ -1, & m>n \end{cases} \tag{21}$$

If the dimensions $\dim X_1 = m$ and $\dim X_2 = n$ coincide and the actions T_1, T_2 are semi-free on X_1, X_2, then formula (21) is reduced to the formula for the calculation modulo k of the degree of the equivariant mapping f:

$$\left[\deg f\right]_k = \sum_{j=1}^{l} \frac{k}{k_j} \cdot \left[\frac{k}{k_j}\right]^{-1}_{k_j} I(j_1) \cdot \left[I(j_2)\right]^{-1}_{k_j} \cdot I(f|S_{1j}) \tag{22}$$

This formula can also be found in [15]. Let us give

Corollary from formula (20) [15]. If two equivariant mappings $f, g: X_1 \to X_2$ are such that $I(f|S_{1j}) = I(g|S_{1j})$ ($j = 1,2\ldots,l$), then $I(f) = I(g)$.

In particular, the equality $I(f) = I(g)$ follows from the assumption that f and g are equivariantly homotopic on the set of all stationary points; in the earlier paper [18] the comparison $\deg f = \deg g \pmod k$ was established. This comparison contains, in particular, the result of M.A.Krasnosel'skiĭ who obtained it by geometrical methods (Theorem 4); note that the generalization of this result by algebraical methods has been previously obtained in paper [13]. The equality I(f) = I(g) which follows from (20) includes (as a consequence) the result of P.P. Zabreiko.

Theorem 9. [22]. Let on the Euclidean sphere S^n the actions T_1, T_2 of the groups Z_k and Z_q be given, with q being the divisor of k. Let S_k, S_q be the sets of stationary points of these actions respectively and let $f, g : S^n \longrightarrow S^n$ be equivariant mappings such that $f \cdot T_1 = T_2 f$ and $g \cdot T_1 = T_2 \cdot g$. Then if $f|S$ and $g|S$ are equivariantly homotopic, then deg f = deg g (mod k).

The proof of a variant of this theorem in [21] appeared to be incorrect; however, its validity followed from more general results [15, 18] obtained by Borel spectral sequence method; in [22] theorem 9 is easier obtained by further development of geometrical methods of M.A.Krasnosel'skiĭ (see also [23]).

A recent paper [24] continues geometrical investigations in this direction. Original constructions, using the principles of equivariant extension of mappings and a further development of methods of [11, 21, 22] allowed the authors to obtain the following generalization of theorem 9.

Theorem 10 (theorem 2 in [24]). Let on the Euclidean sphere S^n two actions of a group G be given. Let $f, g : S^n \longrightarrow S^n$ be equivariant mappings equivariantly homotopic on the set of stationary points of the actions of the group G. Then: the degrees deg f and deg g are comparable modulo k, if G is a finite group of the order $|G| = k$; deg f = deg g (as integers) if G is a compact not quite unconnected topological group.

The next section deals with the application of the cohomological methods described here to the study of the action of the finite (and compact not quite unconnected topological) groups on cohomological spheres.

3. Actions of finite groups *)

Let an arbitrary finite group G of the order $|G| = k$ act (by homeomorphisms) on Z_k- cohomological n-sphere X. Let $K = k_1 \cdot k_2 \cdot \ldots \cdot K_1$ where $k_j = p_j^{\alpha_j}$ be the canonical primary decomposition of the order k of the group G. Consider any Silov subgroup G_j of the group G corresponding to the divisor $k_j = p_j^{\alpha_j}$ of the order k. Then,

*) The results of this section were obtained by T.N.Fomenko.

as it is known [25], there exists a cyclic tower $\{e\} \subset G_{j1} \subset \ldots \subset G_j = G_j$ for G_j such that $G_{j1} = Z_{p_j}$ and $G_{ji+1}/G_{ji} = Z_{p_j}$ $(i = 1,2,\ldots, j)$. The action of the group G on X naturally induces the action of each subgroup G_{ji}. Let F_{ji} be a set of fixed points of this action. It is clear that $F_{j\alpha_j} \subset F_{j\alpha_{j-1}} \subset \ldots \subset F_{j1} \subset X$ where $F_{j\alpha_j}$ is a set of fixed points of the action of the whole subgroup G_j. Consider the greatest subset F_{j1} from this chain of inclusions. It is a set of fixed points of the action of the group $G_{j1} = Z_{p_j}$ and consequently according to Smith theory it is Z_{p_j}-cohomological m_j-sphere with $-1 \leq m_j \leq n$. We shall need the following

Lemma 3. Let the group Z_p act on Z-cohomological n-sphere X with a set F of fixed points being, according to Smith theorem, Z_p-cohomological m-sphere, $-1 \leq m \leq n$. Then F is also a Z_{p^α} - cohomological m-sphere.

Let us give the scheme of the proof. Assume first that $2 \leq m \leq n-2$. The exact coefficient sequence of cohomologies

$$\ldots \to H^q(F;\, Z_{p^t}) \to H^q(F;\, Z_{p^{t+1}}) \to H^q(F;\, Z_p) \to H^{q+1}(F;\, Z_{p^t}) \to \ldots \quad (23)$$

where $1 \leq t \leq \alpha - 1$ with the use of the induction in t gives the equalities: $H^q(F;Z_p) = 0$ at $q \neq 0$; $H^0(F;Z_p) = G^0$; $H^m(F;Z_p) = G^m$, where G^o_α, G^m_α are finite groups of the order p^α. The morphism of exact sequences (23) to their analogues for X induced by the inclusion $F \subset X$ and the five-lemma gives: $H^0(F;Z_p) = Z_p$.
Let us consider now the group $H^m(F;Z_p)$. It follows from the exact sequence of the pair (X,F) that $H^n(X,F;Z_p) = Z_p$, $H^{m+1}(X,F;Z_p) = G^m$, $H^q(X,F;\, Z_{p^\alpha}) = 0$ at $q \neq n,\ m+1$. Then in the spectral sequence of Borel fibration with the fibre (X,F) (and the base B_p) the member $E_2^{**}(X,F)$ has exactly two different nontrivial lines with numbers n and m+1. It is also known [5] that for any group of coefficients Λ there is an isomorphism $H^q(X,F;\, \Lambda) \cong H^q(X/Z_p,\, F;\Lambda)$. Since X is a finite dimensional space the groups $H^q(X/Z_p,\, F;\Lambda)$ and consequently $H^q(X,F;\Lambda)$ are trivial at $q > N$. Therefore the differential d_{n-m} in $E^{**}_{n-m}(X,F)$ must be an isomorphism because otherwise the groups $E_2^{o,n}$ (X,F) and $E_2^{n-m,\, m+1}(X,F)$ (or their subfactorgroups) are preserved till E_∞^{**}, which contradicts the above conditions. But d_{n-m} acts from $E^{o,n}_{n-m} = H^n(X,F;\, Z_{p^\alpha}) = Z_{p^\alpha}$ to the group $E^{n-m,m+1}_{n-m} = H^{m+1}(X,F;\, Z_{p^\alpha}) = G^m_\alpha$. Consequently, $H^{m+1}(X,F;\, Z_{p^\alpha}) = Z_{p^\alpha}$, whence also $H^m(F;\, Z_{p^\alpha}) = Z_{p^\alpha}$, i.e. F is a Z_{p^α} -cohomological m-sphere, QED. If $m < 2$ or $n-2 < m \leq n$, then using a double suspension with identical or free action on S^1 respectively we shall turn from the pair (X,F) to the

pair $(S^1 * X, S^1 * F)$ or $(S^1 * X, F)$ where the dimensions are in the previous relation. The transition from $H^{m+2}(S^1 * F; Z_p\alpha)$ to H^m $H^m(F;Z_p)$ with the use of the suspension homomorphism completes the proof.

Thus, the set F_{j1}, considered above, is a Z_{k_j}-cohomological m_j-sphere. Now let group G and its subgroups $Z_{p_j} = G_{j1}$, considered above, act on Z - cohomological n-spheres X and Y with the sets $F_j \subset X$ and S_j Y of fixed points of the actions $Z_{p_j} = G_{j1}$. Let $\dim F_j = m_j$, $\dim S_j = qj$ for Z_{k_j}-cohomological spheres F_j and S_j and let the mapping $f : X \to Y$ be equivariant (with respect to the actions G). The commutative diagram

$$\begin{array}{ccccc} & X & \xrightarrow{\quad f \quad} & Y & \\ j_1 & \uparrow & & \uparrow & j_2 \\ & F_j & \xrightarrow{\quad f|F_j \quad} & S_j & \end{array} \qquad (24)$$

where j_1, j_2 are inclusions, generates the diagram of homomorphisms of the cohomologies (with the coefficients in $Z_{p_j^{\alpha_j}}$) determining the equivariance indices of the mappings (24) for the action of the subgroup Z_{pj}. Using formula (20) from § 2 we obtain the equality in Z_{k_j}):

$$I(f) = (I(j_1)\ [J(j_2)]^{-1}_{k_j})^{\lambda} \cdot I(f|_{F_j}), \quad \lambda = \begin{cases} 1, & m_j < q_j \\ -1, & m_j > q_j \end{cases} \qquad (25)$$

which involves the following

<u>Lemma 4</u>. Let subgroup G_j of group G have one more cyclic tower besides the one considered above: $\{e\} \subset \bar{G}_{j1} \subset \dots \subset \bar{G}_{j\alpha_j} = G_j$ and $\bar{j}_1 : \bar{F}_j \subset X_1$, $\bar{j}_2 : \bar{S}_j \subset Y$ be the inclusions of the sets of fixed points of the actions of the subgroup $Z_{p_j} = G_{j1}$. Then for the restrictions $(f|F_j)$ and $(f|\bar{F}_j)$ of the equivariant mapping $f : X \to Y$ the following equality takes place: $I(f) = (I(j_1) \cdot [I(j_2)]^{-1}_{k_j})^{\lambda} \cdot I(f|F_j) = (I(\bar{j}_1) \cdot [I(\bar{j}_2)]^{-1}_{k_j})^{\lambda} \cdot I(f|\bar{F}_j)$. $I(f|F_j) = 0$ if the dimensions of Z_{k_j}-cohomological spheres F_j and S_j are different (the same apples to $I(f|\bar{F}_j)$) since Z_{p_j} acts trivially on the corresponding subsets and the spectral sequence is degenerated.

So far we have been considering only one Silov subgroup G_j. Let in group G be one more, different from G_j, Silov subgroup $\hat{G}_j$ of the same order $k_j = p_j^{\alpha_j}$. Then subgroups G_j and $\hat{G}_j$ as is known [25] are conjugated, i.e. $\hat{G}_j = h \cdot G_j \cdot h^{-1}$ for some element $h \in G$. That is why for every cyclic tower $\{e\} \subset G_{j1} \subset \ldots \subset G_{j\,j} = G_j$ of the subgroup G_j there is the conjugated tower : $\{e\} \subset \widehat{G}_{j1} \subset \ldots \subset \widehat{G}_{j\,j} = \widehat{G}_j$ in G_j. In particular, subgroups G_{j1} and $\widehat{G}_{j1}$ are conjugated and, consequently, the action of the element h realizes the homeomorphism between the sets of fixed points of these subgroups in X and Y : $h|F_j : F_j \to \widehat{F}_j$; $h|S_j : S_j \to \widehat{S}_j$, where $\widehat{F}_j \subset X$, $\widehat{S}_j \subset Y$ are the sets of fixed points of the actions of the subgroup $\widehat{G}_{j1}$ in X and Y respectively. From the commutative diagram

$$\begin{array}{ccc} H^{q_j}(\widehat{S}_j) & \xrightarrow{\varphi^*(\widehat{S}_j,\widehat{F}_j)} & H^{m_j}(\widehat{F}_j) \\ (h|S_j)^* \downarrow & & \downarrow (h|F_j)^* \\ H^{q_j}(S_j) & \xrightarrow{\varphi^*(S_j,F_j)} & H^{m_j}(F_j) \end{array}$$

follows the equality: $I(f|F_j) = I(f|F_j)$

Now one can formulate the following main statement on the degree of the mapping equivariant with respect to the actions of the finite group:

<u>Theorem 11</u>. Let X_1, X_2 be Z-cohomological n-spheres ($n>0$) with the actions of the finite group G of the order $k = |G|$. Let $f : X_1 \to X_2$ be an equivariant mapping. Choose for each cofactor $k_j = p_j^{\alpha_j}$ from a canonical primary decomposition of the order k of group G any Silov subgroup G_j and in each subgroup G_j choose any cyclic tower. Thus, for each k_j there is a subgroup $Z_{p_j} = G_{j1}$. Let $F_{ij} \subset X_i$ ($i = 1,2$; $j = 1,2,\ldots,l$) be sets of fixed points of the actions of G_{j1} : $\gamma_{j1} : F_{1j} \to X_1$, $\gamma_{j2} : F_{2j} \to X_2$ be inclusions. Then in Z_k takes place the following equality:

$$\left[\deg f\right]_k = \sum_j \left[\frac{k}{k_j}\right]^{-1}_{k_j} \cdot \frac{k}{k_j} \left[I(\gamma_{j1})\right]^{-1}_{k_j} \cdot I(\gamma_{j2}) \cdot \left[\deg(f|F_{1j})\right]_{k_j} \qquad (26)$$

where $I(\gamma_{j1})$, $I(\gamma_{j2})$ are the equivariance indices calculated over the ring of coefficients Z_{k_j}, $j \in \{ j \mid \dim F_{1j} = \dim F_{2j} \}$.

Indeed, it follows from Lemma 4 and the subsequent reasoning that the right-hand side of the equality(26)is independent of the choice of Silov subgroup corresponding to the given factor k_j and also of the choice of a cyclic tower in Silov subgroup. Solving the system of comparisons (25) for each $j = 1,2,\ldots, l$ we obtain in Z_k equality (26).

Let us give corollaries from Theorem 11.

Corollary 1. If G is a cyclic group of the order k, then formula (26) coincides with formula (22)(see also [15]).

Corollary 2. If in the conditions of Theorem 11 f, g : $X_1 \longrightarrow X_2$ are two equivariant mappings and $I(f|F_{1j}) = I(g|F_{1j})$ (in Z_{k_j}), j = 1, 2,..., l, then deg f = deg g (mod k).

Note that generally speaking the conditions of Corollary 2 are weaker than the requirement that f and g should be equivariantly homotopic on all sets F_{1j} and they are still weaker than the conditions of Theorem 10 on the equivariant homotopy of the mappings f and g on the set S of all stationary points of the action of group G. In this sense Theorem 11 is a strengthening of Theorem 10.

Note also that Lemma 4 and Theorem 11 actually show that the actions of different subgroups of the finite group G are closely associated with each other. A detailed analysis of spectral sequences of Borel fibration constructed for the actions of group G and its different subgroups will probably allow us to study this interdependence in more detail.

In conclusion consider the actions of a compact not quite unconnected topological group G on integer cohomological spheres X_1, X_2. As it is shown in [24] , such group necessarily contains a subgroup of any finite order $k > 0$. As a corollary from Theorem 11 we obtain the following statement:

Theorem 12. Let a compact not quite unconnected topological group G act (by homeomorphism) on Z-cohomological n-spheres X,Y. f, g : $X \longrightarrow Y$ be equivariant mappings, equivariantly homotopic on the set of all stationary points of the action of group G. Then their degrees deg f, deg g coincide in Z.

For the Euclidean spheres the statement of Theorem 12 evidently coincides with the second part of Theorem 2 from [24] .

REFERENCES

1. Smith P.A., Supplement "B" to the book by S.Lefshetz "Algebraic topology", Moscow, 1949 (in Russian).

2. Borel A., Seminar on transformation groups, Ann. of Math. Studies, N.46 (1960).

3. Chernavskiĭ A.V., Transformation groups . The 7th summer mathematical school. IM AN USSR, Kiev, 1970 (in Russian).

4. Izrailevich Ya.A., Index of semi-free periodic mapping. "Mat. zametki", 1973, v.13, N.1. (in Russian).

5. Bredon G., Introduction into the theory of compact transformation groups. Transl. from English, Moscow, 1980 (in Russian).

6. Izrailevich Ya.A., On Lefshetz number of mapping commuting with a periodic one. Tr.matem.f-ta, vyp.4(1973), VGU, Voronezh, 49-54 (in Russian).

7. Bredon G.E. Orientation in generalized manifolds and applications to the theory of transformation groups. Mich.M.J., 7(1960), 35-64.

8. Shchelokova T.N., Floyd-Smith theory and equivariant mappings of manifolds. Sb. rabot asp. po teor. funktsii i diff. uravneniyam, VGU, Voronezh, 1974 (in Russian).

9. Liao S., A theorem on periodic transformation of homology spheres, Ann. of Math., 56 (1952), 68-83.

10. Izrailevich Ya.A., Mukhamadiev E.M., To the theory of periodic mappings of spheres. The 7th summer mathematical school, IM AN USSR, Kiev, 1970 (in Russian).

11. Krasnosel'skiĭ M.A., On the calculation of rotation of vector field on n-dimensional sphere, DAN SSSR, 101, N.3 (1955), 401-404. (in Russian).

12. Fuks D.B., Fomenko A.T., Gutenmakher V.L. Homotopic topology. Izd-vo MGU, Moscow, 1971 (in Russian).

13. Borisovich Yu.G., Izrailevich Ya.A., Calculation of the degree of equivariant mapping by spectral sequences method. Tr.matem.f-ta, vyp.10 (1973), VGU, Voronezh, 1-12. (in Russian).

14. Izrailevich Ya.A . On the calculation of the degree of equivariant mapping by spectral sequences method. Tr.matem.f-ta, vyp.12(1974), VGU, Voronezh, 22-25. (in Russian).

15. Shchelokova T.N., To the problem of calculation of the degree of equivariant mapping. Sibirskii matem. zhurnal, v.XIX, N.2, 1978, 426-435 (in Russian).

16. Shchelokova T.N., To the theory of equivariant mappings of cohomological spheres. In: Methods of solving operator equations. Iz-vo VGU, Voronezh, 1978, 155-158 (in Russian).

17. Shchelokova T.N., To the theory of periodic mappings, Tr.NIIM, vyp.XV (1974), VGU, Voronezh, 75-80 (in Russian).

18. Borisovich Yu.G., Izrailevich Ya.A., Shchelokova T.N., To the method of A.Borel spectral sequence in the theory of equivariant mappings, Uspehi Mat.Nauk, 1977, v.32, No.I (in Russian).

19. Bredon G.E. Equivariant Homotopy. Proc. Conf. Transformation Groups, New Orleans, 1967. Berlin, N.Y. Springer-Verlag, 1968, 281-292.

20. Shchelokova T.N., On the calculation of the degree of mappings , equivariant with respect to the actions of the group Z_k. Tr.NIIM, vyp.XX, 1975, VGU, Voronezh, 51-56 (in Russian).

21. Zabreĭko P.P., To the theory of periodic vector fields. "Vestn. Yarosl. universiteta", vyp.2 (1973), YaGU, Yaroslavl', 24-30 (in Russian).

22. Zabreĭko P.P., To the homotopic theory of periodic vector fields. In: Geometrical methods in the problems of algebra and analysis. Yaroslavskii gos. universitet, 1980, 116-120 (in Russian).

23. Krasnosel'skiĭ M.A., Zabreiko P.P., Geometrical methods of nonlinear analysis, Moscow, "Nauka", 1975 (in Russian).

24. Balanov Z.I., Brodskiĭ S.D., Krasnosel'skiĭ's comparison prin-

ciple and the extension of equivariant mappings. In: Funkts.analiz. Theory of operators, Ul'yanovskiĭ gos.ped.institut, Ul'yanovsk, 1984, 18-31 (in Russian).

25. Leng S., Algebra. Transl. from English, "Mir", Moscow, 1968 (in Russian).

26. Floyd E.E. On periodic maps and Euler characteristics of associated spaces. - Trans. Amer. Math. Soc., 1952, vol.72, 138-147.

THEORY OF OPERATORS AND REAL ALGEBRAIC GEOMETRY

B.A.Dubrovin
Department of Mechanics and Mathematics
Moscow University
119899, Moscow, USSR

Modern applications of the classical algebraic geometry of Riemann surfaces and abelian varieties to the theory of operators and nonlinear equations associated with them are based on the works by S.P.Novikov, the author, V.B.Matveev and A.R.Its [1-5] and P.D.Lax [6,7], published in 1974-1975, in which the class of "finite-gap" periodic and quasiperiodic potentials of Schrödinger (Sturm - Liouville, Hill) operator was introduced and studied. This class is the basis for the construction of a broad class of the solutions to the Korteweg-de Vries (KdV) equation expressed via hyper-elliptic theta-function. (Some results of these investigations were also obtained by McKean and Van Moerbeke in 1975 [8]. As was strictly proved by V.A.Marchenko and I.V.Ostrovskii [9], the set of periodic finite-gap potentials is dense in the periodic function space with a given period). In the cited works the connection was established between the spectral theory of operators with periodic coeeficients and the algebraic geometry, the theory of finite-dimensional completely integrable Hamiltonian systems and the theory of nonlinear equations of KdV type. The generalization of this theory with respect to the spatially two-dimensional (2+1) systems was made by I.M.Krichever [10-12].

The above-mentioned works became the basis for the periodic analogue of the inverse scattering method in the theory of nonlinear equations also called a "finite-gap integration" method or the method of "algebraic-geometric integration". This method consists of three interdependent parts: the theory of nonlinear equations, the spectral theory of operators with periodic and quasiperiodic coefficients and the algebraic geometry of Riemann surfaces and abelian varieties. We illustrate the interaction of these parts using first-order matrix operators and nonlinear equations associated with them. Note that almost all nonlinear equations integrated by inverse scattering method (nonlinear Schrödinger equation, the sine-Gordon equation, equations of nonlinear interaction of wave packets and others) are associated with the spectral theory of matrix linear differential operators, which often are even not self-adjoint. Though it is rather easy to construct complex algebraic-geometric solutions to these equations [11, 13-16] , the attempts to isolate real smooth solutions from them came across great difficulties. The problems of real algebraic geometry, dealt with here, are completely undeveloped (the first serious progress in the solution of these problems as applied to nonlinear Schrödinger equation, two-dimensional Schrödinger operator in a magnetic field and sine-Gordon equation was made by I.V.Cherednik [17] , though the results obtained in his paper are far from being effective Also little was known about spectral properties of non-self-adjoint operators with periodic coefficients, i.e. about the properties of the appearing Riemann surfaces and the analytical properties of Bloch eigenfunctions, meromorphic on these surfaces. The structure of Riemann surface of the spectrum was not investigated even for the self-adjoint case , for which lots of concrete data were accumulated [18, 19] .

A full description of the properties of the spectrum of the first-order matrix finite-gap operators with different conditions of J-self-adjointness type and also the solution of the inverse problem for such operators, i.e. effective theta-functional formulas for their coefficients, and the solution of nonlinear equations associated with them were first obtained by the author [20,21,31] . In the present paper we shall consider $n \times n$-operator sheaves of the form

$$L(\lambda) = i\partial_x + \lambda A - U(x), \quad \text{where} \quad A = \mathrm{diag}\,(a_1,\dots,a_n) \qquad (1)$$

is a real diagonal matrix with various in pairs diagonal elements $a_1,\dots a_n$; λ is a spectral parameter. The condition of "J-self-adjointness of the operator sheaf $L(\lambda)$ has the form

$$U^* = JUJ, \qquad J = \mathrm{diag}(\pm 1,\dots,\pm 1) \qquad (2)$$

or

$$L^*(\bar{\lambda}) = JL(\lambda)J \qquad (2')$$

where the asterisk denotes a Hermitian adjointness.

It is necessary to say a few words about nonlinear equations associated with such sheaves at $n \geqslant 3$. Consider another operator

$$\tilde{L}(\lambda) = i\,\partial_t + \lambda B - \tilde{U}, \text{ where } B = \mathrm{diag}(b_1, \ldots b_n), \quad (3)$$

a real matrix B, also has different in pairs diagonal elements. The condition of the commutation of λ-sheaves

$$[\, L(\lambda), \tilde{L}(\lambda)\,] = 0 \quad (4)$$

is equivalent to a matrix nonlinear equation of the form

$$[A, V_t] - [B, V_x] = i\,\big[[A,V], [B,V]\big] \quad (5)$$

where we assume the diagonal elements of the matrices $U, \tilde{U}$ to be zero and $U = [A,V]$, $\tilde{U} = [B,V]$. The conditions of J-selfadjointness of the operator sheaves $L(\lambda), L(\tilde{\lambda})$ will be reduced to

$$V^{*} = -JVJ. \quad (6)$$

In this case, as was first observed by V.E.Zakharov and S.V.Manakov [22], system (5) describes different types of interaction of n wave packets in the medium with quadratic nonlinearity (the problem of n waves), where the type of interaction depends on the relations between diagonal elements of the matrices A and J.

Another important application of these equations, obtained by S.V. Manakov [23], is the integration of Euler-Arnold equations of the motion of a multidimensional rigid body [24]. These equations have the form

$$\Omega_t = [M, \Omega], \qquad M = I\Omega + \Omega I, \quad (7)$$

where I is an operator of a rigid body inertia, $I = \mathrm{diag}(I_1, \ldots, I_n)$. They follow from (5) if

$$A = I^2, \quad B = I, \quad [I,V] = i\Omega \ ;$$

the matrix Ω is real (and skew-symmetrical) and there is no dependence from x. In a more general sense, the equations

$$[A, V_t] = i\,\big[[A,V], [B,V]\big], \quad (8)$$

which follow from (5), when the solutions independent from x are found, under condition (2) coincide with the equations of geodesic on Lie group SU(p,q), where the numbers p and q are the numbers of plus unities and minus unities in the matrix $J(p + q = n)$ for a right-invariant metrics given on Lie algebra su(p,q) by a quadratic form

$$|\Omega|^2 = \sum_{k,l} \frac{b_k - b_l}{a_k - a_l} |\Omega_{Kl}|^2 , \tag{9}$$

$$\Omega = (\Omega_{kl}) = i\,[A,V] . \tag{10}$$

Imposing on the matrix V an additional condition

$$V^T = JVJ \quad \Leftrightarrow \quad U^T = -JUJ, \tag{11}$$

we obtain the equations of geodesics already on the group SO(p,q) for the metrics of type (9).

More complicated nonlinear equations will be obtained from (4), if the operator sheaf $\tilde{L}(\lambda)$ is taken to be polynomial or even rational. All such equations define isospectral deformations of the operator $L(\lambda)$. For a polynomial case

$$\tilde{L}(\lambda) = \partial_t + M(\lambda), \tag{12}$$

$$M(\lambda) = \lambda^N B + \lambda^{N-1} M_1 + \ldots + M_N, \quad B = \mathrm{diag}(b_1,\ldots,b_n) \tag{12'}$$

we have the following recurrent procedure for the calculation of the coefficients M_k. Let us consider for every diagonal matrix B a formal series of the form

$$m^B(x,\lambda) = B + \frac{m_1^B(x)}{\lambda} + \frac{m_2^B(x)}{\lambda^2} + \ldots , \tag{13}$$

satisfying the equation

$$[L(\lambda), m^B(x,\lambda)] = 0 . \tag{14}$$

<u>Lemma 1</u> ([16] ; see also [25]). The coefficients $m_k^B(x)$ of series (13) are defined uniquely from (14) in the form of polynomials from $V(x), V'(x),\ldots,V^{(k-1)}(x)$ if at $V = 0$ they turn zero.

It is easy to check [16] that for sheaf (12),(12') from the condition of commutation (14) it follows that

$$M(\lambda) = \sum_{k=0}^{N} M_k^{B_k}(x, \lambda) \qquad (15)$$

for some diagonal matrices $B_o, \ldots, B_N = B$, where the polynomials $M_k^{B_k}(x, \lambda)$ have the form

$$M_k^{B_k}(x, \lambda) = \left[\lambda^k m^{B_k}(x, \lambda)\right] \quad (\mathrm{mod} \quad \lambda^{-1}) . \qquad (16)$$

The equation of commutation (4) is written in the form

$$iU_t = \sum_{k=0}^{N} \left[A, m^{B_k}_{k+1}\right] . \qquad (17)$$

The dynamic systems of the form (17) at all N and the collections $B_o, \ldots, B_n$ commute with each other ([16] ; see also [25]). There are many important examples among such systems. For example, at n = 2 and N = 2 (17) transforms into a nonlinear Schrödinger equation [22] with the **attraction** at J = 1 and **repulsion** at $J = \begin{pmatrix} 1 & 0 \\ 0 & -1 \end{pmatrix}$.

<u>Remark 1</u> [25] . The coefficients of the formal series

$$Sp(Am^B(x,\lambda)) = SpAB + \frac{I_1}{\lambda^2} + \ldots + \frac{I_N}{\lambda^{N+1}} + \ldots \qquad (18)$$

at any matrices A,B are integrals of all equations (17).

<u>Remark 2</u>. If the condition of J-self-adjointness (2) is fulfilled, the matrix polynomials $M(\lambda)$ of the form (15) satisfy the relations

$$\left[M_N^B(x, \bar{\lambda})\right]^* = JM_N^B(x, \lambda) J. \qquad (19)$$

Let us come back to the investigation of the spectrum of J-self-adjoint operator sheaf $L(\lambda)$ of the form (1). Let the potential U(x) of this sheaf be periodic with the period $T, U(x + T) = U(x)$. Let $Y = Y(x, x_o, \lambda)$ be a fundamental matrix of the solutions of the equation

$$L(\lambda) Y = 0, \qquad (20)$$

$$Y \Big|_{x = x_o} = 1 . \qquad (21)$$

On the space of the solutions to equation (20) acts the operator of monodromy

$$\hat{T}Y(x) = Y(x+T) . \tag{22}$$

The matrix $\hat{T}(x_0, \lambda)$ of the operator of monodromy T, with respect to the fundamental systems of solutions, consisting of the columns of the matrix $Y(x,x_0,\lambda)$ has the form

$$\hat{T}(x_0, \lambda) = Y(x_0 + T, x_0, \lambda). \tag{23}$$

Its matrix elements are therefore integer functions from the spectral parameter λ. The matrix T is unimodular if SpA = 0. Its eigenvalues do not depend on x_0.

Lemma 2. For the matrix of monodromy (23) the following relation of unitarity is satisfied:

$$\hat{T}^*(\bar{\lambda})J\hat{T}(\lambda) = J \tag{24}$$

This follows from the fact that the operator of monodromy (22) preserves the invariant scalar product

$$\vec{\psi}^*(x,\bar{\lambda})J\vec{\psi}(x,\lambda) = \text{const} \tag{25}$$

which exists on the space of the solutions to the equation $L(\lambda)\vec{\psi} = 0$.

Bloch eigenfunctions of the operator $L(\lambda)$ have the form

$$L(\lambda)\vec{\psi}(x,\lambda) = 0$$

$$\hat{T}\vec{\psi}(x,\lambda) = \vec{\psi}(x+T,\lambda) = \mu(\lambda)\vec{\psi}(x,\lambda) \tag{26}$$

where $\mu = \mu(\lambda)$ is any eigenvalue of the operator of monodromy ("multiplier" of the operator $L(\lambda)$). At arbitrary complex λ there is, generally speaking, n different eigenvalues $\mu_1(\lambda),\ldots,\mu_n(\lambda)$ and, respectively, n linearly independent Bloch functions $\vec{\psi}_1(x,\lambda)$, ..., $\vec{\psi}_n(x,\lambda)$. Naturally, Bloch eigenfunctions $\vec{\psi} = (\psi^1,\ldots,\psi^n)^T$ are defined to within normalization factors. Further the following normalization will be used:

$$(\psi^1 + \ldots + \psi^n)_{x=x_0} = 1. \tag{27}$$

Thus, the normalized Bloch eigenfunction will be denoted by $\vec{\psi}(x,x_0,\lambda)$. It appears that n values $\vec{\psi}_1(x,x_0,\lambda),\dots,\vec{\psi}_n(x,x_0,\lambda)$ of Bloch vector-function are the branches of single-valued function meromorphic on some **n-sheeted** Riemann surface. Namely, holds a simple

Lemma 3. The Bloch eigenfunction $\vec{\psi}(x,x_0,\lambda)$ of the operator $L(\lambda)$ with periodic coefficients, normalized by condition (27), is extended with respect to λ to the function $\vec{\psi}(x,x_0,P)$, meromorphic on Riemann surface Γ of the form

$$F(\lambda,\mu) = \det(\mu\cdot 1 - \hat{T}(x_0,\lambda)) = 0. \tag{28}$$

(Here $P=(\lambda,\mu)$ is a point of the Riemann surface Γ).

Proof. n - valued analytical function $\mu=\mu(\lambda)$, given by equation (28), by the definition is single-valued on Γ. Find $\vec{\psi}$ in the form

$$\vec{\psi}(x,x_0,\lambda) = Y(x,x_0,\lambda)\vec{\xi} \quad , \tag{29}$$

where $\vec{\xi}=(\xi^1,\dots,\xi^n)$ is a vector independent from x. From (26), (27) we have

$$\hat{T}(x_0,\lambda)\vec{\xi} = \mu(\lambda)\vec{\xi},$$

$$\xi^1+\dots+\xi^n = 1. \tag{30}$$

From (30) the coordinates of the vector $\vec{\xi}$, by the rules of linear algebra, take the form of rational combinations of matrix elements of the matrix $\hat{T}(x_0,\lambda)-\mu(\lambda)\cdot 1$, i.e. the vector $\vec{\xi}$ is a single-valued meromorphic function on Riemann surface (28) ($\lambda<\infty$). From (29) we obtain the statement of the lemma, since the matrix elements of the matrix Y are integral functions from λ.

From J-selfadjointness of the operator sheaf $L(\lambda)$ follows the "reality" of the Riemann surface Γ, namely, holds

Lemma 4. Riemann surface Γ of the form (28) allows antiholomorphic involution ("anti-involution") τ of the form

$$\tau(\lambda,\mu) = (\bar{\lambda},\bar{\mu}^{-1}). \tag{31}$$

The proof is evident from (24).

Let us call the constructed **n-sheeted** Riemann surface Γ the spectrum of the operator $L(\lambda)$ with periodic coefficients. This Riemann surface, generally speaking, does not allow compactification (it has an infinite genus). Let us discuss the interpretation of some characteristics of the surface Γ in the language of the problem on eigenvalues for the operator $L(\lambda)$ in $L_2(-\infty, \infty)$.

Lemma 5. This spectral problem is self-adjoint if the condition of positivity

$$JA > 0 \tag{32}$$

is satisfied.

Proof. Multiplying operator $L(\lambda)$ by A^{-1} we obtain $A^{-1}L(\lambda) = L_0 + \lambda$, where $L_0 = iA^{-1}\partial_x - A^{-1}U$. For the operator L_0 from the condition of J-self-adjointness (2) we obtain

$$L_0^* = G^{-1}L_0G, \quad G = JA^{-1}$$

i.e. the condition of self-adjointness with respect to scalar product of the form

$$\|\vec{\Psi}\|^2 = \int_0^T \vec{\Psi}^* G \vec{\Psi}\, dx. \tag{33}$$

This completes the proof of the lemma.

The allowed bands of the spectrum (Lyapunov stability bands) on the Riemann surface Γ are defined by the condition

$$|\mu(\lambda)| = 1 \iff \operatorname{Im} p(\lambda) = 0. \tag{34}$$

(Here we introduced the value $p(\lambda) = (iT)^{-1}\log\mu(\lambda) + \frac{2\pi i m}{T}$, called a quasimomentum. It is easy to show [16], that a multivalued function $p(\lambda)$ is an Abel integral on the Riemann surface Γ.) The Bloch function will be bounded along the entire axis x. Then takes place a simple but important

Lemma 6. Real ovals of the Riemann surface Γ of the form (28) with anti-involution (31) are the allowed bands of the spectrum of the operator $L(\lambda)$ in $L_2(-\infty, \infty)$. If the condition of positivity (32) is satisfied, then there are no other allowed bands.

Proof. The point (λ, μ) lies on a real oval if $\tau(\lambda, \mu) = (\bar{\lambda}, \bar{\mu}^{-1}) = (\lambda, \mu)$. Therefore $|\mu(\lambda)| = 1$. Then, if the condition of positivity is satisfied, the problem

$$L(\lambda)\vec{\Psi} = 0, \quad \vec{\Psi}(T, x_0, \lambda) = \mu\vec{\Psi}(0, x_0, \lambda)$$

at $|\mu| = 1$ is self-adjoint (with respect to the scalar product (33)). Therefore the corresponding eigenvalues $\lambda = \lambda(\mu)$ are real, i.e. the allowed bands can be present only at real values of λ. Therefore the point (λ, μ) lies on a real oval. This completes the proof of the lemma.

Remark 3. At $n = 2$ and $SpA = 0$ the **branch points** of the Riemann surface Γ coincide with the points of the spectrum of a periodic or antiperiodic problem. At $n > 2$ there is no such spectral interpretation of the **branch points**

Definition 1. The operator $L(\lambda)$ with periodic coefficients is called a finite-gap operator if its spectrum, the Riemann surface Γ, has a finite genus.

Let operator $L(\lambda)$ be finite-gap. Denote by g the genus of its Riemann surface Γ. The surface Γ allows an evident compactification by n of infinite points $P_1, \dots, P_n$, where at $P \to P_k$ we have

$$\lambda \to \infty, \quad \mu \sim e^{ia_k \lambda T}.$$

Starting from this moment we shall consider the Riemann surface Γ to be compact.

Let operator $L(\lambda)$ have no point spectrum (e.g. condition (32) is satisfied). In this case it is possible to describe accurately the analytical properties of the Bloch eigenfunctions on the Riemann surface

Lemma 7. The Bloch eigenfunction $\vec{\psi}(x, x_0, P)$, meromorphic on $\Gamma \setminus (P_1 \cup \dots \cup P_n)$ has $g + n - 1$ poles $Q_1, \dots, Q_{g+n-1}$; at $P \to P_k$ it has an exponential **essential singularity** of the form

$$\psi^j(x, x_0, P) = e^{ia_k \lambda (x-x_0)}(\delta^j_k + O(\lambda^{-1})) \qquad (35)$$

For the proof see [20].

Remark 4. The Bloch eigenfunctions of finite-gap operators, having a point spectrum, are meromorphic on the so-called singular Riemann surfaces. In this case it is convenient to describe their analytical properties on some (non-singular) Riemann surface Γ_0 of the genus $g_0 < g$, from which Γ is obtained by "gluing". For example, a singular surface Γ with the simplest double point is a surface Γ_0 with the glued points Q' and Q'', $\lambda(Q') = \lambda(Q'')$. Then the functions on Γ can be presented as the functions on Γ_0, having equal values in points Q' and Q''. Similarly more complex singularities are interpreted. Below we shall consider only the case of the absence of a point spectrum (Sturm-Liouville finite-gap operators with a point spectrum were first studied in [26,27]).

Taking into account the analytical properties of the Bloch eigenfunctions of finite-gap operators with periodic coefficients, described in Lemma 7, let us give a more general definition of a finite-gap operator, now not associated with the condition of periodicity and even smoothness and J-self-adjointness of a potential.

Definition 2. The operator $L(\lambda)$ is called a finite-gap operator if there is a Riemann surface Γ of the genus $g < \infty$ with a meromorphic function $\lambda = \lambda(P)$ of degree n and a family of meromorphic on $\Gamma \setminus \lambda^{-1}(\infty)$ vector-functions $\vec{\Psi}(x,P)$ depending on x and having on Γ the analytical properties described in Lemma 7 with

$$L(\lambda(P)) \vec{\Psi}(x,P) = 0. \tag{36}$$

The surface Γ will be called, as above, the spectrum of a finite-gap operator and a meromorphic function $\lambda(P)$ will be called a spectral parameter: the points $Q_1, \ldots, Q_{g+n-1}$ will be called the points of an auxillary spectrum. The whole collection $(\Gamma, \lambda, (Q_i))$ will be called spectral data of the generic finite-gap operator $L(\lambda)$. It appears [11,16] that by giving arbitrarily a Riemann surface Γ, a spectral parameter $\lambda = \lambda(P)$ and points of an extra spectrum, it is possible to solve effectively the inverse problem, i.e. to restore uniquely a finite-gap operator $L(\lambda)$ with exactly such spectral data. Under the fixed spectrum Γ and spectral parameter $\lambda(P)$ the collection of the corresponding finite-gap operators of the form (1), a complex isospectral class M_Γ, is isomorphic to the generalized Jacobi variety $J(\Gamma; P_1, \ldots, P_n)$ of the Riemann surface Γ with the identified infinite points $(P_1, \ldots, P_n) = \lambda^{-1}(\infty)$. The potential can be expressed in terms of theta-functions of Riemann surface Γ and the exponents. Explicit formulas can be found, for example, in [20,28] . One can see, in particular, that the potential $U(x)$ of a finite-gap operator $L(\lambda)$ is a combination of the exponents and complex quasiperiodic meromorphic functions of x.

The theory of finite-gap operators $L(\lambda)$, as was first shown for the case of the Sturm-Liouville operator by S.P.Novikov [1], is closely connected with nonlinear equations considered above.

Lemma 8. Any stationary solution $U(x)$ of the equations of the form (17) is a potential of a finite-gap operator $L(\lambda)$ (in the sense of Definition 2). In other words, the operator $L(\lambda)$ is a finite-gap operator if and only if there exists a matrix polynomial $M(\lambda)$ (necessarily having the form (15)) commuting with $L(\lambda)$,

$$[L(\lambda), M(\lambda)] = 0. \tag{37}$$

For the proof see [20]. We shall only note that a Riemann surface Γ, a spectrum of the operator $L(\lambda)$, in this case can be given in the form of an algebraic curve

$$R(\lambda, \nu) = \det[\nu \cdot 1 - M(\lambda)] = 0 \tag{38}$$

and the eigenfunction $\vec{\psi}(x,P)$ is an eigenfunction for the pair of commuting operators

$$L(\lambda)\vec{\psi}(x,P) = 0, \ M(\lambda)\vec{\psi}(x,P) = \nu\vec{\psi}(x,P), \ P = (\lambda, \nu) \in \Gamma \tag{39}$$

The coefficients of a characteristic polynomial $R(\lambda, \nu)$ of the matrix $M(\lambda)$ are integrals of equation (37). The fixation of a Riemann surface Γ of the form (38) is thus equivalent to the fixation of the level surface of integrals of an ordinary differential equation (37). That is why the corresponding complex invariant manifold of system (37) coincides with an isospectral class M_Γ of the finite-gap operator $L(\lambda)$.

Example 1. Let the degree of a matrix polynomial $M(\lambda)$ equal 1. In this case equation (37) will take the form of Euler-Arnold equations (5) (with the substitute $x \to t$). A Riemann surface Γ (38) in this case is a flat algebraic curve of degree n (probably, with singularities). The meromorphic function λ is a projection of a curve Γ from some point $P_0 \in CP^2$, where $P_0 \notin \Gamma$. Infinite points $P_1, \ldots, P_n$ of the curve Γ lie in CP^2 on some line l traversing P_0.

Now let us come back to the case of J-self-adjoint operator sheaf $L(\lambda)$. In this case to the collection of spectral data is added anti-involution τ on Γ such that $\tau^* \lambda = \bar{\lambda}$. The collection $(\Gamma, \lambda, \tau, (Q_i))$ will be called spectral data of J-self-adjoint finite-gap sheaf $L(\lambda)$. What are the conditions that spectral data of J-self-adjoint finite-gap sheaf $L(\lambda)$ with a smooth periodic or almost periodic potential should satisfy? The main difficulty in solving the inverse problem for matrix finite-gap J-self-adjoint operators is to obtain such conditions and for $n \geq 3$ the main difficulty is to obtain the conditions which the triplet (Γ, λ, τ) should satisfy.

Theorem 1. The spectral data (Γ, λ, τ) of a finite-gap J-self-adjoint operator $L(\lambda)$ with a smooth periodic or almost periodic potential, if the condition of positivity (32) is fulfilled, satisfy the following conditions:

1°. All infinite points $\lambda^{-1}(\infty)$ of the surface Γ are fixed with respect to the anti-involution τ.

2°. The Riemann surface Γ with anti-involution τ belongs to a separating type, i.e. under the removal of a complete collection of real ovals Γ_{Re} (fixed points of the anti-involution τ) it splits into two components $\Gamma \setminus \Gamma_{Re} = \Gamma^+ \cup \Gamma^-$, $\tau(\Gamma^+) = \Gamma^-$.

3°. Let us orient Γ_{Re} as a boundary of, say, Γ^+. Then the degree of the mapping $\lambda : \Gamma_{Re} \longrightarrow RP^1$ is equal to the signature of a Hermitian form with the matrix J (i.e. to the trace of this matrix).

Remark 5. The theorem is valid without the condition of positivity (32) if the connected component of a real isospectral class M_Γ is compact (under the condition of positivity (32) the compactness is always satisfied). Besides, the theorem is valid without the condition of positivity and other additional restrictions at $J = 1$ (cf. [32]).

Remark 6. The conditions of the theorem are also sufficient in the following sense. If these conditions are satisfied, then the collection of J-self-adjoint finite-gap operators $L(\lambda)$ with a smooth periodic or almost periodic potential given by the triplet (Γ, λ, τ) (a real isospectral class) is non-empty and it is a torus of the dimensionality $g + n - 1$. An explicit form of this torus, a real component of a generalized Jacobi variety $J(\Gamma; P_1, \dots, P_n)$, the corresponding conditions for the auxillary spectrum $Q_1, \dots, Q_{g+n-1}$ and also theta-functional formulas for smooth periodic or almost periodic potentials of the operator $L(\lambda)$ with the condition of J-self-adjointness (2) are obtained in [20] .

Now we shall give the idea for the proof of Theorem 1 for the operator $L(\lambda)$ with a periodic potential. Condition 1° evidently follows from asymptotics $\mu(\lambda) \sim \exp(i \lambda a_k T)$ at $P \longrightarrow P_k$ where $\lambda^{-1}(\infty) = P_1 \cup \dots \cup P_n$. Let us show that the surface belongs to the separating type (condition 2°). We saw above(Lemma 6) that under the condition of positivity (32) the allowed spectral bands, where an absolute value of the multiplier μ is equal to unity, coincide with a complete collection Γ_{Re} of real ovals of the surface Γ. Let us assume $\Gamma^- = \lambda^{-1}\{|\mu| > 1\}$, $\Gamma^+ = \lambda^{-1}\{|\mu| < 1$. We shall have $\Gamma \setminus \Gamma_{Re} = \Gamma^+ \cup \Gamma^-$, i.e. the surface Γ with the anti-involution τ belongs to a separating type. Now let us prove condition 3°. We shall orient Γ_{Re} as a boundary of the domain Γ^+. It means that if the point $P = (\lambda, \mu) \in \Gamma_{Re}$, i.e. $Im\lambda = 0$, $|\mu| = 1$ moves along

Γ_{Re} in the direction of a positive orientation, then the multiplier μ rotates counter-clockwise. Calculate the degree of the mapping $\lambda : \Gamma_{Re} \to RP^1$ in the neighbourhood of the infinite point $\lambda = \infty \in RP^1$. From the asymptotics $\mu_k \sim \exp(i \lambda a_k T)$ at $P \to P_k$ it follows that the degree of the mapping $\lambda : \Gamma_{Re} \to RP^1$ in this point P is equal to the sign of the number a_k, since with the increase of λ the multiplier $\mu_k(\lambda)$ will rotate counter-clockwise at $a_k > 0$ and clockwise - at $a_k < 0$. But the sign of a_k is equal to $\sigma_k = \pm 1$, where $J = \text{diag}(\sigma_1, \dots, \sigma_n)$ by virtue of the condition of positivity (32). By summing with respect to all points $P_1, \dots, P_n$ from $\lambda^{-1}(\infty)$ we obtain that the degree of the mapping $\lambda : \Gamma_{Re} \to RP^1$ is equal to $\sigma_1 + \dots + \sigma_n = \text{sign } J$, Q.E.D.

The orientation on real ovals Γ_{Re}, appearing by Theorem 1, for J-self-adjoint operator $L(\lambda)$ with smooth periodic coefficients and the condition of positivity (32) allows a fine description of the multipliers of the first and second genus for the operators with periodic coefficients in the language of the theory of M.G.Krein. Before giving the corresponding definitions applicable to the language of Riemann surfaces let us prove the subsidiary statement.

Lemma 9. Let the point $P = (\lambda, \mu)$ of the Riemann surface Γ lie on a real part Γ_{Re}(i.e. $\text{Im}\,\lambda = 0$, $|\mu| = 1$) and not be a **branch point**. Then for the Bloch eigenfunction $\vec{\psi}$ of the form (26) we have

$$\vec{\psi}^* J \vec{\psi} \neq 0. \tag{40}$$

Proof. At real λ the matrix of monodromy $\hat{T}(\lambda)$ is pseudounitary (it preserves the form (40)). Its proper subspace satisfying the given eigenvalue μ with $|\mu| = 1$ is one-dimensional and J - orthogonal with respect to another proper subspaces. That is why the restriction on this subspace of the form (40) is non-zero. The lemma is proven.

According to M.G.Krein [19] we shall call the multiplier μ a multiplier of the first genus (the second genus) if for the corresponding Bloch function the inequality $\vec{\psi}^* J \vec{\psi} > 0$ ($\vec{\psi}^* J \vec{\psi} < 0$) is satisfied. By virtue of Lemma 9 the genus of a multiplier is a function correctly defined on each real oval with the discarded **branch points.**

Statement. Let Γ be a spectrum of a finite-gap J-self-adjoint operator sheaf $L(\lambda)$ with a periodic potential and the condition(32).

Let the point $P = (\lambda, \mu) \in \Gamma_{Re}$. The degree of the mapping λ: $: \Gamma_{Re} \longrightarrow RP^1$ in the neighbourhood of the point P is equal to +1 if μ is a multiplier of the first genus and -1 if μ is a multiplier of the second genus.

This statement is in fact a restatement of the known theorem of M.G.Krein on the motion of the multipliers of the first and second genus of the operators with periodic coefficients [19] . By this theorem, with the increase of λ the multiplier of the first genus rotates along the unitary circumference counter-clockwise and the multiplier of the second genus - clockwise, which means the validity of the statement.

Let us consider two examples.

a) J = 1. In this case the degree of the mapping $\lambda: \Gamma_{Re} \longrightarrow RP^1$ equals the number of **sheets** n. That is why Γ_{Re} coincides with a full pre-image of a real axis λ on Γ . This was evident beforehand, however, since on a real axis λ the matrix of monodromy $\hat{T}(x_0, \lambda)$ is unitary and absolute values of all its **eigenvalues** are equal to 1.

b) J = diag(-1, 1,...,1) (one minus unity). In this case the allowed bands are arranged in the following manner: on (n-2) leaves there are the allowed bands from $\lambda = -\infty$ to $\lambda = \infty$, the rest bands are ovals each of which contains exactly two **branch points of the multiplicity two** whose images on a real axis λ do not overlap (they are divided by "lacunas"). All these ovals are situated at $|\lambda| < \infty$ and only one of them traverses infinity. All this easily follows from Theorem 1.

Thus, the problem of classification of J-self-adjoint operators with the condition of positivity (32) is equivalent to the problem of a real algebraic geometry with a topological classification of the triplets (Γ, λ, τ) satisfying the conditions of Theorem 1. A general topological classification of triplets (Γ, λ, τ) was obtained by S.M.Natanzon [29] . It shows that at $J \neq 1$ and a fixed genus g and the degree n the collection of the triplets (Γ, λ, τ) satisfying the conditions of Theorem 1 is unconnected. It would be interesting to find characteristics of finite-gap J-self-adjoint operators $L(\lambda)$ responsible for the corresponding topological invariants of the triplets (Γ, λ, τ).

A much more complicated problem arises when the attempts are made to classify J-self-adjoint finite-gap operators $L(\lambda)$ which are the solutions to equation (37) under the fixed degree N of a matrix polynomial $M(\lambda)$. Already in the simplest case N = 1, where equation (37) is reduced to Euler-Arnold equation (5), we come to the problem of classification of flat real complex-orientable curves Γ of the degree n with the point $P_0 \notin \Gamma$ lying on RP^2 and the line 1 on RP^2 going through P_0 and intersecting Γ in n different points P_1, ..., P_n. The signs $\sigma_i = \deg \lambda |_{P_i}$ of the projection $\Gamma \to RP^1$ from the point P_0 determine the diagonal elements of the matrix J. The triplets (Γ, P_0, 1) satisfying the enumerated properties will be called admissible. The condition of non-singularity of the curve means that the Hamiltonian system (5) has a full-dimensional compact torus. Thus, a continuous deformation of the admissible triplet (Γ, P_0, 1) with nonsingular curves Γ satisfies to continuous deformation of Euler-Arnold equations of the form (5), and the levels of their integrals at which there is no degeneration of the corresponding invariant tori. At J = 1 the classification of the admissible triplets (Γ, P_0, 1) is easily obtained: the curve Γ at n = 2k is a nest of k inserted ovals; at n = 2k + 1 a one-sided component is added; the point P_0 lies inside the most inner oval and the line 1 is arbitrary (going through P_0). The general problem of the classification of the pairs (Γ, 1) is solved at present only at $n \leq 5$ (G.M.Polotovskii; see [30]).

Remark 7. As it follows from Lemma 8, for any admissible triplet (Γ, P_0, 1) the equation of a flat algebraic curve Γ can be written in the form

$$R(\lambda, \nu) = \det(\nu \cdot 1 - \lambda A - U) = 0, \qquad (41)$$

where matrix A is real and diagonal and matrix U satisfies condition (2). Let us consider for the given curve Γ of the form(41) the collection M_Γ of matrices U with condition (2) having this curve as a spectrum. From the main theorem it follows that in this collection there is one compact component which is a torus of the dimensionality n(n - - 1)/2. Inversely, the admissibility of the triplet (Γ, P_0, 1) follows from the presence of such component. Note that a general number of components of the manifold M_Γ is equal to 2^k, where k is a number of ovals of the curve Γ not intersecting improper line $\lambda = \infty$. It would be interesting to find other topological characteristics of the manifold M_Γ which would sense a real isotopic type of the admissible triplet (Γ, P_0, 1).

Remark 8. The theory of Euler-Arnold equations of the form (5) for Lie algebra sl(n,R) leads to flat M-curves. To be more precise, let matrix V in (5) be purely imaginary. In this case equations (5) are Euler-Arnold equations on SL(n,R) satisfying the right-invariant metrics of type (9). Invariant manifolds of these systems are always non-compact, since the systems are invariant with respect to the transformations

$$V \longrightarrow \Lambda^{-1} V \Lambda \quad , \tag{42}$$

where Λ is a diagonal real matrix. It is possible, however, to show that if a factor of a real isospectral class M_{Γ}, with respect to transformations of the form (42), has a compact component , then a flat real curve Γ is an M-curve, the line l intersects it in n real points; in this way all such pairs (Γ, l) are obtained, where Γ is an M-curve.

In conclusion let us consider the properties of the spectrum of matrix finite-gap operator sheaves (1) with condition of J-self-adjointness (2) and a skew symmetry (11). The sheaf $L(\lambda)$ in this case also satisfies, apart from symmetry (2') the symmetry

$$L^{+}(-\lambda) = -JL(\lambda)J, \tag{43}$$

where by $L^{+}(\lambda)$ we denoted a formally adjoint operator sheaf, $L^{+}(\lambda) = -i\partial_x + \lambda A - U^T$. The potential U(x) is purely imaginary. In the periodic case $U(x+T) = U(x)$ the matrix of monodromy $\hat{T}(x_0, \lambda)$ (23) satisfies the condition of orthogonality $[\hat{T}(x_0, -\lambda)]^T = J\hat{T}^{-1}(x_0, \lambda)J$. Then on the Riemann surface Γ of the form (28) the holomorphic involution σ acts by the rule

$$\sigma(\lambda, \mu) = (-\lambda, \mu^{-1}). \tag{44}$$

Involution σ commutes with anti-involution τ. Infinite points $P_1, \ldots, P_n$ are fixed with respect to σ : $\sigma(P_i) = P_i$, $i = 1, \ldots, n$. At odd n on Γ there is at least one more fixed, with respect to σ, point ($\lambda = 0$, $\mu = 1$). (It is clear that finite fixed points of involution σ correspond to eigenvalues of the matrix $\hat{T}(x_0, 0) \in SO(p,q)$). By analogy with Theorem 1 is proven

Theorem 2. Spectral data (Γ, λ, τ, σ) of a finite-gap J-self-adjoint operator $L(\lambda)$ with purely imaginary smooth periodic or almost periodic potential satisfy the following conditions if the condition of positivity (32) is fulfilled:

1°. The properties of the triplet (Γ, λ, τ) are the same as in Theorem 1.

2°. The involution σ commutes with the anti-involution τ .

All infinite points $P_1, \ldots, P_n$ are fixed with respect to σ .

The proof is analogous to Theorem 1.

Conditions of Theorem 2 are also sufficient. The isospectral class of the operators $L(\lambda)$ with spectral data $(\Gamma, \lambda, \tau, \sigma)$ satisfying conditions of Theorem 2 is a covering over a real component of Prym manifold of the Riemann surface Γ with the involution σ (for details see [20]).

As was reported by S.M.Natanzon he obtained a topological classification of the triplets (Γ, τ, σ). As far as the problem of classification of flat real curves with the involution σ in a real algebraic geometry is concerned, it was not yet posed (and it is to such curves that the problem of the construction of Euler-Arnold equations (5) for Lie algebras SO(p,q) leads).

REFERENCES

1. Novikov S.P. Periodic problem of Korteweg-de Vries.1. // Funktz. analiz, 1974. V.8, No.3 (in Russian).

2. Dubrovin B.A., Novikov S.P. Periodic and conditionally periodic analogues of multisoliton solutions of Korteweg - de Vries equation // Zhurn. eksperim. i teoret. fiziki. 1974. V.67, No.12.(In Russian)

3. Dubrovin B.A. Periodic problem for Korteweg - de Vries equation in the class of finite-gap potentials // Funkts. analiz. 1975. V.9, No.3, (in Russian).

4. Its A.R., Matveev V.B. Hill operators with a finite number of lacunas and multisoliton solutions of Korteweg - de Vries equation // Teoret. i mat. fizika. 1975. V.23, No.1. (In Russian)

5. Dubrovin B.A., Matveev V.B., Novikov S.P. Nonlinear equations of a Korteweg - de Vries type, finite-gap linear operators and abelian varieties// Uspekhi mat. nauk. 1976. V.31, No.1. (In Russian)

6. Lax P.D. Periodic solutions of KdV equation // Lect. in Appl. Math. 1974. Vol.15.

7. Lax P.D. Periodic solutions of Korteweg - de Vries equation // Comm. Pure and Appl. Math. 1975. V.23.

8. McKean H.P., Moerbeke P. van. The spectrum of Hill's equation // Invent. Math. 1975. Vol.30.

9. Marchenko V.A. Sturm - Liouville operators and their applications. Kiev, 1977. (In Russian)

10.Krichever I.M. Algebraic-geometric construction of Zakharov-Shabat equations and their periodic solutions // DAN SSSR, 1976. V.227, No.2. (In Russian)

11. Krichever I.M. Algebraic curves and commuting matrix differential operators // Funkts. analiz.1976. V.10, No.2. (In Russian)

12. Krichever I.M. Integration of nonlinear equations by methods of algebraic geometry // Ibid. 1977.V.11, No.2. (In Russian)

13. Its A.R. Transformation of hyperelliptic integrals and integration of nonlinear differential equations // Vestn. LGU.1976, No.7.(In Russian)

14. Its A.R., Kotlyarov V.P. Explicit formulas for the solution of Schrödinger nonlinear equation // DAN USSR. Ser.A. 1976. No.11.(In Russian).

15. Kozel V.A., Kotlyarov V.P. Almost periodic solutions of sine-Gordon equation // Ibid. No.10. (In Russian)

16. Dubrovin B.A. Completely integrable Hamiltonian systems associated with matrix operators and abelian varieties // Funkts. analiz. 1977. V.11, No.4. (In Russian)

17. Cherednik I.V. On conditions of reality in "finite-gap integration" // DAN CCCR. 1980. V.252, No.5. (In Russian)

18. Yakubovich V.A., Starzhinskii V.M. Linear differential equations with periodic coefficients and their applications. Moscow, 1972.(In Russian)

19. Kreĭn M.G. Foundations of the theory of λ-zones of the stability of the canonical system of linear differential equations with periodic coefficients // Amer. Math. Soc. Translations, Ser.2, vol. 120, 1983.

20. Dubrovin B.A. Matrix finite-gap operators // Sovremennye problemy matematiki. Moscow, 1983. V.23. (In Russian)

21. Dubrovin B.A. Spectrum of matrix finite-gap operators // Uspekhi mat. nauk. 1984. V.39 No-12 (In Russian)

22. The theory of solitons :inverse scattering method / Pod red. S.P.Novikova.Moscow, 1980. (In Russian)

23. Manakov S.V. Remark on integration of Euler equation of dynamics of n-dimensional rigid body // Funkts.analiz. 1976. V.10, No.4.(In Russian)

24. Arnold V.I. Mathematical methods of classic mechanics. Moscow, 1974. (In Russian)

25. Gel'fand I.M., Dikii L.A. Resolvent and Hamiltonian systems. // Funkts. analiz. 1977.V.11, No.2. (In Russian)

26. Krichever I.M. Non-reflecting potentials with respect to finite-gap ones // Ibid. 1975. V.9, No.2 (In Russian)

27. Kuznetsov E.A., Mikhailov A.V. Stability of stationary waves in nonlinear media with a weak dispersion // Zhurn. eksperim. i teoret. fiziki. 1974. V.67, No.11. (In Russian)

28. Dubrovin B.A. Theta-functions and nonlinear equations // Uspekhi mat. nauk. 1981. V.36, No.2. (In Russian)

29.Natanzon S.M. Topology of two-dimensional coverings and meromorphic functions on real and complex algebraic curves // Izv. AN SSSR. Ser. mat. 1986. V.50.(In Russian)

30. Polotovskii G.M. Catalogue of M-disintegrating curves of the 6th order // DAN SSSR. 1977. V.236, No.3. (In Russian)

31. Dubrovin B.A. Nonlinear equations connected with matrix finite-gap operators // Nonlinear and turbulent processes in physics. N.Y., 1984. Vol.III.

32. Cherednik I.V. On regularity of "finite-gap" solutions of integrable matrix differential equations // DAN SSSR. 1982. V.266, No.3. (In Russian)

ON THE STRUCTURE OF THE SET OF SOLUTIONS FOR INCLUSIONS WITH MULTIVALUED OPERATORS

B.D. Gel'man

Department of Mathematics, Voronezh
State University, 394693, Voronezh, USSR

In 1959 M.A. Krasnoselsky and A.I. Perov proved the connectedness principle for single-valued continuous operators (see, for example [1]). This principle, which is in fact an operator generalization of Kneser's integral funnel connectedness theorem, has found wide applications in studying the structure of the set of solutions for different classes of equations. In 1942 N. Arronszajn [2] proved that the integral funnel of an ordinary differential equation is an acyclic set.

These topics were developed further in the study of different classes of integral and differential equations. For example, the connectedness of the set of solutions for certain classes of integral inclusions has been considered in [3,4,5] , and theorems on the acyclicity of the set of solutions for certain classes of differential inclusions have been proved in [6,7,8] .

In the present paper the principles of connectedness and acyclicity of the set of solutions are proved for abstract operator inclusions. Note that the connectedness principle, proved in this paper, is a natural development of the Krasnoselsky-Perov scheme, although, as was stated in [5] , this scheme is essentially invalid for inclusions with multivalued operators.

Furthermore, we consider here the dimension of the set of solutions for inclusions. The theorems proved in the paper generalize the results obtained in [9] .

Certain properties of the set of solutions for parameter-dependent inclusions are studied in the last section of the paper.

1. The basic definitions and theorems

Let Y be a topological space, and let C(Y) stand for the collection of all non-empty closed subsets of the space Y, K(Y) for the set of all non-empty compact subsets of Y, and $K_v(Y)$ for the set of all non-empty compact convex subsets of Y.

By a multivalued mapping (M-mapping) of a topological space X into Y we shall mean a correspondence which associates with each point $x \in X$ a closed subset $F(x) \subset Y$, called the image of point x, i.e. this correspondence is a one-to-one mapping $F:X \to C(Y)$. If F has compact (compact-convex) images, we introduce the notation

$$F:X \longrightarrow K(Y) \qquad (F:X \longrightarrow K_v(Y))$$

The set $\Gamma_X(F) \subset X \times Y$, where

$$\Gamma_X(F) = \{(x;y) | \ x \in X, y \in F(x)\}$$

is called the graph of the multivalued mapping $F:X \to C(Y)$.

Definition 1. An M-mapping $F:X \to C(Y)$ is called upper semicontinuous if for any open set $V \subset Y$ the set

$$F_+^{-I}(V) = \{x | \ x \in X, F(x) \subset V\}$$

is open in X.

If for any closed set $G \subset Y$ the set $F_+^{-1}(G)$ is closed, the M-mapping F is called lower semicontinuous.

Definition 2. An M-mapping $F:X \to K(Y)$ is called completely continuous if:

a) F is upper semicontinuous;

b) $\overline{F(X)} = \overline{\bigcup_{x \in X} F(x)}$ is a compact in Y.

Let (X, ρ_X) and (Y, ρ_Y) be metric spaces. The metric ρ in the product $X \times Y$ is defined by

$$\rho((x,y);(x',y'))=\max \{\rho_X(x;x');\ \rho_Y(y;y')\} .$$

Definition 3. Let $F:X \to C(Y)$ be an M-mapping. The multivalued mapping $F_\varepsilon : X \to C(Y)$, where $\varepsilon > 0$, is called a multivalued ε-approximation of the mapping F if

$$\rho_*(\Gamma_X(F_\varepsilon);\ \Gamma_X(F))=\sup_{z\in\Gamma_X(F_\varepsilon)} \rho(z;\Gamma_X(F))<\varepsilon$$

i.e. if the graph $\Gamma_X(F_\varepsilon)$ belongs to an ε-neighbourhood of the graph $\Gamma_X(F)$.

If F_ε is a single-valued continuous mapping, it is said to be a single-valued ε-approximation of the M-mapping F.

Definition 4. An M-mapping $F:X \to C(Y)$ is called proper if for any compact $G \subset Y$ the complete inverse image of F

$$F_-^{-1}(G)=\{x \mid x \in X, F(x) \cap G \neq \emptyset\}$$

is either a compact in X or an empty set.

Note that if X is a closed subset in Y and $F:X \to K(Y)$ is a completely continuous M-mapping, the M-mapping $\varphi : X \to K(Y)$, $\varphi(x) = x - F(x)$, is a proper M-mapping.

Let Y be a Banach space, $U \subset Y$ a bounded open subset, and $F:\overline{U} \to K_v(Y)$ a completely continuous M-mapping. A multivalued vector field, an MV-field, is said to be the M-mapping $\varphi = i - F:\overline{U} \to K_v(Y)$ defined by the condition: $\varphi(x) = x - F(x)$.

The collection of MV-fields $\varphi : \overline{U} \to K_v(Y)$ such that $\forall y \in \partial U$ and $\varphi(y) \not\ni \theta$ will be denoted as $\varphi \in \mathcal{C}(\overline{U}, \partial U)$.

<u>Definition 5</u>. MV-fields Φ_0 , $\Phi_1 \in \mathcal{C}$ $(\bar{U}, \partial U)$ are called homotopic ($\Phi_0 \sim \Phi_1$) if there exists a completely continuous family of MV-fields

$$\Psi : [0;1] \times \bar{U} \longrightarrow K_v(Y)$$

such that $\theta \bar{\in} \Psi([0;1] \times \partial U)$ and $\Psi(\cdot,0) = \Phi_0(\cdot), \Psi(\cdot,1) = \Phi_1(\cdot)$

For an MV-field $\Phi \in \mathcal{C}$ $(\bar{U}, \partial U)$ there is valid an integer characteristic, rotation,

$$\gamma(\Phi, \partial U)$$

which possesses the following properties (see [10,11]):

1. If $F(x) \equiv x_0$ for all $x \in \partial U$, then

$$\gamma(i-F;\partial U) = \begin{cases} 1, & x_0 \in U \\ 0, & x_0 \notin \bar{U}. \end{cases}$$

2. If Φ_0 , Φ_1 $\in \mathcal{C}$ $(\bar{U}, \partial U)$ and $\Phi_0 \sim \Phi_1$, then

$$\gamma(\Phi_0, \partial U) = \gamma(\Phi_1, \partial U).$$

3. Let $\{U_j\}_{j \in \mathcal{Y}}$ be a family of open non-intersecting subsets in U, let $U = \bigcup_{j \in \mathcal{Y}} U_j$, and let $\Phi \in \mathcal{C}$ $(\bar{U}_j, \partial U_j)$, $j \in \mathcal{Y}$. Then the rotations $\gamma(\Phi, \partial U_j)$ are non-zero for finitely many indices j, and the following relation holds true:

$$\gamma(\Phi, \partial U) = \sum_{j \in \mathcal{Y}} \gamma(\Phi, \partial U_j)$$

4. If $\Phi = i - F \in \mathcal{C}$ $(\bar{U}, \partial U)$ and $\gamma(\Phi, \partial U) \neq 0$, then there exists a point $x_o \in U$, called a fixed point, such that $x_o \in F(x_o)$. The main properties of multivalued mappings are considered at greater length in, e.g., [10,11] .

Let $F:\bar{U} \to K_v(Y)$ and $\Phi = i - F \in \mathcal{C}$ $(\bar{U}, \partial U)$. To study the set of fixed points of this mapping, Fix(F), we shall need the following statement.

Lemma 1 [9] . Let X be a metricizable compact space of dimension not greater than $n - 1$, i.e. $\dim X \leqslant n - 1$. Let Y be a Banach space and $T:X \to K_v(Y)$, a lower semicontinuous M-mapping satisfying the conditions:

1. $T(x) \ni \theta$ for any $x \in X$;
2. $\dim T(x) \geqslant n$ for any $x \in X$.

Then there exists a continuous mapping $f:X \to Y$, $f(x) \in T(x)$ for any $x \in X$, which does not vanish at any point $x \in X$.

Theorem 1. Let $F:U \to K_v(Y)$ be an upper semicontinuous M-mapping, $\Phi = i - F \in \mathcal{C}$ $(\bar{U}, \partial U)$. If:
1. $\gamma(\Phi, \partial U) \neq 0$; and
2. there exists a lower semicontinuous M-mapping $G:\bar{U} \to K_v(Y)$ such that $\dim G(x) \geqslant n$ and $G(x) \subset F(x)$ for any $x \in X$,
then the set of fixed points of the M-mapping F is not empty and its dimension is greater than or equal to n, i.e. $\mathrm{Fix}(F) \neq \emptyset$ and $\dim \mathrm{Fix}(F) \geqslant n$.

Proof. Owing to the properties of rotation of a multivalued vector field, the set $\mathrm{Fix}(F) \neq \emptyset$ and is a compact. We denote it by X. Let us assume the converse, i.e. that $\dim X \leqslant n - 1$. Consider the multivalued mapping $\Phi_1 = i - G:X \to K_v(Y)$. According to Lemma 1, this mapping has a non-vanishing selection f. Then, by virtue of Michail's theorem [12] , there exists the selection $\tilde{f}:\bar{U} \to Y$ of the M-mapping $i - G$ which coincides with f on the set X. Apparently, this selection is also a selection of the M-mapping Φ . It can easily be seen that $i-\tilde{f}$ is a completely continuous non-degenerate vector field on ∂U, and its rotation $\gamma(i-\tilde{f}, \partial U) = \gamma(\Phi, \partial U)$. Thus, according to the property of rotation, there exists a point x_o such that $\varphi(x_o) = 0$ and $\varphi(x_o) = x_o - \tilde{f}(x_o)$, which contradicts the assumption. The theorem is proved.

Let us consider the following operation over the M-mapping F. For each $x_0 \in \bar{U}$ we define the set $L(F)(x_0)$ according to the rule

$$L(F)(x_0) = \bigcap_{\varepsilon>0} \overline{\left(\bigcup_{\delta>0} \left(\bigcap_{x\in U_\delta(x_0)} F^{\varepsilon}(x)\right)\right)}$$

where $U_\delta(x_0)$ is a δ-neighbourhood of the point x_0, and $F^{\varepsilon}(x) = \bigcup_{y\in F(x)} U_\varepsilon(y)$ is an ε-inflation of the set F(x). Apparently, the set $L(F)(x_0)$ can be empty for certain points x_0.

Let us consider iterations of the operation L:

$$L^0(F)=F, L^n(F)=L(L^{n-1}(F)), \ n \geq 1.$$

We continue this process for each transfinite number of type 1, and for each transfinite number of type 2 we set

$$L^{\alpha}(F)(x) = \bigcap_{\beta<\alpha} L^{\beta}(F)(x).$$

The sequence $\{L^{\alpha}(F)\}$ is said to be stabilized at the step α_0 if

$$L^{\alpha_0}(F)(x) = L^{\alpha_0+1}(F)(x).$$

for any $x \in X$. The stabilized term of the sequence will be denoted by $\hat{L}(F)$.

<u>Theorem 2 [13]</u>. Once the sequence $\{L^{\alpha}(F)\}$ has been stabilized and the set $\hat{L}(F)(x) \neq \emptyset$ for any $x \in \bar{U}$, the M-mapping $\hat{L}(F):\bar{U} \to K_v(Y)$ is lower semicontinuous and satisfies the conditions:

1) $\hat{L}(F)(x) \subset F(x)$ for any $x \in \bar{U}$;
2) if $\Psi:\bar{U} \to K_v(Y)$ is a lower semicontinuous mapping and $\Psi(x) \subset F(x)$ for any $x \in \bar{U}$, then $\Psi(x) \subset \hat{L}(F)(x)$ for any $x \in \bar{U}$.

If the M-mapping F contains a continuous selector, the sequence

$\{L^{\alpha}(F)\}$ is stabilized and $\hat{L}(F)(x) \neq \emptyset$ for any $x \in \bar{U}$.

Thus, the M-mapping $\hat{L}(F)$ is a maximal lower semicontinuous mapping which selects the M-mapping F.

The following statement is a consequence of Theorems 1 and 2:

Corollary 1. Let $F:\bar{U} \to K_V(Y)$ be a completely continuous M-mapping and $\Phi = i - F \in \mathcal{C}(\bar{U}, \partial U)$. If:

1) $\gamma(\Phi, \partial U) \neq 0$ and
2) the sequence $\{L^{\alpha}(F)\}$ is stabilized, $\hat{L}(F)(x) \neq \emptyset$, and $\dim \hat{L}(F)(x) \geq n$ for any $x \in \bar{U}$,

then the set $Fix(F) \neq \emptyset$ and its dimension is equal to or greater than n.

2. Connectedness principle for the set of solutions to an inclusion with a multivalued operator

Let Y be a Banach space, U a bounded open set in Y, and $F:\bar{U} \to K_V(Y)$ a completely continuous M-mapping.

Theorem 3. Let $\gamma(i - F, \partial U) \neq 0$. If for any $\varepsilon > 0$ and for any point $x_1 \in Fix(F)$ there exists a completely continuous M-mapping $F_{\varepsilon, x_1} : \bar{U} \to K_V(Y)$ such that

1) F_{ε, x_1} is a multivalued ε-approximation of F;
2) the set $Fix(F_{\varepsilon, x_1})$ is either empty or belongs to an ε-neighbourhood of the point x_1,

then the set Fix(F) is connected.

Proof. That the set Fix(F) is non-empty follows from Property 4 of the rotation of a multivalued vector field. We now prove that this set is connected. Assume the converse. Then the set Fix(F) can be represented as a union of two non-empty non-intersecting closed sets N_0 and N_1, i.e. $Fix(F) = N_0 \cup N_1$. Let U_0 and U_1 denote non-intersecting neighbourhoods of these sets, which lie in U. According to Property 3 of the rotation of a multivalued vector field, we have

$$\gamma(i-F, \partial U) = \gamma(i-F, \partial U_0) + \gamma(i-F, \partial U_I)$$

Hence, one of the numbers $\gamma(i - F, \partial U_0)$ and $\gamma(i - F, \partial U_1)$ is not equal to zero. Let, for certainty, $\gamma(i - F, \partial U_0) \neq 0$. Consider an arbitrary point $x_1 \in N_1$. It is a simple matter to demonstrate that there exists an $\varepsilon_0 > 0$, such that $\min_{x \in \partial U_0} \rho(x, F(x)) \geq \varepsilon_0$.

Let $0 < \varepsilon < \varepsilon_0/3$. Consider an M-mapping F_{ε, x_1} which satisfies the conditions of the theorem. We can assume, without loss of generality, that $Fix(F_{\varepsilon, x_1}) \cap U_0 = \emptyset$, otherwise we could choose a smaller ε. However, it is easy to prove that the fields $i - F_0$ and $i - F_{\varepsilon, x_1}$ are pathwise homotopic on ∂U_0. Hence, $\gamma(i - F, \partial U_0) = \gamma(i - F_{\varepsilon, x_1}, \partial U_0) \neq 0$, and we come to contradiction, since the field $i - F_{\varepsilon, x_1}$ must not have singular points in U_0. On the other hand, the properties of the rotation do imply the existence of a singular point belonging to U_0. The contradiction proves the theorem.

The proof of this theorem is an extension of the Krasnoselsky-Perov connectedness principle to the multivalued case. Yet, even for a single-valued completely continuous mapping this theorem is more general than the well-known previous theorem.

Corollary 2. Let $f: \bar{U} \to Y$ be a completely continuous single-valued mapping such that $\gamma(i - f, \partial U) \neq 0$. If for any $\varepsilon > 0$ and for any point $x_1 \in Fix(f)$ there exists a completely continuous mapping $f_{\varepsilon, x_1}: \bar{U} \to Y$ such that $\| f(x) - f_{\varepsilon, x_1}(x) \| < \varepsilon$ for any $x \in \bar{U}$ and $Fix(f_{\varepsilon, x_1}) \in U_\varepsilon(x_1)$, where $U_\varepsilon(x_1) = \{x' \mid \| x_1 - x' \| < \varepsilon\}$, then $Fix(f)$ is a non-empty connected set.

The following theorem may also prove useful for studying the structure of the set of solutions to an M-mapping F.

Theorem 4. Let $F: \bar{U} \to K_v(Y)$ be a completely continuous M-mapping, and let there exist a sequence $\{F_n\}$ of completely continuous M-mappings $F_n: \bar{U} \to K_v(Y)$ which satisfy the following conditions:

1. $F_n(x) \supset F_{n+1}(x)$ for any $x \in \bar{U}$;
2. $\bigcap_{n=1}^{\infty} F_n(x) = F(x)$ for any $x \in \bar{U}$;
3. the set $Fix(F_n)$ is connected for any $n = 1, 2, \ldots$.

Then the set $Fix(F)$ is connected.

Proof. Complete continuity of the M-mappings F and F_n implies that

$Fix(F)$ and $Fix(F_n)$ are both compacts, and also that $Fix(F) \subset Fix(F_{n+1}) \subset Fix(F_n)$ for any $n = 1,2, \ldots$. Assume the converse, i.e. let the set $Fix(F)$ be non-connected. Then there exist in Y two open, non-intersecting sets, U_0 and U_1, such that $Fix(F) \subset (U_0 \cup U_1)$and $Fix(F) \cap U_i \neq \emptyset$, $i = 0,1$. Since the sets $Fix(F_n)$ are connected, there exists, for any n, a point $x_n \in Fix(F_n)$ such that $x_n \bar{\in} (U_0 \cup U_1)$. One may assume, without loss of generality, that $\{x_n\} \longrightarrow x_0$, and, as can easily be seen, $x_0 \in \bigcap_{n=1}^{\infty} Fix(F_n)$, i.e. $x_0 \in Fix(F)$. But this contradicts the condition $Fix(F) \subset (U_0 \cup U_1)$. Hence, the set $Fix(F)$ is connected.

As an application of these theorems, we now prove the connectedness of the set of fixed points for a certain class of M-mappings.

Let U be a bounded open domain in a Banach space Y and $F:\bar{U} \longrightarrow K_v(Y)$, a completely continuous M-mapping. Let also F be a multivalued Lipschitz mapping, i.e.

$$h(F(x_1);F(x_2)) \leqslant k \| x_1 - x_2 \| ,$$

where h is the Hausdorff metric (see, for example, [11]).

Lemma 2. Suppose F satisfies the above conditions, and the number $k < 1$. Then, if $Fix(F) \neq \emptyset$, for any point $x_0 \in Fix(F)$ and the number k_1, $k < k_1 < 1$, there exists a continuous mapping $f_{x_0}:\bar{U} \longrightarrow Y$ such that:

1. $f_{x_0}(x) \in F(x)$ for any $x \in \bar{U}$;
2. $\| x_0 - f_{x_0}(x)\| \leqslant k_1 \| x_0 - x\|$ for any $x \in \bar{U}$.

Proof. Let $x_0 \in Fix(F)$ and k_1 satisfy the inequality $k < k_1 < 1$. Consider the function $\alpha(x) = \rho(x_0, F(x))$. It can easily be seen that this function is continuous. Let $\varphi(x) = (1 + \beta)\alpha(x)$, where $0 < \beta \leqslant k_1 - k/k$. Introduce the notation $F_1(x) = F(x) \cap U_{\varphi(x)}(x_0)$. It can easily be seen that $F_1(x) \neq \emptyset$ for any $x \in \bar{U}$, and is a convex closed set. It is a simple matter to prove that $F_1(x)$ is a lower semicontinuous M-mapping (see [11]). Hence, by Michail's theorem [12], there exists for F_1 a continuous selection, i.e. a continuous mapping $f_{x_0}:\bar{U} \longrightarrow Y$ such that $f_{x_0}(x) \in F_1(x)$ for any $x \in \bar{U}$. Since $F_1(x) \subset F(x)$, this just means that Condition 1 is satisfied. We now demonstrate that Condition 2 is also satisfied, namely:

$$\rho(x_0;f_{x_0}(x)) < \varphi(x)=(I+\beta)\alpha(x)=(I+\beta)\rho(x_0;F(x))$$

Since

$$x_0 \in \mathrm{Fix}(F),\ \rho(x_0;F(x)) \leq h(F(x_0);F(x)) \leqslant k\rho(x_0;x)$$

we have

$$\rho(x_0;f_{x_0}(x)) \leqslant (I+\beta)k\rho(x_0,x) \leq k_I\rho(x_0;x).$$

The lemma is proved.

Theorem 5. Let $F:\overline{U} \longrightarrow K_V(Y)$ be a completely continuous M-mapping satisfying the conditions:

1. $h(F(x_1);F(x_2)) \leq \|x_1 - x_2\|$ for any $x_1,x_2 \in \overline{U}$;
2. $\gamma(i - F, \partial U) \neq 0$.

Then the set $\mathrm{Fix}(F)$ is connected and non-empty.

Proof. That $\mathrm{Fix}(F)$ is non-empty follows from the properties of the rotation of a multivalued vector field. Let $x_1 \in \mathrm{Fix}(F)$. Consider the M-mapping $F_\varepsilon(x) = (1 - \varepsilon)F(x) + \varepsilon x_1$, $0 < \varepsilon < 1$. Apparently, $x_1 \in \mathrm{Fix}(F_\varepsilon)$ and F_ε is a completely continuous M-mapping. We now demonstrate that F_ε is a contraction operator. We have

$$h(F_\varepsilon(x_I);F_\varepsilon(x_2))=h((I-\varepsilon)F(x_I);(I-\varepsilon)F(x_2)) \leqslant (I-\varepsilon)\|x_I-x_2\|$$

Hence, according to Lemma 2, there exists $f_{x_1}:\overline{U} \longrightarrow Y$, $f_{x_1}(x) \in F_\varepsilon(x)$ for any $x \in \overline{U}$, and x_1 is the only fixed point of this mapping. Then,

$$f_{x_I}(x) \in F_\varepsilon(x) \subset U_{\varepsilon_I}(F(x)) + U_{\varepsilon\|x_I\|}(\theta) \subset U_{\varepsilon_I+\varepsilon\|x_I\|}(F(x)),$$

where $\varepsilon_1 = \varepsilon(N + 1)$, $N = \max_{y \in F(\bar{U})} \|y\|$.

Thus,

$$f_{x_I}(x) \in U_{\varepsilon(N + \|x_I\| + I)}(F(x)).$$

Suppose $\varepsilon_0 > 0$ is an arbitrary number and consider $0 < \varepsilon < \varepsilon_0 / N + \|x_1\| + 1$. Then the mapping f_{x_1} constructed for this ε is just the mapping f_{ε, x_1} appearing in Theorem 5, which proves our statement.

Corollary 3. Let an M-mapping F satisfy the conditions of Theorem 5 and let for any point $x \in \bar{U}$ the dimension $\dim F(x) \geqslant n$. Then the set Fix(F) is a non-empty connected compact and $\dim(\mathrm{Fix}(F)) \geqslant n$.

Since F is a continuous M-mapping, the assertion directly follows from Theorems 5 and 1.

3. On the acyclicity of the set of solutions

In this section we consider reduced Čech cohomologies with integer coefficients.

Definition 6. A subset A of a topological space X is called acyclic if $\bar{H}^i(X, \mathbb{Z}) = 0$, $i = 0, 1, 2, \dots$.

Let Y be a Banach space, X a metric space, and $\Phi : X \longrightarrow C(Y)$ an M-mapping. We shall consider the following inclusion:

$$\theta \in \Phi(x),$$

where θ is zero of the space Y.

The set of solutions for this inclusion is denoted by $N(\Phi, \theta)$.

Theorem 6. Let there exist an acyclic paracompact topological space V and an M-mapping $S: X \times V \longrightarrow C(Y)$ such that the following conditions are satisfied:

1) $\Phi(x) = \bigcup_{v \in V} S(x,v)$ for any $x \in X$;
2) the inclusion $S(x,v) \ni \theta$ has only one solution for any $v \in V$;
3) the mapping $\alpha: V \longrightarrow X$, for which $S(\alpha(v); v) \ni \theta$, is a continuous closed mapping;
4) the set $\{v \mid S(x,v) \ni \theta\}$ is acyclic for any $x \in N(\Phi; \theta)$.

Then the set $N(\Phi, \theta)$ is acyclic.

Proof. Let us consider the single-valued mapping $\alpha: V \longrightarrow X$. According to the conditions of the theorem, this mapping is continuous and closed, and, furthermore, the set

$$\alpha^{-1}(x) = \{v \mid S(x;v) \ni \theta\}$$

is non-empty and acyclic for any $x \in N(\Phi, \theta)$. Since $\alpha(V) = N(\Phi, \theta)$, all the conditions of the Vietorisse theorem are satisfied (see, for example, [14]), and therefore the homomorphism $\alpha^*: \bar{H}^i(N(\Phi, \theta); \mathbb{Z}) \longrightarrow \bar{H}^i(V, \mathbb{Z})$ is an isomorphism for $i = 0,1,2, \ldots$. The acyclicity of the set $N(\Phi, \theta)$ follows then from the acyclicity of the space V.

Just as in Section 2, we now prove an approximation theorem which is convenient for studying particular classes of inclusions.

Lemma 3. Let X be a paracompact Hausdorff topological space and $\{N_i\}_{i=0}^{\infty}$, closed subsets in X which satisfy the following conditions:

1) $N_0 \subset N_i$, $i \geqslant 1$;
2) for any neighbourhood U of the set N_0 there exists a number $n_0 = n_0(U)$ such that $N_n \subset U$ for $n \geqslant n_0$;
3) the sets N_i, $i \geqslant 1$, are acyclic.

Then, the set N_0 is also acyclic.

Proof. Let us consider the confinal sequence $\{U_\alpha\}_{\alpha \in \mathcal{Y}}$ of neighbourhoods of the set N_0. Then, one can define the homomorphism of cohomology groups

$$\varinjlim \bar{H}^n(U_\alpha : \mathbb{Z}) \xrightarrow{j^*} \bar{H}^n(N_0; \mathbb{Z}),$$

induced by the embedding mapping $j_\alpha : N_0 \longrightarrow U_\alpha$
According to rigidness property of the Čech cohomology theory (see, for example, [15]), the homomorphism j^* is an isomorphism. Then there exist the set N_i satisfying $U_\alpha \overset{j_1}{\supset} N_i \overset{j_2}{\supset} N_0$ and the homomorphism $j^*_\alpha = j^*_2 \circ j^*_1$, where $H^n(U_\alpha, \mathbb{Z}) \xrightarrow{j^*_1} \bar{H}^n(N_i, \mathbb{Z}) \xrightarrow{j^*_2} \bar{H}^n(N_0, \mathbb{Z})$.

Since the set N_i is acyclic, j^*_1 is a zero homomorphism. Hence, j^*_α is a zero homomorphism as well, and therefore the limiting homomorphism j^* is also a zero one. Since j^* is an isomorphism, $\bar{H}^n(N_0, \mathbb{Z}) = 0$, i.e. the set N_0 is acyclic.

Theorem 7. Let X be a metric space and $\Phi : X \longrightarrow C(Y)$, a proper upper semicontinuous M-mapping. If there exists the sequence $\{\Phi_n\}$ of upper semicontinuous M-mappings $\Phi_n : X \longrightarrow C(Y)$ satisfying the conditions

1) $\Phi_n(x) \supset \Phi(x)$ for any $x \in X$;
2) for any number $\varepsilon > 0$ there exists a number $n_0 = n_0(\varepsilon)$ such that for $n \geq n_0$ the graph $\Gamma_X(\Phi_n) \subset U_\varepsilon(\Gamma_X(\Phi))$, where $U_\varepsilon(\Gamma_X(\Phi))$ is an ε-neighbourhood of the graph $\Gamma_X(\Phi)$; and
3) the set $N(\Phi_n, \theta)$ is acyclic for any $n = 1, 2, \ldots$,

then the set $N(\Phi, \theta)$ is acyclic.

Proof. Since the M-mappings Φ and Φ_n are upper semicontinuous, it follows that the sets $N(\Phi, \theta)$ and $N(\Phi_n, \theta)$ are closed, and also $N(\Phi, \theta)$ $N(\Phi_n, \theta)$ for any $n = 1, 2, \ldots$.

It is a simple matter to demonstrate that $N(\Phi, \theta) = \bigcap_{n=1}^{\infty} N(\Phi_n, \theta)$. Indeed, let $x_0 \in \bigcap_{n=1}^{\infty} N(\Phi_n, \theta)$, then $\Phi_n(x_0) \ni \theta$ for any $n = 1, 2, \ldots$. Let us consider the sequence $\{\varepsilon_m\} \longrightarrow 0$, $\varepsilon_m > 0$. By virtue of Condition 2, there exist points $x_m, x_m \longrightarrow x_0$ and y_m; $y_m \in \Phi(x_m)$, $\|y_m\| < \varepsilon_m$, i.e. $\{y_m\} \longrightarrow \theta$.

Since Φ is an upper semicontinuous M-mapping, then $\theta \in \Phi(x_0)$ and therefore $x_0 \in N(\Phi, \theta)$.

Let U be an arbitrary open neighbourhood of the set $N(\Phi, \Theta)$. Consider the graph $\Gamma_{X\setminus U}(\Phi)$ of the M-mapping Φ over the set $X \setminus U$. We now demonstrate that there exists a positive number ε_0 such that $U_{\varepsilon_0}(\Gamma_{X\setminus U}(\Phi)) \cap (X \times \Theta)$ is an empty set. Assume the converse, i.e. let there exist a sequence of positive numbers ε_n, $\varepsilon_n \to 0$, and a sequence of points $\{x_n\} \subset X$ such that $\rho(x_n; X\setminus U) < \varepsilon_n$ and $\Phi(x_n) \ni y_n$, where $\|y_n\| < \varepsilon_n$. Then the sequence $\{y_n\} \to \Theta$ and the set $\Phi^{-1}(\{y_n\})$ is a compact in X because the M-mapping Φ is proper. We may therefore assume, without loss of generality, that $\{x_n\} \to x_0$. Since the set $X \setminus U$ is closed, $x_0 \in X \setminus U$, but because the M-mapping Φ is upper semicontinuous we have $\Phi(x_0) \ni \Theta$, which is contrary to the condition $N(\Phi, \Theta) \subset U$.

Let $\varepsilon_0 > 0$ be this number. Condition 2 implies then that there exists a number n_0 such that for $n \geqslant n_0$ we have $\Gamma_{X\setminus U}(\Phi_n) \subset \subset U_{\varepsilon_0}(\Gamma_{X\setminus U}(\Phi))$. Hence, $\Gamma_{X\setminus U}(\Phi_n) \cap (X \times \Theta) = \emptyset$, i.e. $N(\Phi_n, \Theta) \subset U$, and the proof follows from Lemma 3.

As an application of Theorems 6 and 7, we now consider the following statement.

<u>Theorem 8</u>. Let X be a complete metric space and $\Phi: X \to K_v(Y)$, an upper semicontinuous proper M-mapping. If for any $\varepsilon > 0$ there exists a completely continuous M-mapping $G_\varepsilon: X \to K_v(Y)$ satisfying the conditions

1) $\max_{y \in G_\varepsilon(x)} \|y\| \leqslant \varepsilon$ for any $x \in X$ and
2) the inclusion $\Theta \in \Phi(x) + G_\varepsilon(x) + v$ has only one solution for any v, $\|v\| \leqslant \varepsilon$,

then the set $N(\Phi, \Theta)$ is non-empty and acyclic.

<u>Proof</u>. We first prove that the set $N(\Phi, \Theta)$ is non-empty. Let $\{\varepsilon_n\}$, $\varepsilon_n > 0$, be a vanishing sequence of numbers. Then there exist sequences $\{G_n\}$ and $\{v_n\}$ satisfying the conditions of the theorem and such that the inclusion $\Theta \in \Phi(x) + G_n(x) + v_n$ has only one solution x_n. Let $\Theta = y_n + z_n + v_n$, where $y_n \in \Phi(x_n)$ and $z_n \in G_n(x_n)$. In this case $\|z_n\| \to 0$ and $\|v_n\| \to 0$, and therefore $\|y_n\| \to 0$.

Let us consider the set $\Phi^{-1}(\{y_n\})$. Since the M-mapping Φ is proper, this set is a compact and, consequently, one may assume, without loss of generality, that the sequence $\{x_n\}$ converges to the point x_o. Furthermore, since the M-mapping Φ is upper semicontinuous, we have $\Phi(x_o) \ni \theta$, which just means that the set $N(\Phi, \theta)$ is non-empty.

We now prove the acyclicity of the set $N(\Phi, \theta)$. Again, let us consider an arbitrary sequence of positive numbers $\{\varepsilon_n\}$, $\varepsilon_n \to 0$, and denote $B_n \subset Y$, $B_n = \{x \mid \|x\| \leqslant \varepsilon_n\}$. Thus, there is defined the M-mapping $\Phi_n : X \to C(Y)$, $\Phi_n(x) = \Phi(x) + G_n(x) + B_n$, where the M-mapping G_n (constructed for ε_n) satisfies the conditions of the theorem. We now demonstrate that the sequence Φ_n satisfies the conditions of Theorem 7.

To prove the acyclicity of the sets $N(\Phi_n, \theta)$, we use Theorem 6. Consider the M-mapping $S_n : X \times B_n \to C(Y)$ defined by the relation

$$S_n(x,v) = \Phi(x) + G_n(x) + v$$

Apparently, $\Phi_n(x) = \bigcup_{v \in B_n} S_n(x,v)$, and the inclusion $S_n(x,v) \ni \theta$ has only one solution for any $v \in B_n$. Let $\alpha_n : B_n \to X$ denote the mapping through which the solution of this inclusion is associated with $v \in B_n$. Let us demonstrate that the mapping α is continuous. Consider sequences $\{v_m\} \subset B_n$, $v_m \to v_o$, and let $x_m = \alpha_n(v_m)$. Then there exist $z_m \in G_n(x_m)$ and $y_m \in \Phi(x_m)$ such that $\theta = y_m + z_m + v_m$. Since G_n is completely continuous, one may assume, without loss of generality, that $z_m \to z_o$. In this case the sequence $-y_m = z_m + v_m$, i.e. the sequence z_m is convergent and, therefore, the set $\Phi_-^{-1}(\{y_m\})$ is a compact because the M-mapping Φ is proper. Hence, we may assume, without loss of generality, that the sequence $\{x_m\}$ is convergent and $x_o = \lim x_m$. Since the M-mappings Φ and G_n are upper semicontinuous, we obtain $z_o \in G_n(x_o)$, $-z_o - v_o \in \Phi(x_o)$, i.e. the inclusion $\theta \in \Phi(x_o) + G_n(x_o) + v_o$ is valid and, consequently, $\alpha_n(v_o) = x_o$, which means that the mapping α is continuous.

Let us now prove that the mapping α_n is closed. To this end, suffice it to verify that if the sequence $\{v_m\} \subset B_n$ and $x_m = \alpha_n(v_m)$ converges to x_o, then a convergent subsequence can be found in $\{v_m\} \subset B_n$.

Let us introduce $y_m \in \Phi(x_m)$ and $z_m \in G_n(x_m)$ such that $\theta = y_m + z_m + v_m$. Since under the action of an upper semicontinuous M-mapping with compact images the image of a compact is again a compact, then the sets $\Phi(\{x_m\})$ and $G_n(\{x_m\})$ are compacts.

Hence, the sequences $\{y_m\}$ and $\{z_m\}$ include convergent subsequences, and, therefore, the sequence $\{v_m\}$ also includes a convergent subsequence.

The set $\{v \mid S_n(x_o, v) \ni \theta\}$ is convex for any $x_o \in N(\Phi_n, \theta)$, since $\Phi(x_o)$, $G_n(x_o)$, and B_n are convex sets. Thus, all the conditions of Theorem 6 are satisfied, i.e. the set $N(\Phi_n, \theta)$ is acyclic.

We now demonstrate that the M-mapping Φ and the sequence $\{\Phi_n\}$ satisfy the conditions of Theorem 7. Condition 1 is obvious, Condition 3 has already been proved, so it remains to verify Condition 2. This condition follows from the inclusion

$$\Phi_n(x) = \Phi(x) + G_n(x) + B_n \subset \Phi(x) + \overline{U_{2\varepsilon_n}(\theta)} \subset U_\varepsilon(\Phi(x))$$

where $\varepsilon > 2\varepsilon_n$. The theorem is proved.

<u>Corollary 4</u>. Let U be a bounded open domain in a Banach space Y, and let $f: \overline{U} \to Y$ be a completely continuous single-valued operator satisfying the conditions:

1) the operator f does not have fixed points on the set ∂U, and the rotation $\gamma(i - f; \partial U) \neq 0$;
2) for any $\varepsilon > 0$ there exists a completely continuous operator $f_\varepsilon : \overline{U} \to Y$ such that $\|f(x) - f_\varepsilon(x)\| \leq \varepsilon$ for any $x \in \overline{U}$, and the equation $x = f_\varepsilon(x) + v$ does not have in $\overline{U}$ more than one solution for each v, $\|v\| \leq \varepsilon$.

Then the set of fixed points of the operator f is an acyclic set.

The proof of this statement follows from Theorem 8 if the mapping $i - f$ is taken as Φ and the mapping $f_\varepsilon - f$, as G_ε.

Thus, if the Krasnoselsky-Perov connectedness principle is valid (see, for example, [1]), the set of fixed points is acyclic. This statement was proved in [6] by other methods.

4. Certain properties of the set of solutions for parameter-dependent inclusions

Let X be a compact metric space, M a closed subset of a Banach space Y, and $F:X \times M \longrightarrow K(Y)$ an upper semicontinuous M-mapping. Let also the set $\overline{F(X \times M)}$ be a compact in Y. For a fixed $x \in X$ we obtain the mapping $F_x = F(x, \cdot):M \longrightarrow K(Y)$. Let us consider the inclusion

$$F_x(y) \ni y$$

Suppose that this inclusion has a solution for any x. Then the M-mapping $\Phi :X \longrightarrow K(M)$ is valid, where $\Phi(x) = \{y \mid y \in Fix(F_x)\}$. That images of this mapping are compact follows from the fact that the M-mapping F is upper semicontinuous and compact.

<u>Theorem 9</u>. Let F satisfy the above assumptions. If there exists a sequence $\{f_n\}$ of completely continuous single-valued mappings $f_n:X \times M \longrightarrow K$, where K is a compact in Y, which satisfy the conditions

1) for any point $(x_0, y_0) \in X \times M$, any sequence $\{(x_n, y_n)\}$, $\{(x_n, y_n)\} \longrightarrow (x_0, y_0)$, and any subsequence $\{f_{n_k}\}$ the following limiting process is valid: $\lim\limits_{k \to \infty} \rho(f_{n_k}(x_k, y_k); F(x_0, y_0)) = 0$, and

2) for any $x \in X$ the equation $f_n(x, y) = y$ has only one solution, then the M-mapping Φ admits single-valued ε-approximations for any $\varepsilon > 0$.

<u>Proof</u>. For any mapping f_n one can naturally define the mapping $\varphi_n:X \longrightarrow M$ associating with a point $x \in X$ a certain solution of the equation $f_n(x, y) = y$. The continuity of this mapping follows from complete continuity of the mapping f_n. We now demonstrate that the mappings φ_n are ε-approximations of the M-mapping Φ.

Assume the converse. Then there exist a positive number ε_0 and a sequence of points $\{x_n\}$ such that $\rho(\varphi_n(x_n); \Phi(x)) \geqslant \varepsilon_0$, provided $\rho(x_n; x) < \varepsilon_0$. Since X is a compact metric space, one can choose from the sequence $\{x_n\}$ a convergent subsequence $\{x_{n_k}\}$, $\{x_{n_k}\} \longrightarrow x_0$. Let us denote $\varphi_{n_k}(x_{n_k}) = y_{n_k}$, then $f_{n_k}(x_{n_k}; y_{n_k}) = y_{n_k}$. Since the sequence $\{y_{n_k}\}$ belongs to the compact K, we shall consider it convergent (otherwise, we choose a convergent subsequence).

Let us denote $y_o = \lim y_n$. Then,

$$\rho(y_o;F(x_o;y_o)) \leqslant \| y_o - f_{n_k}(x_{n_k};y_{n_k}) \| + \rho(f_{n_k}(x_{n_k};y_{n_k});F(x_o,y_o))$$

i.e. for sufficiently large n we can make ρ less than an arbitrary positive number. Since $F(x_o,y_o)$ is a compact, $y_o \in F(x_o,y_o)$, and therefore $y_o \in \Phi(x_o)$. We have come to contradiction, since there exists a number n_k such that $\rho(x_{n_k};x_o) < \varepsilon_o$ and $\rho(\varphi_{n_k}(x_{n_k});\Phi(x_o)) < \varepsilon_o$. The theorem is proved.

The author is indebted to A.A. Talashev, translation editor.

REFERENCES

1. Krasnoselsky M.A. and Zabreiko P.P., Geometric methods in nonlinear analysis, Moscow, "Nauka", 1975 (in Russian).
2. Arronszajn N., Le correspondant topologique de l'unicite dans la théorie des équations différentielles, Annls. Math., v. 43, No. 4, 1972, pp. 730-738.
3. Bulgakov A.I., Kneser's theorem for a certain class of integral inclusions, Differentsialnye Uravneniya, 1980, v. 16, No. 5, pp. 894-900 (in Russian).
4. Bulgakov A.I. and Lyapin L.N., Certain properties of the set of solutions for the Volterra-Hammerstein integral inclusion, Differentsialnye Uravneniya, 1978, v. 14, No. 8, pp. 1465-1472 (in Russian).
5. Bulgakov A.I. and Lyapin L.N., On the connectedness of solutions for functional inclusions, Matem. Sbornik, 1982, v.119, No.2 (in Russian).
6. Lasry G.M. and Robert R., Acyclicité de l'ensemble des solutions de certains équations fonctionnelles, C.r. Acad. Sci., v.282, No.22, 1976, pp. 1283-1286.
7. Himmielberg C.G. and Van Vleek F.S., On the topological triviality of solutions sets, Rocky. Monnt. J. Math., v.10, No.1, 1980, pp. 247-252.
8. Haddad G., Topological properties of the sets of solutions for functional differential inclusions, Nonlinear Analysis Theory, Methods and Appl., v.5, No.12, 1981, pp. 1349-1366.
9. Raymond G.S., Points fixes des multiapplications à valeurs convexes, C.R. Acad. Sci., t.298, No.4, 1984, pp. 71-74.
10. Borisovich Yu.G., Gel'man B.D., Myshkis A.D., and Obukhovsky V.V., Topological methods in the theory of fixed points of multivalued mappings, Usp. Matem. Nauk, 1980, v. 35, No. 1, pp. 59-126 (in Russian).

11. Borisovich Yu.G., Gel'man B.D., Myshkis A.D., and Obukhovsky V.V., Multivalued mappings, Itogi Nauki i Tekhniki, Ser. Mathematical analysis, 1982, v. 19, pp. 127-230 (in Russian).

12. Michail E.A., Continuous selections. I, Ann. Math., v. 63, No. 2, 1956, pp. 362-381.

13. Gel'man B.D., On certain classes of selections of many-valued mappings, Lect. Notes Math., 1986, vol.1214.

14. Sklyarenko E.G., On certain applications of sheaf theory in general topology, Usp. Matem. Nauk, v. 19, No. 6, 1964, pp. 47-70 (in Russian).

15. Spanier E.H., Algebraic Topology, N.Y., 1966.

SCHOUTEN BRACKET AND CANONICAL ALGEBRAS

I.S.Krasil'shchik
All-Union Institute for
Scientific and Technical Information
Baltiyskaya ul.,14
125219, Moscow, USSR

In [1] , when studying an algebraic model of the classical Hamiltonian mechanics, we have introduced the notion of the canonical ring (algebra), which is analogous to the ring of smooth functions on a symplectic manifold. In [2] , for any canonical algebra the natural complex was constructed, the cohomologies of which were called the Hamiltonian cohomologies of a canonical algebra. In fact, this construction is closely related to the existence of a specific algebraic operation (the Schouten bracket) in modules $D_*(A)$ of polyderivations of an algebra A (cf. [3,4]). The Schouten bracket together with the exterior multiplication defined in $D_*(A)$ induces in $D_*(A)$ a structure which is natural to be called a canonical super-algebra and may be viewed as an analogue of the Poisson structure in commutative canonical algebras. This interpretation gives rise to various parallels, such as:

the differential of the Hamiltonian complex of a canonical algebra $(A;h)$	$\Longleftrightarrow$	a "super-Hamiltonian" vector field on $D_*(A)$ with the Hamiltonian $h \in D_2(A)$
commuting Hamiltonian structures*)	$\Longleftrightarrow$	Hamiltonians in involution

etc.

The Schouten bracket construction can be extended onto A-modules P with connections. If the curvature of the connection in question vanishes on Hamiltonian (with respect to h) derivations (we call such connections Hamiltonian ones) then the Hamiltonian cohomologies

*) See, for example, [5] .

with the coefficients in P can be constructed.

The invariant approach adopted here allows to clarify the algebraic nature of main concepts which is, as a rule, obscured by analytical calculations. In consequence, all the proofs are reduced to simple manipulations with basic definitions. For example, it is shown (see Theorem 4.4 below) that the reason for the existence of infinite series of conservation laws for commuting canonical structures is purely cohomological.

The material is exposed in the following way. The first section includes the review of the basic definitions and facts from the algebra of logic of linear differential operators (DO) over commutative rings (see [6,7]) which is necessary for the understanding of the paper. In the second section the definitions of the exterior multiplication and the Schouten bracket in modules with connections are given and the main properties of these constructions are investigated. Section 3 concerns the study of the properties of the Hamiltonian cohomologies of canonical algebras and the discussion of some important examples. Finally, the last section includes the results concerning commuting canonical structures.

I. The basic facts from the calculus over commutative algebras.

I.I. Suppose K is a commutative ring with identity, A is a commutative K-algebra, P and Q are modules over A. Consider the set $\mathrm{Hom}_K(P,Q)$ and for any element $a \in A$ define the operations r_a, $l_a : \mathrm{Hom}_K(P,Q) \longrightarrow \mathrm{Hom}_K(P,Q)$ setting

$$r_a(\Delta)(p) = \Delta(ap), \quad l_a(\Delta)(p) = a \cdot \Delta(p),$$

where $\Delta \in \mathrm{Hom}_K(P,Q)$, $p \in P$. Let $\delta_a = r_a - l_a$.

Definition. An element $\Delta \in \mathrm{Hom}_K(P,Q)$ is said to be a linear differential operator of the order $\leq s, s \geq 0$, over A acting from P into Q, if for any elements $a_0,\ldots,a_s$ A the equality

$$(\delta_{a_0} \circ \ldots \circ \delta_{a_s})(\Delta) = 0$$

holds.

The set of all DOs of order $\leq s$ acting from the module P into the module Q is additively closed and for each a A is stable with respect to the action of r_a and l_a. Hence, two A-module structures are determined in this set which are denoted by $\mathrm{Diff}_s(P,Q)$ (the structure induced by l_a) and by $\mathrm{Diff}_s^+(P,Q)$ (the structure induced by r_a). Since for any $a,b \in A$ the equality $r_a \cdot l_b = l_b \cdot r_a$ holds it follows that a bi-module $\mathrm{Diff}_s^{(+)}(P,Q)$ is determined as well. Obviously, for any s t the embeddings of A-bi-modules $\mathrm{Diff}_s^{(+)}(P,Q) \subsetneq \mathrm{Diff}_t^{(+)}(P,Q)$ take place and, besides, $\mathrm{Diff}_o^{(+)}(P,Q) = \mathrm{Hom}_A(P,Q)$. In the sequel for any $\mathrm{Diff}_s^{(+)}(P,Q)$, $a \in A$ we shall set

$$l_a(\Delta) \overset{\text{def}}{=} a\,\Delta\ , \quad r_a(\Delta) \overset{\text{def}}{=} a^+\Delta\ .$$

Denote by $\mathrm{Diff}_s^{(+)}(P)$ bi-modules $\mathrm{Diff}_s^{(+)}(A,P)$ and by $\mathrm{Diff}_*^{(+)}(P,Q)$ bi-modules $\bigcup_{s\geq 0} \mathrm{Diff}_s^{(+)}(P,Q)$.

Suppose that R is also an A-module, $\Delta_I \in \mathrm{Diff}_s^{(+)}(P,Q)$ and $\Delta_2 \in \mathrm{Diff}_t^{(+)}(Q,R)$. Then the composition $\Delta_2 \cdot \Delta_I$ is a DO acting from P into R and its order $\leq s + t$. For any $a \in A$ the following equalities take place:

$$(a\,\Delta_2)\cdot\Delta_I = a(\Delta_2\cdot\Delta_I),\ (a^+\Delta_2)\cdot\Delta_I = \Delta_2\cdot(a\,\Delta_I),$$

$$\Delta_2\cdot(a^+\Delta_I) = a^+\,(\Delta_2\cdot\Delta_I).$$

<u>I.2</u>. Let P be an A-module as before. Define the special class of DOs of order ≤ 1 acting from A into P.

<u>Definition</u>. An element $\Delta \in \mathrm{Hom}_K(A,P)$ is said to be a derivation of A into P if for any $a,b \in A$ the equality

$$\Delta(ab) = a\,\Delta(b) + b\,\Delta(a)$$

holds. The set of all such elements is denoted by D(P).

Obviously, D(P) is a submodule in $\mathrm{Diff}_I(P)$. Moreover, D(A) becomes a Lie algebra over K if we set $[X,Y] = X\cdot Y - Y\cdot X$ for any $X,Y \in D(A)$.

Suppose S is a subset in P. Denote by $D(S \subset P)$ the set of such derivations from D(P) images of which lie in S:

$$D(S\subset P) \overset{\text{def}}{=} \{\Delta \in D(P)\ |\ \ \Delta(a) \in S,\ \forall a \in A\}.$$

If S is a submodule then $D(S\subset P) = D(S)$. By induction let us define now the sequence of A-modules $D_i(P)$ together with embeddings $D_i(P) \subset (\mathrm{Diff}_I^+)^i(P)$, $i \geq 0$. Namely, set $D_o(P) = P$, $D_I(P) = D(P)$, and

$$D_{i+I}(P) = D(D_i(P) \subset (\text{Diff}_I^+)^i(P)),$$

where $(\text{Diff}_I^+)^i(P) = \text{Diff}_I^+(\ldots(\text{Diff}_I^+)^i(P)\ldots)$. Obviously, this con-
$\underbrace{}_{i\ times}$
struction is well defined while the correspondence $P \mapsto D_i(P)$ is a
functor of the category of A-modules into itself. The elements of
$D_i(P)$ are called polyderivations of A into P.

From the definition it immediately follows that for any $a \in A$ and
$\Delta \in D_i(P)$ the substitution operation

$$a \lrcorner \Delta \quad = \quad \Delta(a) \in D_{i_1}(A), \ i \geqslant I,$$

can be defined. Moreover, for any $a_1,\ldots,a_j \in A$, $j \leqslant i$, polyderivations

$$\Delta(a_1,\ldots,a_j) = \Delta(a_1)(a_2,\ldots,a_j) \in D_{i-j}(P)$$

are determined. If $j > i$ we shall put $\Delta(a_1,\ldots,a_j) = 0$.

Let us give now a clearer description of the modules $D_i(P)$ which
will be useful for further considerations.

Proposition. For $i \geqslant 2$ an element $\Delta \in \text{Hom}_K(A, D_{i-I}(P))$ belongs to
$D_i(P)$ iff for all $a,b \in A$ the following equalities hold:

(i) $\Delta(ab) = a\Delta(b) + b\Delta(a)$

(ii) $\Delta(a,b) + \Delta(b,a) = 0.$

I.3. Introduce now the modules of exterior forms over A. For this we
set $\wedge^0 = \wedge^0(A) \overset{\text{def}}{=} A$ and define the module $\wedge^I = \wedge^I(A)$ as the A-
module generated by the elements $\{d_a \mid a \in A\}$ together with the re-
lations

$$d_{ab} = ad_b + bd_a, \quad d_{a+b} = d_a + d_b, \quad d_{ka} = kd_a, \quad a,b \in A, \ k \in K.$$

Let us set also $\wedge^i = \wedge^i(A) = \underbrace{\wedge^I \wedge \ldots \wedge \wedge^I.}_{i\ times}$

Suppose $\omega = ad_{a_1} \wedge \ldots \wedge d_{a_i} \in \wedge^i$, $a,a_1,\ldots,a_i \in A$. Set $d\omega = d_a \wedge d_{a_I} \wedge \ldots \wedge d_{a_i}$. Obviously, it defines the sequence

$$0 \longrightarrow A \overset{d}{\longrightarrow} \wedge^I(A) \longrightarrow \ldots \longrightarrow \wedge^i(A) \overset{d}{\longrightarrow} \wedge^{i+I}(A) \to \ldots (I)$$

which is a complex of DOs of order $\leqslant$ I. This complex is called the de Rham complex of the algebra A, its cohomologies being the de Rham cohomologies of A.

Proposition. For all $i \geqslant 0$ the functors D_i are representable. Namely, for every A-module P there are natural isomorphisms

$$D_i(P) \simeq \mathrm{Hom}_A(\Lambda^i, P).$$

In particular, for $i = I$ such an isomorphism is defined in the following way: for each derivation $\Delta \in D_I(P)$ there exists the unique homomorphism $\varphi_\Delta : \Lambda^I \to P$ for which the diagram

$$\begin{array}{ccc} A & \xrightarrow{d} & \Lambda^I \\ & \underset{\Delta}{\searrow} & \downarrow \varphi_\Delta \\ & & P \end{array} \qquad (2)$$

is commutative.

For all $i, j \geqslant 0$ define the homomorphism of A-modules

$$\varkappa : D_i(P) \otimes_A \Lambda^j \longrightarrow D_{i-j}(P)$$

by setting

$$\varkappa(\Delta \otimes \omega) = \langle \Delta, \omega \rangle = \begin{cases} 0, & \text{when } i < j, \\ \Delta(\omega), & \text{when } i = j, \end{cases}$$

and $\langle \Delta, \omega \rangle (a) = \langle \Delta(a), \omega \rangle$ for all $a \in A$ when $i > j$. Obviously, the operation $\langle\ ,\ \rangle$ is well defined. Note that the operations of substitution $\lrcorner\, a : \Delta \mapsto \Delta(a)$ and of exterior multiplication $\omega \mapsto da \wedge \omega$ are conjugated in the following sense:

$$\langle \Delta(a), \omega \rangle = (-1)^{i+j+I} \langle \Delta, da \wedge \omega \rangle .$$

I.4. Let us interpret the concepts introduced above in the situation which we call classical. Let M be a smooth manifold, $A = C^\infty(M)$, ξ and η be smooth vector bundles over M and $P = \Gamma(\xi)$, $Q = \Gamma(\eta)$ be corresponding A-modules of smooth sections of these bundles. Then the modules $\mathrm{Diff}_i(P,Q)$ coincide with the sets of all linear DOs of order $\leqslant i$ on the manifold M acting from the sections of the bundle

ξ into the sections of η. Elements of the module D(A) can be viewed as vector fields on M while the modules $D_i(A)$ are isomorphic to $C^\infty(M)$-modules of polyvector fields on M. The elements of $D_i(P)$ may be undestood as "P-valued" vector fields on M.

To establish a correspondence between the modules $\Lambda^i(A)$ and classical exterior forms on M let us introduce one more notion. We call an A-module R a geometrical one if the equality $\bigcap_\mu \mu R = 0$ holds, where μ lies in the set Specm(A)of all maximal ideals of the algebra A. Obviously, all the modules of the form $\Gamma(\xi)$ are geometrical ones. Consider the category $\mathcal{M}_g(A)$ of all the geometrical A-modules. It is clear that the correspondence

$$R \Rightarrow R_g = R \Big/ \bigcap_{\mu \in \mathrm{Specm}(A)} \mu R$$

is the functor from the category (A) of all A-modules into $\mathcal{M}_g(A)$.

Proposition. For all $i \geq 0$ the functors D_i are representable in $\mathcal{M}_g(A)$, where $A = C^\infty(M)$. Namely, for every geometrical A-module P the following isomorphisms take place:

$$D_i(P) = \mathrm{Hom}_A(\Lambda^i_g, P).$$

The modules Λ^i_g coincide with the modules of i-th exterior forms on M while the complex (I) constructed in the geometrical category coincides with the de Rham complex of the manifold M.

I.5. Definition. An element $\bar{X} \in \mathrm{Hom}_K(P,P)$ is said to be a derivation of the module P covering the derivation $X \in D(A)$ if for all $a \in A$ and $p \in P$ the equality

$$\bar{X}(ap) = X(a)p + a\bar{X}(p)$$

holds.

The set of all such derivations is denoted by Der(P). Obviously, Der(P) is a submodule in $\mathrm{Diff}_1(P,P)$. It will become a Lie algebra over K if we set $[\bar{X},\bar{Y}] = \bar{X} \cdot \bar{Y} - \bar{Y} \cdot \bar{X}$ for all $\bar{X},\bar{Y} \in \mathrm{Der}(P)$. Note that to every element $\bar{X} \in \mathrm{Der}(P)$ a derivation $X \in D(A)$ corresponds which is covered by $\bar{X}$. Such a correspondence defines an A-algebra homomorphism $\pi : \mathrm{Der}(P) \to D(A)$ which is a homomorphism of Lie algebras over K as well.

A homomorphism $\nabla \in \mathrm{Hom}_A(D(A),\mathrm{Der}(P))$, $X \mapsto \nabla_X$, is said to be a connection in the module P, if $\pi \circ \nabla = id_{D(A)}$. Thus, a connection should satisfy the following conditions:

(i) $\nabla_{X+Y} = \nabla_X + \nabla_Y$,

(ii) $\nabla_{aX} = a \nabla_X$,

(iii) $\nabla_X (ap) = X(a)p + a \nabla_X (p)$,

where $a \in A$, $p \in P$, $X, Y \in D(A)$. We say that the curvature of a connection ∇ vanishes if $\nabla_{[X,Y]} = [\nabla_X, \nabla_Y]$ for any $X, Y \in D(A)$. In other words, connections with the vanishing curvature are Lie algebra homomorphisms.

Examples. I. If P = A then the natural embedding $D(A) \subsetneq Diff_I(A)$ determines the connection with the vanishing. This curvature in the algebra A considered as the A-module. This follows from the obvious fact that $Der(A) = Diff_I(A)$.

2. If P and Q are A-modules and ∇ is a connection in Q then the connection Diff ∇ in the A-module $Diff_*(P,Q)$ is determined by

$$(Diff \nabla)_X (\Delta) = \nabla_X \cdot \Delta ,$$

where $X \in D(A)$, $\Delta \in Diff_*(P,Q)$. In the classical situation, i.e. when $A = C^\infty(M)$, $P = \Gamma(\xi)$, $Q = \Gamma(\eta)$ while ∇ is the connection from example I, the connection Diff ∇ coincides with the Cartan connection in the bundle $\xi_\infty : J^\infty(\xi) \longrightarrow M$ of infinite jets of the bundle ξ (see [7]).

2. Exterior multiplication in the modules $D_i(P)$ and the Schouten bracket.

2.I. For any pair of elements $X \in D_i(A)$ and $\Delta \in D_j(P)$ define by induction their exterior product $X \wedge \Delta \in D_{i+j}(P)$ by setting $X \wedge \Delta = X \cdot \Delta$ when $i + j = 0$ and

$$(X \wedge \Delta)(a) = X(a) \wedge \Delta + (-I)^i X \wedge \Delta(a) \qquad (3)$$

when $i+j > 0$ and a is an arbitrary element of A.

Proposition. By (3) the exterior multiplication $\wedge$:

$D_i(A) \otimes_A D_j(P) \longrightarrow D_{i+j}(P)$ is well defined.

Proof. It needs to show that for any $X \in D_i(A)$ and $\Delta \in D_j(P)$ the element $X \wedge \Delta$ defined by (3) lies in $D_{i+j}(P)$. By the proposition I.2 it suffices to show that

(i) $X \wedge \Delta \in \mathrm{Hom}_K(A, D_{i+j-I}(P))$,

(ii) $(X \wedge \Delta)(ab) = a(X \wedge \Delta)(b) + b(X \wedge \Delta)(a)$,

(iii) $(X \wedge \Delta)(a,b) + (X \wedge \Delta)(b,a) = 0$,

where a,b are arbitrary elements of A. The first is obvious by the definition of $X \wedge \Delta$.

For the following considerations we shall need the

Lemma. The operation $\wedge$ is distributive with respect to summation:

$$(X+Y) \wedge \Delta = X \wedge \Delta + Y \wedge \Delta,\ X \wedge (\Delta + \Gamma) = X \wedge \Delta + X \wedge \Gamma ,$$

where $X, Y \in D_i(A)$ and $\Gamma, \Delta \in D_j(P)$.

Lemma's proof. Induction by i+j. When i+j =o both equalities are obvious. Let i+j > o and the equalities are correct for i+j-I. Then for any $a \in A$ we have

$$((X+Y) \wedge \Delta)(a) = (X+Y)(a) \wedge \Delta + (-I)^i (X+Y) \wedge \Delta(a) = X(a) \wedge \Delta + Y(a) \wedge \Delta + (-I)^i X \wedge \Delta(a) + (-I)^i Y \wedge \Delta(a) = (X \wedge \Delta)(a) + (Y \wedge \Delta)(a) = (X \wedge \Delta + Y \wedge \Delta)(a).$$

The second equality is proved in the same way.

Let us prove (ii) now. This is done by the induction with respect to i+j also. For i+j=o the equality (ii) is obvious. Suppose i+j > o and for i+j -I (ii) is correct. Then, by (3) and the lemma, we have

$$(X \wedge \Delta)(ab) = X(ab) \wedge \Delta + (-I)^i X \wedge \Delta(ab) = (aX(b) + bX(a)) \wedge \Delta$$

$$+ (-I)^i X \wedge (a\,\Delta(b) + b\,\Delta(a)) = a(X(b) \wedge \Delta + (-I)^i X \wedge \Delta(b)) +$$

$$+ b\,(X(a) \wedge \Delta + (-I)^i X \wedge \Delta(a)) = a\,(X \wedge \Delta)(b) + b\,(X \wedge \Delta)(a).$$

Prove the equality (iii):

$(X \wedge \Delta)(a,b) = (X(a) \wedge \Delta + (-I)^i X \wedge \Delta(a))(b) = X(a,b) \wedge \Delta + (-I)^{i-I}$

$X(a) \wedge \Delta(b) + (-I)^i (X(b) \wedge \Delta(a) + (-I)^i X \wedge \Delta(a,b)) = -(X(b,a) \wedge \Delta$

$+ (-I)^{i-I} X(b) \wedge \Delta(a) + (-I)^i X(a) \wedge \Delta(b) + X \wedge \Delta(b,a)) =$

$-(X \wedge \Delta)(b,a).$

Q.E.D.

Remark. This proof is typical for the most of the results formulated below. For this reason we, as a rule, either omit such proofs or confine ourselves to brief comments.

2.2. Suppose $\Delta \in D_k(P)$ and $P \overset{\varphi}{\dashrightarrow} Q$ is a homomorphism of A-modules. Then we can define an element $D_k(\varphi)(\Delta) \in D_k(Q)$ by setting $D_k(\varphi)(\Delta) = \varphi \circ \Delta$. The main properties of the exterior multiplication are described by the following

Proposition. $D_*(A) = \sum_{i \geq 0} D_i(A)$ is the Grassmann algebra with respect to the operation $\wedge$ while $D_* = \sum_{i \geq 0} D_i$ is the functor from the category of A-modules into the category of $D_*(A)$-modules.

For the proof it suffices to show that

(i) $X \wedge Y = (-I)^{ij} Y \wedge X$,

(ii) $X \wedge (Y \wedge \Delta) = (X \wedge Y) \wedge \Delta$,

(iii) $D_{i+k}(\varphi)(X \wedge \Delta) = X \wedge D_k(\varphi)(\Delta)$,

for all $X \in D_i(A)$, $Y \in D_j(A)$, $\Delta \in D_k(P)$, $\varphi \in Hom_A(P,Q)$.

2.3. Let, as above, P be an A-module and ∇ be a connection in P. For any pair of elements $X \in D_i(A)$ and $Y \in D_j(P)$ define the element $[\![X, \Delta]\!]_\nabla \in D_{i+j-I}(P)$ by setting

$$[\![a,p]\!]_\nabla = o, \quad [\![X,p]\!]_\nabla = \nabla_X(p), \quad [\![a, \Delta]\!]_\nabla = \Delta(a),$$

when $X \in D_i(A)$, $p \in P$, $\Delta \in D_i(P)$, $a \in A$, and

$$[\![X, \Delta]\!]_\nabla(a) = [\![X(a), \Delta]\!]_\nabla + (-I)^i [\![X, \Delta(a)]\!]_\nabla , \qquad (4)$$

when $i+j > I$.

Proposition. By (4) the operation

$$[\![\ ,\ ]\!]_\nabla : D_i(A) \otimes_k D_j(P) \dashrightarrow D_{i+j-I}(P)$$

is well defined.

For the proof proposition I.2 and the following lemma are used.

Lemma. If $X \in D_i(A)$, $\Delta \in D_i(P)$ and $a \in A$ then the equalities

$$[\![X, a\Delta]\!]_\nabla = (-I)^{i+j+I} X(a) \wedge \Delta + a [\![X, \Delta]\!]_\nabla ,$$

$$[\![aX, \Delta]\!]_\nabla = (-I)^{j+I} X \wedge \Delta(a) + a [\![X, \Delta]\!]_\nabla$$

hold.

Definition. The element $[\![X, \Delta]\!]_\nabla \in D_{i+j-I}(P)$ is called the Schouten bracket of X and Δ constructed with respect to the connection ∇.

In particular, the canonical connection $\nabla : D(A) \dashrightarrow Der(A)$ introduced in example I of subsection I.5 allows to construct the bracket $[\![X, Y]\!]_\nabla \overset{def}{=} [\![X, Y]\!]$ for any elements $X, Y \in D_*(A)$.

2.4. Before passing to the description of the Schouten bracket basic properties let us introduce one more construction generalizing the notion of curvature. Suppose ∇ is a connection in an A-module P. For any triad of elements $X \in D_i(A)$, $Y \in D_j(A)$, $\Delta \in D_k(P)$ define the element $R_\nabla(X, Y, \Delta) \in D_{i+j+k-2}(P)$ in the following way. When $i+j+k \leqslant 2$ we set

$$R_\nabla(X, Y, \Delta) = \begin{cases} ([\nabla_X, \nabla_Y] - \nabla_{[X,Y]})(\Delta), & \text{if } i=j=I,\ k=0, \\ 0 & \text{eitherwise}, \end{cases}$$

and when $i + j + k > 2$ we set

$$R_\nabla(X, Y, \Delta)(a) = R_\nabla(X(a), Y, \Delta) + (-I)^i R_\nabla(X, Y(a), \Delta) + (-I)^{i+j} R_\nabla(X, Y, \Delta(a)) \qquad (5)$$

for any $a \in A$.

Proposition. By (5) the element $R_\nabla(X, Y, \Delta) \in D_{i+j+k-2}(P)$ is well defined.

For the proof proposition I.2 and the following lemma are used.

Lemma. The element $R_\nabla(X, Y, \Delta)$ is A-linear with respect to all the arguments.

2.5. From the last lemma it follows that the correspondence $R_\nabla : (X, Y, \Delta) \longmapsto R_\nabla(X, Y, \Delta)$ determines the element

$$R_\nabla \in \mathrm{Hom}_A(D_*(A)_A \otimes D_*(A) \otimes_A D_*(P), D_*(P))$$

the basic properties of which are described by the

Proposition. (i) if $X \in D_i(A)$, $Y \in D_j(A)$, then

$$R_\nabla(X, Y, \Delta) = (-1)^{ij} R(Y, X, \Delta)$$

for any $\Delta \in D_*(P)$;

(ii) the curvature of ∇ vanishes iff $R_\nabla = 0$.

2.6. Proposition. The Schouten bracket $[\![\ ,\ ]\!]_\nabla$ is K-linear with respect to both arguments and possesses the following properties

(i) $[\![X, Y]\!] = (-1)^{ij} [\![Y, X]\!]$,

(ii) $[\![X, Y \wedge \Delta]\!]_\nabla = (-1)^k [\![X, Y]\!] \wedge \Delta + (-1)^{ij} Y \wedge [\![X, \Delta]\!]_\nabla$,

(iii) $[\![X \wedge Y, \Delta]\!]_\nabla = (-1)^{ij} Y \wedge [\![X, \Delta]\!]_\nabla + X \wedge [\![Y, \Delta]\!]_\nabla$,

(iv) $[\![X, [\![Y, \Delta]\!]_\nabla]\!]_\nabla + (-1)^k [\![[\![X, Y]\!], \Delta]\!]_\nabla +$

$+ (-1)^{ij} [\![Y, [\![X, \Delta]\!]_\nabla]\!]_\nabla = R_\nabla(X, Y, \Delta)$,

where $X \in D_i(A)$, $Y \in D_j(A)$, $\Delta \in D_k(P)$.

Corollary. If the curvature of ∇ vanishes then for any $X \in D_i(A)$, $Y \in D_j(A)$, $\Delta \in D_k(P)$ the equality

(iv′) $[\![X, [\![Y, \Delta]\!]_\nabla]\!]_\nabla = (-1)^{k+1} [\![[\![X, Y]\!], \Delta]\!]_\nabla +$

$+ (-1)^{ij+1} [\![Y, [\![X, \Delta]\!]_\nabla]\!]_\nabla$

holds.

2.7. In conclusion we shall write out the explicit representation for the Schouten bracket of two polyderivations $X \in D_i(A)$, $Y \in D_j(P)$. Let $a_1, \ldots, a_n \in A$. Denote by T_k^n the set of all sequences of natural numbers of the form $\tau=(\tau_1,\ldots,\tau_k)$ where $1 \le \tau_1 < \ldots < \tau_k \le n$, and set

$$\tau^n = (1,\ldots,\tau_1 -1, \tau_1 +1,\ldots, \tau_k -1, \tau_k +1,\ldots, n).$$

Let $|\tau| = \tau_1 + \ldots + \tau_k$. For any $\square \in D_l(Q)$, where Q is an A-module, $l \ge k$, we shall denote by $\square(a_\tau)$ the element $\square(a_{\tau_1}, \ldots, a_{\tau_k}) \in D_{l-k}(Q)$.

Proposition. Let $X \in D_i(A)$, $\Delta \in D_j(P)$ and ∇ be a connection in P. Then for any elements $a_1, \ldots, a_{i+j-1} \in A$ the equality

$$[\![X, \Delta]\!]_\nabla (a_1,\ldots, a_{i+j-1}) = (-1)^{ij+\frac{j(j+1)}{2}} \sum_{\tau \in T_j^{i+j-1}} (-1)^{|\tau|} \nabla_{X(a_{\tau^{i+j-1}})}$$

$$(\Delta(a_\tau)) + (-1)^{\frac{i(i+1)}{2}} \sum_{\tau \in T_i^{i+j-1}} (-1)^{|\tau|} \Delta(a_{\tau^{i+j-1}},$$

$$X(a_\tau)). \tag{6}$$

is valid.

Proof. Induction with respect to $i+j$. When $i+j=1$ (6) is obvious. Suppose now that $i+j > 1$ and for $i+j-1$ the equality (6) is correct. Then we have

$$[\![X, \Delta]\!]_\nabla (a_1,\ldots, a_{i+j-1}) = (-1)^{i+j} \left([\![X(a_{i+j-1}), \Delta]\!]_\nabla + (-1)^i [\![X,\right.$$

$$\Delta(a_{i+j-1})]\!]_\nabla)(a_1,\ldots, a_{i+j-2}) = (-1)^{i+j} \left((-1)^{(i-1)j+\frac{j(i+1)}{2}} \cdot \right.$$

$$\sum_{\tau \in T_j^{i+j-2}} (-1)^{|\tau|} \nabla_{X(a_{i+j-1}, a_{\tau^{i+j-2}})} (\Delta(a_\tau)) +$$

$$+ (-1)^{\frac{(i-1)i}{2}} \sum_{\tau \in T_{i-1}^{i+j-2}} (-1)^{|\tau|} \Delta(a_{\tau^{i+j-2}}, X(a_{i+j-1}, a_\tau)) +$$

$$+ (-1)^j \left((-1)^{i(j-1)+\frac{(j-1)i}{2}} \sum_{\tau \in T_{j-1}^{i+j-2}} (-1)^{|\tau|} \nabla_{X(a_{\tau^{i+j-2}}}\right.$$

$$(\Delta(a_{i+j-1}, a_\tau) + (-1)^{\frac{i(i+1)}{2}} \sum_{\tau \in T_i^{i+j-2}} (-1)^{|\tau|} \Delta(a_{i+j-1}, a_{i+j-2},$$

$X(a_{\tau})))$.

Note now, that for any k 0 the following set equalities hold:

$$T_k^{i+j-2} \cup \Big(\bigcup_{\tau \in T_{k-1}^{i+j-2}} \{(\tau,\ i+j-1)\} \Big) = T_k^{i+j-1},$$

$$T_k^{i+j-2} \cap \Big(\bigcup_{\tau \in T_{k-1}^{i+j-2}} \{(\tau,\ i+j-1)\} \Big) = \emptyset ,$$

and if $\tau \in T_k^{i+j-2}$, then $\tau^{i+j-1} = (\tau^{i+j-2},\ i+j-1)$, while from $\tau \in T_{k-1}^{i+j-2}$ it follows that $(\tau,\ i+j-1)^{i+j-1} = \tau^{i+j-2}$. Hence, after necessary algebraic transformations the sum of the first and the third addenda in the previous expression will take the form

$$(-1)^{ij+\frac{j(j+1)}{2}} \sum_{\tau \in T_j^{i+j-1}} (-1)^{|\tau|} X(a_{\tau^{i+j-1}}) \, (\Delta(a_{\tau}))$$

while the sum of the second and the forth ones will be

$$(-1)^{\frac{i(i+1)}{2}} \sum_{\tau \in T_i^{i+j-1}} (-1)^{|\tau|} \Delta(a_{\tau^{i+j-1}},\ X(a_{\tau})).$$

Q.E.D.

<u>Corollary</u>. For any $X \in D_i(A)$, $Y \in D_j(A)$ and $a_1, \ldots, a_{i+j-1} \in A$ we have

$$[\![X, Y]\!](a_1, \ldots, a_{i+j-1}) = \tag{7}$$

$$= (-1)^{ij+\frac{i(j+2)}{2}} \sum_{\tau \in T_j^{i+j-1}} (-1)^{|\tau|} X(a_{\tau^{i+j-1}},\ Y(a_{\tau})) + (-1)^{\frac{i(i+1)}{2}}$$

$$\sum_{\tau \in T_i^{i+j-1}} (-1)^{|\tau|} Y(a_{\tau^{i+j-1}}\ X(a_{\tau})).$$

3. Lie derivatives, canonical algebras, and Hamiltonian cohomologies

3.I. Suppose P is an A-module with a connection ∇ and $X \in D_I(A)$.

Definition. The Lie derivative of the element $\Delta \in D_j(P)$ along the derivation X with respect to ∇ is the element

$$L_X^{\nabla}(\Delta) = (-I)^i [\![X, \Delta]\!]_{\nabla} \in D_j(P). \tag{8}$$

In particular, for any $Y \in D_j(A)$ there exists its Lie derivative $L_X(Y) = (-I)^j [\![X, Y]\!]$.

Proposition. The Lie derivative can be defined by the induction as follows:

(i) $L_X^{\nabla}(p) = \nabla_X(p)$, $p \in P$,

(ii) $L_X^{\nabla}(\Delta)(a) = L_X(\Delta(a)) - \Delta(X(a))$, $j > o$,

$\Delta \in D_j(P)$, $a \in A$.

3.2. The basic properties of the Lie derivative are described by the following.

Proposition. For any $X \in D_I(A)$ the mapping $L_X: D_*(P) \longrightarrow D_*(P)$ is K-linear. If $a \in A$, $Y \in D_j(A)$, $\Delta \in D_k(P)$, then we have:

(i) $L_{aX}^{\nabla}(\Delta) = aL_X^{\nabla}(\Delta) - X \wedge \Delta(a)$,

(ii) $L_X^{\nabla}(Y \wedge \Delta) = L_X(Y) \wedge \Delta + Y \wedge L_X^{\nabla}(\Delta)$,

(iii) $L_X^{\nabla}[\![Y, \Delta]\!]_{\nabla} = [\![L_X(Y), \Delta]\!]_{\nabla} + [\![Y, L_X^{\nabla}(\Delta)]\!]_{\nabla} +$

$+ (-I)^{k+j-I} R_{\nabla}(X, Y, \Delta)$.

In particular, if $Y \in D_I(A)$, then, in addition, we have

(iv) $L_X^{\nabla} L_Y^{\nabla}(\Delta) - L_Y^{\nabla} L_X(\Delta) = L_{[X,Y]}(\Delta) + R_{\nabla}(X, Y, \Delta)$,

(v) $L_X(Y) = [X, Y]$.

The proof follows from proposition 2.6.

3.3. From proposition 2.7 the explicit expression for the Lie derivative immediately follows

Proposition. Let $X \in D_I(A)$, $Y \in D_j(P)$ and ∇ be a connection in P. Then for any $a_I, \ldots, a_j \in A$ we have

$$L_x^\nabla (\Delta)(a_I, \ldots, a_j) = \nabla_x(\Delta(a_I, \ldots, a_j)) - \Delta(X(a_I), a_2,$$

$$\ldots, a_j) - \Delta(a_I, X(a_2), \ldots, a_j) - \Delta(a_I, a_2, \ldots, X(aj)). \quad (9)$$

In particular, if $Y \in D_j(A)$, then

$$L_x(Y)(a_I, \ldots, a_j) = X\, Y(a_I, \ldots, a_j) - Y(X(a_I), a_2, \ldots, a_j) -$$

$$- Y(a_I, X(a_2), \ldots, a_j) - Y(a_I, a_2, \ldots, X(a_j)). \quad (IO)$$

3.4. Consider now an element $X \in D_2(A)$ and for any $j \geqslant 0$ define the mappings $\partial_x^\nabla : D_j(P) \longrightarrow D_{j+I}(P)$ by

$$\partial_x^\nabla (\Delta) = (-I)^j [\![X, \Delta]\!]_\nabla, \quad (II)$$

$\Delta \in D_j(P)$. In particular, for any $Y \in D_j(A)$ the element $\partial_x(Y) =$
$= (-I)^j [\![X, Y]\!] \in D_{j+I}(A)$ is defined.

Proposition. The mapping ∂_x^∇ can be defined by the induction as follows:

(i) $\partial_x^\nabla(p)(a) = \nabla_{X(a)}(p)$, $p \in P$, $a \in A$,

(ii) $\partial_x^\nabla (\Delta)(a) = L_{X(a)}^\nabla(\Delta) - \partial_x^\nabla(\Delta(a))$, $j > 0$,

$\quad \Delta \in D_j(P)$, $a \in A$.

3.5. The basic properties of ∂_x^∇ are described by the following

Proposition. For any $X \in D_2(A)$ the mapping $\partial_x^\nabla : D_*(P) \longrightarrow$
$D_*(P)$ is K-linear. If $a \in A$, $Y \in D_j(A)$, $\Delta \in D_k(P)$, then we have:

(i) $\partial_{ax}^\nabla(\Delta) = a\, \partial_x^\nabla(\Delta) - X \wedge \Delta(a)$,

(ii) $\partial_x^\nabla(Y \wedge \Delta) = \partial_x(Y) \wedge \Delta + (-I)^j Y \wedge \partial_x^\nabla(\Delta)$,

(iii) $\partial^\nabla_X [\![Y, \Delta]\!]_\nabla = [\![\partial_X(Y), \Delta]\!]_\nabla + (-I)^j [\![Y, \partial^\nabla_X(\Delta)]\!]_\nabla +$

$(-I)^{j+k-I} R_\nabla(X, Y, \Delta)$.

In particular, when $Y \in D_2(A)$, then

(iv) $\partial^\nabla_X \partial^\nabla_Y(\Delta) + \partial^\nabla_Y \partial^\nabla_X(\Delta) = (-I)^k [\![[\![X, Y]\!], \Delta]\!]_\nabla -$

$R_\nabla(X, Y, \Delta)$,

and when $Y \in D_I(A)$, then

(v) $L^\nabla_Y \partial^\nabla_X(\Delta) - \partial^\nabla_X L^\nabla_Y(\Delta) = \partial^\nabla_{L_Y(X)}(\Delta) - R_\nabla(X, Y, \Delta)$.

If $X = X_I \wedge X_2$, $X_i \in D_I(A)$, then, in addition, we have

(vi) $\partial^\nabla_X(\Delta) = X_I \wedge L^\nabla_{X_2}(\Delta) - X_2 \wedge L^\nabla_{X_I}(\Delta)$.

The proof follows from proposition 2.6.

From (ii) we see that for any $a \in A$ the equality

$$\partial^\nabla_X(a\Delta) = a\partial^\nabla_X(\Delta) - X(a) \wedge \Delta$$

is valid.

<u>Corollary</u>. For any $X \in D_2(A)$ the sequence

$$0 \longrightarrow P \xrightarrow{\partial^\nabla_X} D_I(P) \longrightarrow \ldots \longrightarrow D_j(P) \xrightarrow{\partial^\nabla_X} D_{j+I}(P) \longrightarrow \ldots \qquad (I2)$$

is the sequence of DO's of order $\leq I$. The mappings ∂^∇: $D_2(A) \dashrightarrow \mathrm{Diff}_I(D_j(P), D_j(P))$ are DO's of order $\leq I$ as well.

From proposition 2.7 the explicit expression for ∂^∇_X follows Namely, if $X \in D_2(A)$, $\Delta \in D_j(P)$ and ∇ is a connection in P, then for any $a_I, \ldots, a_{j+I} \in A$ we have

$$\partial^\nabla_X(\Delta)(a_I, \ldots, A_{j+I}) = \sum_{k=I}^{j+I} (-I)^{k-I} \nabla_{X(a_k)}(\Delta(a_I, \ldots$$

$$\ldots, a_{k-I}, a_{k+I}, \ldots, a_{j+I})) + \sum_{k<l} (-I)^{k+l} \Delta(X(a_k, a_l), a_I,$$

$$\ldots, a_{k-I}, a_{k+I}, \ldots, a_{l-I}, a_{l+I}, \ldots, a_{j+I}). \qquad (I3)$$

In particular, if $Y \in D_j(A)$, then

$$\partial^{\nabla}_{X}(Y)(a_I,\dots,a_{j+I}) = \sum_{k=I}^{j+I} (-I)^{k-I}X(a_k, Y(a_I,\dots,a_{k-I},$$

$$a_{k+I},\dots,a_{j+I}) \sum_{k<l} (-I)^{k+l}\, Y(X(a_k,a_l), a_I,\dots,a_{k-I}, a_{k+I},$$

$$\dots, a_{l-I}, a_{l+I}, \dots, a_{j+I}). \qquad (I4)$$

<u>3.6.</u> Consider now an important class of situations in which the sequence (I2) is a complex.

<u>Definition</u>. Let K be a commutative ring with identity, A be a commutative unitary K-algebra and P be a module over A.

(i) A pair (A;h), $h \in D_2(A)$, is said to be a canonical algebra if $[\![h,h]\!] = o$. In this case h is called the canonical structure in A while the element $\{a_I, a_2\}_h \overset{\text{def}}{=} h(a_I, a_2) \in A$ is called the Poisson bracket of the elements $a_I, a_2 \in A$ with respect to h.

(ii) A connection ∇ in P is said to be a Hamiltonian one with respect to h if for any $\Delta \in D_*(P)$ the equality $R_{\nabla}(h,h,\Delta) = o$ holds.

In the sequel we shall suppose that the equation 2x=0 has the unique solution both in A and in P.

<u>Proposition</u>. An element $h \in D_2(A)$ is a canonical structure iff one of the following equivalent conditions is satisfied:

(i) $\partial_h \circ \partial_h = o$,

(ii) $\partial_h(\partial_h(a)) = o$ for any $a \in A$,

(iii) $[\![h, h]\!] = \partial_h(h) = o$,

(iv) $h(a_I, h(a_2, a_3)) + h(a_2, h(a_I, a_3)) + h(a_3, h(a_I, a_2))$

$= o$ for any $a_I, a_2, a_3 \in A$,

(v) the Poisson bracket $\{\ ,\ \}_h$ defines in A the structure of a Lie algebra over K.

<u>Proof</u>. Let $h \in D_2(A)$ be a canonical structure. Then, since the curvature of the connection constructed in example I of subsection I.5 vanishes, from proposition 3.5 (iv) it follows that $2\partial_h(\partial_h(\Delta)) = o$ for any $\Delta \in D_*(A)$. Hence, $\partial_h \circ \partial_h = o$.

The implication (i) => (ii) is obvious.

From propositions 3.I and 3.4 it immediately follows that for any $a \quad A$ the equality $(\partial_h(h))(a) = 2\partial_h(\partial_h(a))$ holds. Hence, (ii) is equivalent to (iii).

From equality (II) we see that $[\![h,h]\!] = \partial_h(h)$. Therefore, (iii) is equivalent to the fact that h is a canonical structure.

Suppose a_I, a_2, a_3 are arbitrary elements of A. Then, by (I4),

$$[\![h, h]\!](a_I, a_2, a_3) = 2\,(h(a_I, h(a_2, a_3)) + h(a_2, h(a_3, a_I)) +$$

$$+ h(a_3, h(a_I, a_2))),$$ i.e. $[\![h,h]\!] = 0$ iff (iv) holds.

Finally, the condition (iv) is obviously equivalent to (v).

Q.E.D.

Now, we can give the following important

<u>Definition.</u> K-modules

$$\gamma_h^i(A) = \ker(\partial_h : D_i(A) \longrightarrow D_{i+I}(A)) \Big/ \operatorname{im}(\partial_h : D_{i-I}(A) \longrightarrow D_i(A))$$

are said to be the Hamiltonian cohomologies of the canonical algebra (A; h).

<u>3.7</u>. <u>Proposition</u>. A connection ∇ in an A-module P is Hamiltonian with respect to a canonical structure $h \in D_2(A)$ iff one of the following equivalent conditions is satisfied:

(i) $\partial_h^\nabla \circ \partial_h^\nabla = 0$,

(ii) $\partial_h^\nabla(\partial_h^\nabla(p)) = 0$ for any $p \in P$,

(iii) $R_\nabla(h(a_I), h(a_2), p) = 0$ for any $a_I, a_2 \in A$, $p \in P$; in particular, ∇ is Hamiltonian of $R_\nabla = 0$.

Next definition generalizes definition 3.6.

<u>Definition.</u> K-modules

$$\gamma_h^i(P;\nabla) = \ker(\partial_h^\nabla : D_i(P) \longrightarrow D_{i+I}(P)) \Big/ \operatorname{im}(\partial_h^\nabla : D_{i-I}(P) \longrightarrow D_i(P))$$

are said to be the Hamiltonian cohomologies of an A-module P constructed with respect to the Hamiltonian connection ∇ in P.

3.8. Suppose $(A;h)$ is a canonical algebra, P is an A-module with the Hamiltonian connection ∇. Define $\varphi : A \dashrightarrow Der(P)$ from the commutativity of the diagram

$$\begin{array}{ccc} A & \xrightarrow{\varphi} & Der(P) \\ & \partial_h \searrow \quad \nearrow \nabla & \\ & D_I(A) & \end{array}$$

Thus, $\varphi(a)(p) = \nabla_{h(a)}(p)$ for any $a \in A$, $p \in P$. Since ∇ is Hamiltonian and h is the canonical structure, we have the following properties of φ :

(i) φ is K-linear,

for any a_I, a_2 A the equalities

(ii) $\varphi(h(a_I, a_2)) = [\varphi(a_I), \varphi(a_2)]$,

(iii) $\varphi(a_I a_2) = a_I \varphi(a_2) + a_2 \varphi(a_I)$,

(iv) $\delta_{a_2}(\varphi(a_I)) = h(a_I, a_2)$

are valid. In the last equality $h(a_I, a_2)$ is understood as the element from $Hom_A(P,P)$: $h(a_I, a_2)(p) = h(a_I, a_2) \cdot p$, $p \in P$ (the definition of δ_a see in subsection I.I).

Definition. Let $(A;h)$ be a canonical algebra and P be an A-module. The pair $(P;\varphi)$, where $\varphi : A \longrightarrow Der(P)$ is a mapping satisfying the conditions (i-iv) above, is said to be a canonical $(A;h)$-module, while φ is said to be a canonical structure in P.

For any cannonical structure φ we can construct the Lie derivatives of the form $L^{\varphi}_{h(a)}: D_*(P) \longrightarrow D_*(P)$ by setting

$$L^{\varphi}_{h(a)}(p) = \varphi(a)(p), \; p \in P, \; a \in A,$$

and

$$L^{\varphi}_{h(a)}\Delta(b) = L^{\varphi}_{h(a)}(\Delta(b)) - \Delta(h(a,b)),$$

if $\Delta \in D_j(P)$, $b \in A$, $j > o$ (cf. proposition 3.I).

In the same way as in proposition 3.4 we shall define the

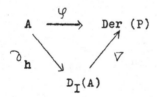

operators ∂^{φ}_h: $D_j(P) \dashrightarrow D_{j+I}(P)$ by setting

$$\partial^{\varphi}_h (p)(a) = \varphi(a)(p),$$

when $a \in A$, $p \in P$, and

$$\partial^{\varphi}_h (\Delta)(a) = L^{\varphi}_{h(a)}(\Delta) - \partial^{\varphi}_h(\Delta(a)),$$

when $\Delta \in D_j(P)$, $j > o$.

We can prove that $L^{\varphi}_{h(a)}$ and ∂^{φ}_h are well defined and formulate their basic properties in the same way as it was done in propositions 3.I, 3.2, 3.4 and 3.5. In particular, $\partial^{\varphi}_h \circ \partial^{\varphi}_h = o$ and therefore K-modules.

$$\gamma^i_h(P;\varphi) \overset{\text{def}}{=} \ker(\partial^{\varphi}_h: D_i(P) \dashrightarrow D_{i+I}(P)) \Big/ \operatorname{im}(\partial^{\varphi}_h: D_{i-I}(P) \dashrightarrow D_i(P))$$

are defined which are called the Hamiltonian cohomologies of the canonical module $(P;\varphi)$.

<u>3.9</u>. Let us describe the modules $\gamma^o_h(A)$. Let $a \in A$; from $\gamma^o_h = \ker(\partial_h): A \dashrightarrow D_I(A)$ it follows that a belongs to $\gamma^o_h(A)$ iff $\{a, a\}_h = o$ for all $a \in A$. In other words, a lies in the center of the Lie algebra which is determined in A by the Poisson bracket. This center is called the Poisson center of the canonical algebra (A;h) and obviously is a subalgebra in A.

Further, the set $\operatorname{im}(\partial_h: D_o(A) \dashrightarrow D_I(A))$ consists of the derivations of the form $h(a) = \{a, \}_h$, $a \in A$. Such a derivation is said to be a Hamiltonian one (with the Hamiltonian a). The set of these derivations is denoted by Ham (A;h). On the other hand a derivation $X \in D_I(A)$ lies in $\ker(\partial_h: D_I(A) \dashrightarrow D_2(A))$ iff for any $a_I, a_2 \in A$ the equality.

$$X\{a_I, a_2\}_h = \{Xa_I, a_2\}_h + \{a_I, Xa_2\}_h \tag{I5}$$

holds. Derivations satisfying the condition (I5) are called canonical and the set of such derivations is denoted by Can (A;h). Thus, $\gamma^1_h(A)$ coincides with the factor Can $(A;h)/_{\text{Ham}(A;h)}$. It should be noted that the so-called main theorem of mechanics (cf. [8]) postulates identity between Can (A,h) and Ham (A;h) when $A = C^{\infty}(T^*(M))$, where M is a simply connected smooth menifold while the canonical structure in A is induced by the symplectic structure on $T^*(M)$. In other words the main theorem of mechanics is equivalent to the I-acyclicity of the complex

$$0 \longrightarrow A \xrightarrow{\partial_h} D_1(A) \longrightarrow \dots \longrightarrow D_i(A) \xrightarrow{\partial_h} D_{i+1}(A) \longrightarrow \dots$$

in the classical case.

In conclusion note that the proposition 3.7 can be formulated in the following form.

<u>Proposition</u>. A connection ∇ in an A-module P is Hamiltonian with respect to the canonical structure $h \in D_2(A)$ iff the mapping

$$R_\nabla : \mathrm{Ham}(A;h) \otimes_K \mathrm{Ham}(A;h) \longrightarrow \mathrm{Hom}_A(P,P),$$

where $R_\nabla(h(a_1) \otimes h(a_2))(p) = R_\nabla(h(a_1), h(a_2), p)$, $a_1, a_2 \in A$, $p \in P$, is trivial.

<u>3.10</u>. Consider the general situation again. Suppose, as above, (A;h) is a canonical algebra and ∇ is a Hamiltonian connection in an A-module P. From proposition 3.5(ii) it follows that

$$\ker\partial_h \wedge \ker\partial_h^\nabla \subset \ker\partial_h^\nabla, \quad \ker\partial_h \wedge \mathrm{im}\,\partial_h^\nabla \subset \ker\partial_h^\nabla, \tag{16}$$

while from 3.5(iii) we have

$$[\![\ker\partial_h, \ker\partial_h^\nabla]\!]_\nabla \subset \ker\partial_h^\nabla, \quad [\![\ker\partial_h, \mathrm{im}\,\partial_h^\nabla]\!]_\nabla \subset \ker\partial_h^\nabla. \tag{17}$$

Thus, the exterior multiplication and the Schouten bracket induce K-linear mappings

$$\gamma_h^i(A) \otimes_K \gamma_h^j(P;\nabla) \longrightarrow \gamma_h^{i+j}(P;\nabla)$$

and

$$\gamma_h^i(A) \otimes_K \gamma_h^j(P;\nabla) \longrightarrow \gamma_h^{i+j-1}(P;\nabla),$$

which are also denoted by $\wedge$ and $[\![\ ,\ ]\!]_\nabla$ respectively. The properties of these operations are described by the following

<u>Proposition.</u> With respect to the exterior multiplication $\gamma_h^*(A) = \sum_{i\geq 0}\gamma_h^i(A)$ is a graded $\gamma_h^0(A)$-algebra, while $\gamma_h^*(P;\nabla)$ =

$= \sum_{i\geq 0} \gamma_h^i (P; \nabla)$ is a graded $\gamma_h^*(A)$-algebra. For any elements $x \in \gamma_h^i(A)$, $y \in \gamma_h^j(A)$, $\delta \in \gamma_h^k(P;\nabla)$ the equalities

(i) $[\![x,y]\!] = (-I)^{ij} [\![y, x]\!]$,

(ii) $[\![x, y \wedge \delta]\!]_\nabla = (-I)^k [\![x,y]\!] \wedge \delta + (-I)^{ij} y \wedge [\![x, \delta]\!]_\nabla$,

(iii) $[\![x \wedge y, \delta]\!]_\nabla = (-I)^{ij} y \wedge [\![x, \delta]\!]_\nabla + x \wedge [\![y, \delta]\!]_\nabla$,

(iv) $[\![x, [\![y, \delta]\!]_\nabla]\!]_\nabla + (-I)^k [\![[\![x,y]\!] , \delta]\!]_\nabla + (-I)^{ij} [\![y, [\![x, \delta]\!]_\nabla]\!]_\nabla$

are valid. In particular, $\gamma_h^*(A)$ is the graded Lie elgebra with respect to the Schouten bracket, which is represented in $\gamma_h^*(P;\nabla)$.

The proof follows from embeddings (I6) and (I7) and from propositions 2.2 and 2.6.

3.II. Consider some examples.

I. For any commutative K-algebra A its symbolic algebra $S_*(A) = \sum_{i\geq 0} \mathrm{Diff}_i(A)/\mathrm{Diff}_{i-I}(A)$ is determined (see [I]). Denote by $S_i(A)$ the A-module $\mathrm{Diff}_i(A)/\mathrm{Diff}_{i-I}(A)$. Let $s_{i_k} = \Delta_{i_k} + \mathrm{Diff}_{i_k - I}(A) \in S_{i_k}(A)$, $\Delta_{i_k} \in \mathrm{Diff}_{i_k}(A)$, k = I,2. Set

$$\{ s_{i_I}, s_{i_2} \} = [\Delta_{i_I}, \Delta_{i_2}] + \mathrm{Diff}_{i_I + i_2 - I}(A) \in S_{i_I + i_2 - I}(A).$$

As it was shown in [1] , the operation { , } (the Poisson bracket) determines the structure of the canonical A-algebra in $S_*(A)$.

2. If (A;h) is a canonical algebra and P is an A-module, then (see subsection I.5, example 2) there is the connection ∇ : $D_I(A) \longrightarrow \mathrm{Der}(\mathrm{Diff}_*(P,A))$ whose curvature vanishes. Hence, ∇ is the Hamiltonian connection and for any A-module P the module $\mathrm{Diff}_*(P,A)$ is canonical.

3. Let G be a Lie algebra over K, U_G be its enveloping algebra and $K = U_G^0 \subset U_G^1 \subset \ldots \subset U_G^i \subset \ldots \subset U_G$ be the natural filtration in U_G . Then the graded K-algebra $S_G = \sum_{i\geq 0} S_G^i = \sum_{i\geq 0} U_G^i / U_G^{i-I}$ is

commutative and has the natural structure of the canonical algebra:

if $s_{i_k} \in S_G^{i_k}$, $s_{i_k} = u_{i_k} + S_G^{i_k - I}$, $u_{i_k} \in U_G^{i_k}$, $i = I,2$, then we set

$$h_G(s_{i_I}, s_{i_2}) = \{ s_{i_I}, s_{i_2} \}_G \overset{\text{def}}{=} (u_{i_I} u_{i_2} - u_{i_2} u_{i_I}) + U_g^{i_I + i_2 - I}$$

(see [I]). Note that S_G consists of symbols of left-invariant differential operators on the Lie group corresponding to G.

4. If H is also a K-algebra Lie and Φ: G $\longrightarrow$ H is a homomorphism of Lie algebras, then we have the commutative diagram

$$\begin{array}{ccc} G & \xrightarrow{\ \Phi\ } & H \\ {\scriptstyle i_G}\downarrow & & \downarrow{\scriptstyle i_H} \\ U_G & \xrightarrow{\ U(\Phi)\ } & U_H, \end{array}$$

where i_G and i_H are the natural embeddings, while $U(\Phi)$ is the homomorphism of filtered algebras which exists due to the functorial nature of the correspondence $G \longrightarrow U_G$. Therefore the homomorphism of graded algebras $S(\Phi)$: $S_G \longrightarrow S_H$ is defined by which S_H becomes the S_G-module. Define the mapping $\varphi : S_G \longrightarrow \operatorname{Der} S_H$ by setting

$$\varphi(s)\,(s') = \{ S(\Phi)\,(s),\, s' \}_H, \quad s \in S_G,\ s' \in S_H. \tag{I8}$$

Obviously, (I8) determines the structure of a canonical S_G-module in S_H.

In particular, if V is a K-module and

$$\Phi : G \longrightarrow \operatorname{End}(V) = gl(V)$$

is a representation of the Lie algebra G in V, then the structure of the canonical S_G-module in $S_{gl(v)}$ is defined.

5. Let, as above, G be a K-algebra Lie and V be a K-module. Consider the S_G-module $S_G \otimes_K V \overset{\text{def}}{=} S_G(V)$. If Φ is a representation of G in V, then the structure of a canonical S_G-module is $S_G(V)$ can be defined. Namely, when $s \in S_G^I = G$ we set

$$\varphi(s)\,(s' \otimes v) = \{s, s'\}_G \otimes v + s' \otimes \Phi(s)v, \quad s' \in S_G,\ v \in V.$$

If $s \in S_G^i$, then s can be represented as the sum of elements of the form $s_I \cdot s_2$, where $s_k \in S_G^{i_k}$, $i_k < i$, k=I,2. Hence, for any $w \in S_G(V)$ we can set by the induction

$$\varphi(s)\,(w) = \sum (s_I\, \varphi(s_2)\,(w) + s_2\, \varphi(s_I)\,(w)).$$

3.I2. In this subsection we consider the relations between the Hamiltonian cohomologies and other cohomological theories.

Let (A;h) be a canonical K-algebra. Then, by proposition I.3, we have the commutative diagram

$$\begin{array}{ccc} A & \xrightarrow{\partial_h} & D_I(A) \\ & {\scriptstyle d}\searrow \quad \nearrow{\scriptstyle f_h} & \\ & \Lambda^I(A) & \end{array} \qquad (I9)$$

where $f_h \in \mathrm{Hom}_A(\Lambda^I, D_I(A))$. Since Λ^i is the i-th exterior power of Λ^I, we can introduce homomorphisms $f_h\colon \Lambda^i \longrightarrow D_i(A)$ by setting

$$f_h(\omega_I \wedge \omega_2) = f_h(\omega_I) \wedge f_h(\omega_2) \in D_i(A)$$

for $\omega_1 \in \Lambda^{i_I}$, $\omega_2 \in \Lambda^{i_2}$, $i_I+i_2=i$, $i_I, i_2 < i$. By the definitions we have $\partial_h \circ f_h = f_h \circ d$, where d is the exterioir derivation.

Thus, we have the commutative diagram

$$\begin{array}{ccccccccccc} 0 & \longrightarrow & A & \xrightarrow{\partial_h} & D_I(A) & \longrightarrow & \cdots & \longrightarrow & D_i(A) & \xrightarrow{\partial_h} & D_{i+I}(A) \longrightarrow \cdots \\ & & \uparrow{\scriptstyle id} & & \uparrow{\scriptstyle f_h} & & & & \uparrow{\scriptstyle f_h} & & \uparrow{\scriptstyle f_h} \\ 0 & \longrightarrow & A & \xrightarrow{d} & \Lambda^I & \longrightarrow & \cdots & \longrightarrow & \Lambda^i & \xrightarrow{d} & \Lambda^{i+I} \longrightarrow \cdots\cdots \end{array}$$

Let $H^*(A) = \sum\limits_{i \geq 0} H^i(A)$ denote the de Rham cohomologies of the algebra A. Then we have K-module homomorphism $f_h^*\colon H^*(A) \dashrightarrow \gamma_h^*(A)$. Call the canonical structure $h \in D_2(A)$ non-degenerate if $f_h\colon \Lambda^I \longrightarrow D_I(A)$ is the isomorphism.

Theorem. If K is an algebra over the field of rational numbers and the algebra A is such that $\Lambda^I(A)$ is a projective finitely generated A-module, then for all nondegenerate canonical structures

$h \in D_2(A)$ the cohomologies $\gamma^*_h(A)$ are mutually isomorphic and coincide with the de Rham cohomologies of A.

Proof. If Λ^I is the projective module of finite type then from the equality $D_i(A) = \mathrm{Hom}_A(\Lambda^i, A)$ it follows that for all $i \geq o$ the isomorphisms

$$D_i(A) \simeq \underbrace{D_I(A) \wedge \ldots \wedge D_I(A)}_{i \text{ times}}$$

take place. Moreover, $f_h: \Lambda^i \dashrightarrow D_i(A)$ in this case is the i-th exterior power of $f_h: \Lambda^I \dashrightarrow D_I(A)$. So, if the latter is an isomorphism then f^*_h is an isomorphism as well. Now, if $g \in D_2(A)$ is another nondegenerate canonical structure then we have

$$f^*_g \circ (f^*_h)^{-I} : \gamma^*_h(A) \dashrightarrow \gamma^*_g(A). \text{ Q.E.D.}$$

In particular, the conditions of the theorem hold if $A = C^\infty(M)$ is the function algebra on a symplectic manifold $(M; \Omega)$, $\Omega \in \Lambda^2(M)$ while the Rham complex for A is constructed in the geometrical category. In this context the relation between the canonical structure h in $C^\infty(M)$ and the symplectic form Ω is expressed by the equality $h = f_h(\Omega)$.

3.I3. Let $h \in D_2(A)$ be a nondegenerate canonival structure. Consider the form $\Omega_h = f_h^{-I}(h) \in \Lambda^2$ corresponding to h. In this case, provided the conditions of theorem 3.I2 hold, the Schouten bracket of arbitrary elements $X, Y \in D_*(A)$ can be expressed through the Hamiltonian differentiation ∂_h and the pairing $\langle \ , \ \rangle$ (see subsection I.3). Namely, the following result takes place.

Proposition. If the canonical structure $h \in D_2(A)$ is nondegenerate, K is an algebra over the field of rational numbers and $\Lambda^I(A)$ is a projective module of finite type then for any $X \in D_i(A)$ and $Y \in D_j(A)$ we have

$$[\![X, Y]\!] = (-I)^{i+j}\partial_h \langle X \wedge Y, \Omega_h\rangle + (-I)^{i+j+I}\langle (\partial_h X), \Omega_h\rangle + (-I)^{j+I}\langle X \wedge \partial_h(Y), \Omega_h\rangle + (-I)^{i+j}\langle \partial_h X, \Omega_h\rangle \wedge Y +$$
$$+ (-I)^j X \wedge \langle \partial_h Y, \Omega_h\rangle + (-I)^{i+j+I}(\partial_h\langle X, \Omega_h\rangle) \wedge Y +$$
$$+ (-I)^{j+I} X \wedge \partial_h \langle Y, \Omega_h\rangle .$$

Proof. Since the modules $\Lambda^i(A)$ are generated by the forms $da_I \wedge \ldots \wedge da_i$, $a_\alpha \in A$, and $f_h(da) = h(a)$, then, f_h: $\Lambda^i \longrightarrow D_i(A)$ being isomorphisms, any element from $D_i(A)$ can be represented as the sum of the elements $a_o h(a_I) \wedge \ldots \wedge h(a_i)$, $a_\alpha \in A$. The result to be proved follows from the basic properties of the Schouten bracket and the equality.

$$\langle h(a_I) \wedge \ldots \wedge h(a_i), \Omega_h \rangle = \sum_{\alpha_1 < \alpha_2} (-I)^{\alpha_I + \alpha_2 + I} \{ a_{\alpha_I}, a_{\alpha_2} \} \cdot$$

$$\cdot h(a_I) \wedge \ldots h(a_{\alpha_I - I}) \wedge h(a_{\alpha_I + I}) \wedge \ldots \wedge h(a_{\alpha_2 - I}) \wedge$$

$$\wedge h(a_{\alpha_2 + I}) \wedge \ldots \wedge h(a_i)$$

which is easy to verify by the induction. Q.E.D.

Corollary. When the conditions of theorem 3.I3 hold the Schouten bracket in $\gamma_h^*(A)$ is trivial.

Proof. In fact, we have

$$[\![X, Y]\!] = (-I)^{i+j} \partial_h \langle X \wedge Y, \Omega_h \rangle + (-I)^{i+j+I} (\partial_h \langle X, \Omega_h \rangle) \wedge Y +$$

$$+ (-I)^{j+I} X \wedge \partial_h \langle Y, \Omega_h \rangle \in \operatorname{im} \partial_h$$

when $X, Y \in \ker \partial_h$. Q.E.D.

Remark. Nondegeneratedness of h is essential. For example, if $h = 0$ then $\gamma_h^*(A) = D_*(A)$ where the Schouten bracket is, in general, nontrivial.

3.I4. Consider again example 5 from subsection 3.II. Suppose in addition that G is a free K-module and V is a projective K-module of finite type. Then the modules $D_i(S_G(V)$ can be represented in the form

$$D_i(S_G(V)) \simeq S_G \otimes_K \operatorname{Hom}(\Lambda^i G, V).$$

Using this representation equality (I3) can be rewritten as

$$\partial_h(s \otimes f)(a_I, \ldots, a_{i+I}) = \sum_{k<l} (-I)^{k+I} s \bullet f([a_k, a_l], a_I, \ldots,$$

$$a_{k-1}, a_{k+1}, \ldots, a_{l-1}, a_{l+1}, \ldots, a_{i+1}) + \sum_{k<1}^{i+1} (-1)^{k+1}.$$

$$(\{a_k, s\} f(a_1, \ldots, a_{k-1}, a_{k+1}, \ldots, a_{i+1}) + s \otimes \Phi(a_k)(f(a_1,$$

$$\ldots, a_{k-1}, a_{k+1}, \ldots, a_{i+1}))).$$

where $s \otimes f \in D_i(S_G(V))$, $s \in S_G$, $f \in \mathrm{Hom}_K(\wedge^i G, V)$, $a_1, \ldots, a_{i+1} \in G$.

Hence, the following commutative diagram

$$\begin{array}{ccccccccc} 0 \longrightarrow & S_G(V) & \longrightarrow \ldots \longrightarrow & D_i(S_G(V)) & \xrightarrow{\gamma^{\varphi}_{h_G}} & D_{i+1}(S_G(V)) & \longrightarrow \ldots \\ & \uparrow {\scriptstyle 1\otimes id} & & \uparrow {\scriptstyle 1\otimes id} & & \uparrow {\scriptstyle 1\otimes id} & \\ 0 \longrightarrow & V & \longrightarrow \ldots \longrightarrow & \mathrm{Hom}_K(\wedge^i G, V) & \longrightarrow & \mathrm{Hom}_K(\wedge^{i+1} G, V) & \longrightarrow \ldots, \end{array}$$

takes place where the lower complex determines the cohomologies of the Lie algebra G with coefficients in the representation Φ, while $(1 \otimes id)(f) = 1 \otimes f$ for any $f \in \mathrm{Hom}_K(\wedge^i G, V)$.

From the above said we get the following

<u>Theorem</u>. Let K be a field, G be a finite dimensional Lie algebra over K, V be a finite dimensional vector space over K and Φ: $G \longrightarrow \mathrm{End}(V)$ be a representation of G in V. Then for all $i \geq 0$ the isomorphisms

$$\gamma^i_{h_G}(S_G(V); \varphi) \simeq H^i(G; \Phi) \otimes_K S_G$$

take place, where φ is the structure of the canonical S_G-module in $S_G(V)$ defined in example 5 of subsection 3.II, while $H^i(G;\Phi)$ are the cohomologies of G with the coefficients in the representation Φ.

The proof follows from the above mentioned , the Poincaré-Birkhoff-Witt theorem and the universal coefficients theorem (see [9,10]).

4. Commuting canonical structures.

We, as before, suppose, that the equation $2x = 0$ has the unique solution in the algebras which we consider and in all the modules over them.

4.1. Definition. Canonical structures $g, h \in D_2(A)$ are called commuting if the equality $[\![h, g]\!] = 0$ takes place.

Proposition. Suppose $g, h \in D_2(A)$ are canonical structures. Then the following statesments are equivalent:

(i) g and h commute,

(ii) g + h is a canonical structure,

(iii) $\partial_g(h) = 0$,

(iv) $\partial_h(g) = 0$,

(v) $\partial_g \circ \partial_h + \partial_h \circ \partial_g = 0$.

Proof. This immediately follows from proposition 3.5.

4.2. The equality (v) from the previous proposition holds in a more general situation hamely, we have

Proposition. Suppose g and h are commuting canonical structures in A, P is an A-module and ∇ is a connection in P which is Hamiltonian with respect to both structures. Then the equality

$$\partial_g^\nabla \circ \partial_h^\nabla + \partial_h^\nabla \circ \partial_g^\nabla = 0$$

holds if for any $\Delta \in D_*(P)$ we have $R_\nabla(g, h, \Delta) = 0$.

4.3. Example. Let V be a module over the ring K. Consider two structures, G and H. of Lie algebras in V. Suppose further that $[v_1, v_2]_G$ and $[v_1, v_2]_H$ are the commutators of elements $v_1, v_2 \in V$ with respect to G and H, while ad_G and ad_H are the corresponding adjoint actions. Then the canonical structures h_G and h_H commute iff the equality

$$[ad_G v_1, ad_H v_2] + [ad_H v_1, ad_G v_2] = ad_G [v_1, v_2]_H + ad_H [v_1, v_2]_G$$

holds (note, that in the left-hand side of the equality $[\ ,\ ]$ denotes the commutator of operators in V).

4.4. The role of commuting canonical structures is illustrated by the following results.

Lemma. Suppose $g, h \in D_2(A)$ are commuting canonical structures, P is an A-module and ∇ is a connection in P which satisfies the conditions of proposition 4.2. If $\gamma_g^{k+1}(P; \nabla) = 0$ and $\Delta_1, \Delta_2 \in$

$D_k(P)$ satisfy $\partial^{\nabla}_h(\Delta_I) = \partial^{\nabla}_g(\Delta_2)$ then there exist such elements $\Delta_3, \Delta_4, \dots, \Delta_\alpha, \dots \in D_k(P)$ that

$$\partial^{\nabla}_h(\Delta_\alpha) = \partial^{\nabla}_g(\Delta_{\alpha+I}).$$

Proof. Induction with respect to α. Suppose that the elemets $\Delta_I, \dots, \Delta_\alpha$ have been constructed. Then

$$\partial^{\nabla}_g(\partial^{\nabla}_h(\Delta_\alpha)) = -\partial^{\nabla}_h(\partial^{\nabla}_g(\Delta_\alpha)) = -\partial^{\nabla}_h(\partial^{\nabla}_h(\Delta_{\alpha-I}) = o.$$

Hence, $\partial^{\nabla}_h(\Delta_\alpha) \in \ker(\partial^{\nabla}_g : D_{k+I}(P) \longrightarrow D_{k+2}(P))$. This, together with the triviality of $\gamma^{k+I}_g(P;\nabla)$, implies the existence of such $\Delta_{\alpha+I} \in D_k(P)$ that $\partial^{\nabla}_h(\Delta_\alpha) = \partial^{\nabla}_g(\Delta_{\alpha+I})$ Q.E.D.

Theorem(cf. [5]). Suppose $g, h \in D_2(A)$ are commuting canonical structures and $\gamma_g^I(A) = o$. If $a_I, a_2 \in A$ are such elements, that $h(a_I) = g(a_2)$, then:

(i) there exist such elements $a_3, a_4, \dots, a_\alpha, \dots \in A$, that

$$h(a_\alpha) = g(a_{\alpha+I}),$$

(ii) elements $a_I, \dots, a_\alpha, \dots$ are in involution i.e.

$$\{a_\alpha, a_\beta\}_h = \{a_\alpha, a_\beta\}_g = o$$

for any $\alpha, \beta \geq I$.

Proof. The first statement is a particular case of the previous lemma. The second one can be proved just in the same way as it was done in [5] and is omitted here.

4.5. Let $h \in D_2(A)$ be a canonical structure. Denote by Com(h) $\subset D_2(A)$ the set of all canonical structures which commute with h. From proposition 4.I(iii) it follows that $\mathrm{Com}(h) \subset \ker \partial_h$ and thus the mapping

$$\tau : \mathrm{Com}(h) \dashrightarrow \gamma^2_h(A)$$

is defined. An element $g \in \mathrm{Com}(h)$ lies in the kernel of τ iff it has the form $g = \partial_h(X) = -L_X(h)$, $X \in D_I(A)$. Since g is a canonical structure, then the equality $[\![L_xh, L_xh]\!] = o$ should hold, which is equivalent (as it follows from the proposition 3.5(v)) to $L_X^2(h) \in \ker \partial_h$. The elements of the form $L_X(h)$, $X \in D_I(A)$, are natural to be called infinitesimal deformations of the structure h along the derivations $X \in D_I(A)$. Thus, we have the following result.

Proposition. If $h \in D_2(A)$ is a canonical structure and $\gamma_h^2(A)=0$ then all the structures commuting with h are the infinitesimal deformations of h along such derivations $X \in D_I(A)$, that $L_X^2(h) = L_Y(h)$ for some $Y \in D_I(A)$.

4.6. Describe now the image of τ. By the results of subsections 3.5 and 3.I0 the element $\delta \in \gamma_h^2(A)$ lies in $\operatorname{im}\tau$ iff the equality

$$[\![\Delta, \Delta]\!] = \partial_h(L_X^2(h) - 2L_X(\Delta)) \tag{20}$$

hold, where Δ is a representative of the cohomological class δ, while $X \in D_I(A)$ is some derivation. When the conditions of subsection 3.I3 are satisfied, equality (20) can be rewritten in a more convenient form. Namely, $\delta \in \operatorname{im}\tau$ if

$$\langle \Delta \wedge \Delta, \Omega_h \rangle - 2\langle \Delta, \Omega_h \rangle \Delta + 2L_X(\Delta) - L_X^2(h) \in \ker \partial_h$$

for some $X \in D_I(A)$.

4.7. Let us give the coordinate form of the condition $g \in \operatorname{Com}(h)$. Suppose $D_I(A)$ is a free A-module with the generators $\partial_I, \ldots, \partial_k, \ldots$ and $D_i(A) = \Lambda^i D_I(A)$. Let

$$h = \sum_{i<j} h_{ij}\, \partial_i \wedge \partial_j, \quad g = \sum_{i<j} g_{ij}\, \partial_i \wedge \partial_j, \quad h_{ij}, g_{ij} \in A.$$

Then, by proposition 2.6 (ii,iii) and by collorary 2.7, we have

$$[\![g,g]\!] = 2 \sum_{i<j<k} \sum_{l} (g_{il}\partial_l(g_{jk}) - g_{jl}\partial_l(g_{ik}) + g_{kl}\partial_l(g_{ij})) \cdot \partial_i \wedge \partial_j \wedge \partial_k$$

and

$$[\![g,h]\!] = \sum_{i<j<k} \sum_{l} (g_{il}\partial_l(h_{jk}) - g_{jl}\partial_l(h_{ik}) + g_{kl}\partial_l(h_{ij}) +$$

$$+ h_{il}\partial_l(g_{jk}) - h_{jl}\partial_l(g_{ik}) + h_{kl}\partial_l(g_{ij}))\partial_i\wedge\partial_j\wedge\partial_k .$$

Hence, $g \in \mathrm{Com}(h)$ iff the coeficients $g_{\alpha\beta}$ satisfy the following system of nonlinear differential equations

$$\sum_l (g_{il}\partial_l(g_{jk}) - g_{jl}\partial_l(g_{ik}) + g_k\partial_l(g_{ij})) = 0,$$

$$\sum_l (g_{il}\partial_l(h_{jk}) - g_{jl}\partial_l(h_{ik}) + g_{kl}\partial_l(h_{ij}) +$$

$$+ h_{il}\partial_l(g_{jk}) - h_{jl}\partial_l(g_{ik}) + h_{kl}\partial_l(g_{ij})) = 0,$$

$$i < j < k.$$

4.8. Suppose $g,h \in D_2(A)$ are commuting canonical structures, ∇ is a connection in an A-module P satisfing the conditions of proposition 4.2. Then the family of modules $D_{ij}(P) \overset{\mathrm{def}}{=} D_{i+j}(P)$, $i,j \geq 0$, together with the operators

$$\partial_g^\nabla : D_{ij}(P) \longrightarrow D_{i+1,j}(P), \quad \partial_h^\nabla : D_{ij}(P) \longrightarrow D_{i,j+1}(P), \quad i,j \geq 0,$$

forms a bi-complex. Denote by $H_h^{i,j}(P;\nabla)$ the cohomologies of the complex

$$0 \longrightarrow D_i(P) \xrightarrow{\partial_h^\nabla} D_{i+1}(P) \longrightarrow \ldots \xrightarrow{\partial_h} D_{i+j}(P) \xrightarrow{\partial_h} \ldots$$

in the member $D_{i+j}(P)$. Thus,

$$H_h^{i,j}(P;\ \ ) = \begin{cases} \ker(\partial_h^\nabla : D_i(P) \longrightarrow D_{i+1}(P)), & \text{when } j = 0, \\ \mathcal{H}_h^{i+j}(P;\nabla), & \text{when } j > 0 \end{cases}$$

As it follows from proposition 4.2, ∂_g^∇ determines the family of complexes

$$0 \longrightarrow H_h^{0,j}(P;\nabla) \longrightarrow H_h^{1,j}(P;\nabla) \longrightarrow \ldots \longrightarrow H_h^{j,i}(P;\nabla) \longrightarrow \ldots \qquad (21.j)$$

Denote by $H_g^i H_h^j(P;\nabla)$ the cohomologies of (21.j) in the i-th member. Then (see [9]) the spectral sequence of the bi-complex converges to the cohomologies of the complex $(\sum_{i+j=n} D_{i,j}(P), \partial_h^\nabla + \partial_g^\nabla)$ while the member E_2 of this spectral sequence is of the form $E_2^{i,j} = H_g^i H_h^j(P; \nabla)$.

Note that the canonical structure g determines a filtration

$$F_g^r D_*(P) = \underbrace{g \wedge \ldots \wedge g}_{r \text{ times}} \wedge D_*(P)$$

in $D_*(P)$ which is in agreement with the differentials ∂_P^∇ (when the latter are constructed for any structure h commuting with g). The corresponding spectral sequance converges to $\gamma_h^*(P; \nabla)$.

References

I. Vinogradov,A.M., and Krasil'shchik,I.S., What is the Hamiltonian formalism? Uspekhi Mat. Nauk 30 (I975), English transl. in Russian Math. Surveys 30 (I975), I77-202.

2. Krasil'shchik,I.S, Hamiltonian cohomology of canonical algebras, Dokl.Akad. Nauk SSSR 25I(I980), English transl. in Soviet Math. Dokl. 2I (I980), 625-629.

3. Frolicher, A., Nijenhuis, A., Theory of vector-valued differential forms I,Indag. Math. I8(I956), 338-359; II, Indag.Math. 20 (I958), 4I4-429.

4. Lichnerowicz, A., New geometrical dynamics, Lecture Notes in Math. 570(I975), 377-394.

5. Gel'fand, I.M., and Dorfman I.Ya., Schouten bracket and Hamiltonian operators, Funct. Anal. and Appl. I4(I980), 7I-74.

6. Vinogradov, A.M., The logic algebra for the theory of linear differential operators, Dokl. Akad. Nauk SSSR 205(I972); English transl. in Soviet Math. Dokl. I3(I972), I058-I062.

7. Krasil'shchik,I.S., Vinogradov, A.M. and Lychagin, V.V. Geometry of Jet Spaces and Nonlinear Differential Equations, Gordon and Breach Sci. Publ, New York, I986.

8. Arnol'd, V,I, Mathematical Methods of Classical Mechanics, Nauka, Moscow, I974.

9. Maclane, S., Homology, Springer-Verlag, Berlin, I963.

I0. Serre, J.-P., Lie Algebras and Lie Groups, Benjamin, New York-Amsterdam, I965.

MULTIDIMENSIONAL SLEEPING TOPS

Yu.I. Sapronov

Department of Mathematics,
Voronezh State University,
394693, Voronezh, USSR

Recent advances in the study of solid state dynamics are caused, to a great extent, by the application of the Lie group co-adjoint representation orbit method [4,7,13,15,16] , and the related reduced phase space method [5,6,7,9,10,15] . The reduced phase space is a level manifold of first integrals factorized with respect to the orbits of action of the subgroup of symmetries preserving the values of these integrals. In papers [6,9,10] the above methods were used to study bifurcation of level manifolds for the mappings of energy and momentum. In particular, these papers also contain information concerning the branching of stable stationary rotations.

The stability of stationary rotations of a heavy rigid body has been studied by many authors (see, e.g., [1-3,7,8,17-20]). It is known that the stability of a sleeping top is determined by the condition $C^2\omega^2 - 4Amg\ell > 0$ in the case of dynamical symmetry and by the condition $\min\{C-A, C-B\}\omega^2 > mg\ell$, for an asymmetric top. Once stability has been lost, a sleeping top "awakes": stationary motion turns into a more complicated motion.

In the present paper we consider stationary rotations of a multidimensional rigid body in the gravity field. The motion of an n-dimensional body is described by the Euler-Poisson generalized equations on the SO(n) group [4,11,12,14] . The multidimensional case is specific in that the factorization of the effective potential does not lead to a function on a sphere. The reduced effective potential also depends on some additional variables responsible for the location of the subgroup of level manifold symmetries relative to the kinetic moment in the group of rotations which preserve the vertical direction. If the kinetic moment has a regular value (an annulator or a Cartan subalgebra), this subgroup is a maximal torus in the group of rotations "about" the vertical axis. The additional variables can be eliminated (using, for example, the Lyapunov-Schmidt modified scheme [21]). After the elimination, we arrive at a key function on a sphere whose critical points correspond to stationary rotations. To analyse qualitatively the branching of critical points of this function and to find the asymptotic behaviour of the bifurcating critical points with respect to small increments of the bifurcation parameter, one has to calculate the leading part of the key function.

In the present paper we report some preliminary results. It is found that no stable standing stationary rotations (sleeping tops) can exist in an even-dimensional space, and that degenerate standing rotations cannot be stable in an odd-dimensional space. Sufficient conditions for the stability of stationary rotation of an asymmetric top have been obtained, which generalize the classical conditions (the inequality presented above). The stability conditions are derived from the assumption that the reduced effective potential is minimal.

1. The relative kinetic moment and separation of motions

An extension of the Euler-Poisson classical equations [14] to an n-dimensional case leads to the following equations:

$$\left.\begin{aligned} M\dot{\Omega} + \dot{\Omega}M + [\Omega, M\Omega + \Omega M] + \lambda r \wedge \gamma &= 0, \\ \dot{\gamma} + \Omega\gamma &= 0, \end{aligned}\right\} \tag{1}$$

where $\Omega = f^{-1}df/dt$ is angular velocity (in moving coordinates) of the motion defined by the function $f(t)$ having values in $SO(n)$; $M = \mathrm{diag}(m_1, \ldots, m_n)$, $m_j > 0$; $\lambda > 0$ is the load parameter which is proportional to the body mass; $[x,y] = xy - yx$; $e_n \in \mathbb{R}^n$ is a vertical unit vector; $Me_n = m_n e_n$; $r \wedge \gamma$ is a skew-symmetric matrix defined by $(r\wedge\gamma)x = (r,x)\gamma - (\gamma,x)r$, $\forall x \in \mathbb{R}^n$; $\gamma = f^{-1}e_n$ is the Poisson vector; $r \in \mathbb{R}^n$ is centre of gravity. System (1) is equivalent to the following equation:

$$M\dot{\Omega} + \dot{\Omega}M + [\Omega^2, M] + \lambda r \wedge f^{-1} e_n = 0 \tag{2}$$

(we have taken into account that $[\Omega, M\Omega + \Omega M] = [\Omega^2, M]$). The kinetic and potential energy of a body can be written as: $T = K(\Omega, \Omega)$ and $U = \lambda(r, \gamma) = \lambda(r, f^{-1}e_n)$, where $K(x,y) = \frac{1}{2}\langle Mx + xM, y\rangle$; $\langle\cdot,\cdot\rangle$ is the canonical scalar product (the Cartan-Killing form) defined by $\langle x,y\rangle = 1/2 \operatorname{tr} x^* y$. As usual, the total energy, $E = T + U$, is a first integral:
$\dot E = 2K(\dot\Omega, \Omega) + \lambda(r, \dot\gamma) = \langle -[\Omega, M\Omega + \Omega M] - \lambda(r\wedge\gamma), \Omega\rangle - \lambda(\Omega\gamma, r) = -\lambda\langle r\wedge\gamma, \Omega\rangle - \lambda(\Omega\gamma, r) = 0$.
Here we have used the obvious relations

$$\langle[\Omega, x], \Omega\rangle = 0, \quad \langle r\wedge\gamma, \Omega\rangle = \langle\Omega r, \gamma\rangle. \tag{3}$$

Let $\mathcal{M}(\Omega) = M\Omega + \Omega M$ be the kinetic moment (in moving coordinates). Consider the mapping of the relative kinetic moment $P_f = \pi\cdot J_f$, $J_f = f\mathcal{M}(\Omega)f^{-1}$, where π is an orthoprojector onto $\mathfrak{N} = \operatorname{Ann}(e_n) = \{y \in so(n) \mid ye_n = 0\}$. The mapping of the relative kinetic moment is an integral of equation (2) (see [5]). Indeed, differentiating $P_f(\Omega)$ with respect to time, we obtain: $d/dt\, P_f(\Omega) = \pi f[\Omega, \mathcal{M}(\Omega)]f^{-1} + \pi f\mathcal{M}(\dot\Omega)f^{-1} = \pi f([\Omega, \mathcal{M}(\Omega)] + [\mathcal{M}(\Omega), \Omega] - \lambda r\wedge f^{-1}e_n)f^{-1} = \lambda\pi(fr\wedge e_n) = 0$ (see Proposition 1). We have taken into account the well-known relation (see [5,7]):

$$\frac{d}{dt}(fxf^{-1}) = f[\Omega, x]f^{-1}, \quad \forall x \in so(n). \tag{4}$$

In what follows by $e_1, \ldots, e_n$ we shall mean a standard orthonormal basis in R^n ; $Me_j = m_j e_j$. Since $\langle e_j\wedge e_i, e_p\wedge e_q\rangle = ((e_p\wedge e_q)e_j, e_i) = (e_p, e_j)(e_q, e_i)$, then $\{\mathcal{E}_{ij} = e_i\wedge e_j\}$ is an orthonormal basis in $so(n)$. We assume that M is a simple matrix: $m_j \neq m_k$, $j \neq k$.

<u>Proposition 1</u>. The mappings $\operatorname{Ann}(\gamma)^\perp \to \operatorname{Lin}(\gamma)^\perp$ and $\operatorname{Lin}(\gamma)^\perp \longrightarrow \operatorname{Ann}(\gamma)^\perp$ defined by the correspondences $\Omega \longrightarrow \Omega\cdot\gamma$ and $x \longrightarrow \gamma\wedge x$ are mutually inverse isometries.

Proof. Without loss of generality, we may put $\gamma = e_n$. Since $\{e_i\wedge e_j\}$ is a basis in $so(n)$, the mapping $\Omega \longrightarrow \Omega\cdot e_n$ is a linear isomorphism ($\Omega \in \mathrm{Ann}(c_n)$). By definition of the exterior product,

$$(\gamma \wedge x)\gamma = (\gamma, \gamma)x - (x, \gamma)\gamma = x , \quad \forall x \in \mathrm{Lin}(\gamma)^{\perp}.$$

Hence, the mappings in question are mutually inverse. And their orthogonality follows from the relations $\langle \mathcal{E}_{j,n} , \mathcal{E}_{k,n} \rangle =$ $= (e_j, e_k)(e_n, e_n) = \delta_{j,k}$.

In accordance with direct decomposition $so(n) = \mathfrak{n}_\gamma + \mathfrak{n}_\gamma^{\perp}$ ($\mathfrak{n}_\gamma = \mathrm{Ann}(\gamma)$) , the angular velocity can be represented as the sum

$$\Omega = X + Y , \quad X \in \mathfrak{n}_\gamma^{\perp} , \quad Y \in \mathfrak{n}_\gamma . \tag{5}$$

This decomposition suggests that the motion of a rigid body is "separated" into motions in a fiber and in the base of the canonical bundle $SO(n) \longrightarrow S^{n-1}$. It follows from (5) that

$$\Omega \cdot \gamma = X \cdot \gamma \tag{6}$$

and, therefore,

$$\dot{\gamma} = - X \cdot \gamma$$

i.e. the evolution of γ is determined by the component X.

Proposition 2. The following relations are valid:

$$X = \dot{\gamma} \wedge \gamma , \quad |X| = |\dot{\gamma}| . \tag{7}$$

Proof. From (1) and (6) we have $\gamma = -X \cdot \gamma$, and from Proposition 1 it follows $\dot{\gamma} \wedge \gamma = -(X \cdot \gamma) \wedge \gamma = X$. Hence, $|X| = |\dot{\gamma}|$, and by virtue of Proposition 1 we obtain relation (7).

In what follows we shall identify (according to the Cartan-Killing form) coalgebras $SO(n)^*$ and $\mathfrak{n}^*$ with $SO(n)$ and $\mathfrak{n}$. Let Ad_f and Ad_f^* be adjoint and coadjoint representations: $Ad_f(\Omega)=f\Omega f^{-1}$, $Ad_f^*(\Omega)=f^{-1}\Omega f$; and let $\mathcal{T}: \mathfrak{so}(n) \longrightarrow \mathfrak{n}$ be an orthoprojector in the metric $\langle \cdot , \cdot \rangle$, $\mathcal{T}(\Omega)=\sum_{1\le i<j\le n-1}\langle\Omega,\mathcal{E}_{ij}\rangle\mathcal{E}_{ij}$. We now define the following operators:

$$B_f : \mathfrak{n}_\gamma \longrightarrow \mathfrak{n} \quad , \qquad \tilde{B}_f : \mathfrak{n} \longrightarrow \mathfrak{n} \, ,$$

$$B_f = P_f|_{\mathfrak{n}_\gamma} \quad , \qquad \tilde{B}_f = B_f \cdot Ad_f^* \, .$$

Lemma 1. The operators B_f and $\tilde{B}_f$ are isomorphisms.

Proof. Suffice it to demonstrate that B_f is an isomorphism. For $\forall u,v \in \mathfrak{n}$ we have $\langle \tilde{B}_f u, v\rangle = \langle \mathcal{T}J_f Ad_f^*(u), v\rangle = \langle J_f Ad_f^*(u), v\rangle =$
$= \langle Ad_f(\mathcal{M}(Ad_f^*(u))), v\rangle = \langle \mathcal{M}(\tilde{u}), \tilde{v}\rangle$, $\tilde{u} = Ad_f^*(u)$, $\tilde{v} = Ad_f^*(v)$.
Since $\mathcal{M}: \mathfrak{so}(n) \longrightarrow \mathfrak{so}(n)$ is a positive self-adjoint operator, $\tilde{B}_f$ is also positive and self-adjoint.

Lemma 2. The operator $Q_f : \mathfrak{so}(n) \longrightarrow \mathfrak{n}_\gamma$, $Q_f = B_f^{-1}\cdot P_f$, is an orthoprojector in the metric $K(\cdot , \cdot)$.

Proof. Let $u \in \mathfrak{n}_\gamma$ and $v \perp \mathfrak{n}_\gamma$ in the metric $K(\cdot , \cdot)$. Then, by construction, $Q_f(u) = u$, and it remains to prove that $Q_f(v) = 0$. The latter equality follows from the relations:

$$2K(Q_f(v), u) = \langle \mathcal{M}(Q_f(v)), u\rangle = \langle \mathcal{M}(B_f^{-1}\mathcal{T}Ad_f(\mathcal{M}(v)), u\rangle =$$

$$\langle P_f B_f \mathcal{T} Ad_f(\mathcal{M}(v)), Ad_f(u)\rangle = \langle Ad_f(\mathcal{M}(v)), Ad_f(u)\rangle = 2K(u,v) \, .$$

By virtue of Lemma 2 we have for $\Omega \in \mathfrak{so}(n)$ and $f \in SO(n)$:

$$K(\Omega, \Omega) = K((I-Q_f)(\Omega), (I-Q_f)(\Omega)) + K(Q_f(\Omega), Q_f(\Omega)) \quad (8)$$

Since $(I - Q_f)(Y_f) = 0$, where $Y_f = \mathcal{T}(\Omega)$, the expression

$$K((I-Q_f)(\Omega), (I-Q_f)(\Omega)) = K((I-Q_f)(X_f), (I-Q_f)(X_f))$$

is the distance from $X_f = \Omega - Y_f$ to $\mathfrak{n}_\gamma$ in the metric $K(\cdot , \cdot)$.

Let us introduce the notation

$$\widetilde{K}_f(X_f, X_f) = K((I-Q_f)(X_f), (I-Q_f)(X_f)) . \qquad (9)$$

Below we shall follow the procedure described in [5,7] . To this end, the second term on the right-hand side of (8) is represented as

$$K(Q_f(\Omega), Q_f(\Omega)) = K(B_f^{-1} \cdot P_f(\Omega), B_f^{-1} P_f(\Omega)) . \qquad (10)$$

Let G be an **isotropy** group at point e_n. For $\forall g \in G$ the restrictions of $\overline{Ad}_g$ and $\overline{Ad}_g^*$ onto $\mathfrak{N}$ will be denoted by Ad_g and Ad_g^*. By construction, $P_{gf} = \overline{Ad}_g P_f$. Let us consider an integral submanifold $F_p = P_f^{-1}(p)$, which can be represented in the form $F_p = \{(f, \widetilde{x}_f + B_f^{-1}(p))\}$, where $\widetilde{x}_f = (I - Q_f)x_f$, $\widetilde{x}_f \in \widetilde{\mathfrak{N}}_\gamma^\perp$, and $\widetilde{\mathfrak{N}}_\gamma^\perp = Im(I - Q_f)$ is orthogonal complement to $\mathfrak{N}_\gamma$ in the metric $K(\cdot , \cdot)$. It can be seen from the construction that $\dim F_p = 2\dim SO(n) - \dim \mathfrak{N}_\gamma = \frac{1}{2}(n-1)(n+2)$. The submanifold F_p is a subbundle of the trivial bundle $SO(n) \times SO(n) \longrightarrow SO(n)$ with the fiber $\widetilde{\mathfrak{N}}_\gamma^\perp$.

Hereinafter we shall always assume that $p \in Reg\, \mathfrak{N}$, i.e. Ann(p) is a Cartan subalgebra. Let $\mathcal{H}_p = Ann(p)$ and let H_p be a maximal torus in G which corresponds to $\mathcal{H}_p$. To describe the reduced phase space $\widetilde{F}_p = F_p/H_p$, we note first of all that $B_{gf}^{-1} = B_f^{-1}(p)$, $\forall g \in H_p$, which follows from the relation $B_{gf}^{-1} = B_f^{-1}\, \overline{Ad}_g^*$. Hence, the action of H_p on F_p is reduced to the transformation of the first component

$$(g, (f, \widetilde{x}_p)) \longrightarrow (gf, \widetilde{x}_p) , \quad \widetilde{x}_p = \widetilde{x} + B_f^{-1}(p) .$$

Thus, the reduced phase space F_p is diffeomorphic to

$$SO(n)/H_p \times \mathfrak{N}^\perp \quad \text{and} \quad \dim F_p = \begin{cases} 2\ell(\ell+1) , & n = 2\ell+1 , \\ 2\ell^2 , & n = 2\ell . \end{cases}$$

If a submanifold S in SO(n) is a section of the bundle SO(n) $\longrightarrow$

$\longrightarrow SO(n)/H_p$, the reduced effective potential, $U_{eff} = E - K_f(x_f, x_f)$, can be written (with an allowance for (9) and (10)) in the following form:

$$U_{eff} = K(B_\sigma^{-1}(p), B_\sigma^{-1}(p)) + \lambda(r, \gamma), \ \sigma \in S. \tag{11}$$

The minima of the reduced effective potential are known [5,7] to correspond to stable stationary rotations. To study local behaviour of U_{eff}, one should introduce coordinates which are compatible with the bundle mentioned above. Let $\mathcal{O}$ be a neighbourhood of zero in R^{n-1}, and let $\Theta_{f_0} : \mathcal{O} \longrightarrow SO(n)$ be a mapping defined by

$$\Theta_{f_0}(\xi) = f_0 \exp\left(\sum_{j=1}^{n-1} \xi_j \, \varepsilon_{j,n}\right), \quad \xi = (\xi_1, \ldots, \xi_{n-1})^T. \tag{12}$$

The mapping Θ_{f_0} is transversal at point f_0 to the fiber of the canonical bundle $\rho : SO(n) \longrightarrow S^{n-1}$. The relation

$$\gamma(\xi_1, \ldots, \xi_{n-1}) = \Theta_I^{-1}(\xi)\, e_n \tag{13}$$

defines coordinates in a certain neighbourhood U of the point e_n on S^{n-1}. By virtue of (12) and (13) we can define a trivialization chart

$$\mathcal{O} \times G \longrightarrow \rho^{-1}(U), \quad (\xi, g) \longrightarrow g\, \Theta_I(\xi), \tag{14}$$

on U for the bundle $\rho : SO(n) \longrightarrow S^{n-1}$. The mapping $g(v) = \exp(v)$, $v \in \mathcal{H}_p^\perp \cap \mathfrak{n}$, is transversal to H_p at point I, and, therefore, the mapping

$$v \longrightarrow \overline{Ad}^*_{\exp(v)}(p) \tag{15}$$

specifies parametrization of the orbit of coadjoint action $G \times \mathfrak{n} \longrightarrow \mathfrak{n}$ and parametrization of the factor G/H_p. The parametrization

$$\sigma(u,v,w) = \exp(w)\exp(v)\exp(u),\; w\in \mathcal{H}_p,\; v\in \mathcal{H}_p^{\perp}\cap \mathfrak{n},\; u\in \mathfrak{n}^{\perp}, \quad (16)$$

is compatible with all the aforementioned bundles, i.e. (16) confirms the validity of the following commutation diagram (see (13),(14),(15)) composed of embedding and factorization mappings:

$$\begin{array}{ccccc} G & \longrightarrow & SO(n) & & \\ \downarrow & & \downarrow & \searrow & \\ & & & & S^{n-1} \\ & & & \nearrow & \\ G/H_p & \longrightarrow & SO(n)/H_p & & \end{array}$$

Let $k(u,v,p)$ be an expression for $K(B_\sigma^{-1}(p),\; B_\sigma^{-1}(p))$ in the variables (u,v):

$$k(u,v,p) = K(B^{-1}_{\exp(v)\exp(u)}(p),\; B^{-1}_{\exp(v)\exp(u)}(p)) \;. \quad (17)$$

Since the transition from parametrization $\exp(v)\exp(u)$ to parametrization $g\exp(v)\exp(u)$ is reduced only to the change $p \longrightarrow Ad^*_g(p)$, it is sufficient, in what follows, to consider motions in the neighbourhood of the subgroup H_p belonging to chart (16). Of primary interest to us will be rotations occurring under the condition $\gamma = r = e_n$

Here are a number of relations (we present them without proof) that will be used in calculations to follow:

$$\left.\begin{aligned} &\exp(u) = u + \left(1 - \frac{|u|^2}{2}\right) I + o(|u|^2) + h,\quad h\in\mathfrak{n}, \\ &\exp(w\,\mathcal{E}_{i,j}) = I + \sin w\;\mathcal{E}_{i,j} + (1-\cos w)\,\mathcal{E}_{i,j}^2\,, \\ &Ad_{\exp(\Omega)} = \exp(ad_\Omega)\,, \end{aligned}\right\} \quad (18)$$

where $ad_\Omega(x) = [\Omega, x]$; and

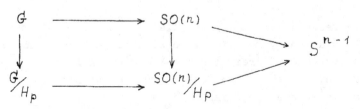

$$\left.\begin{aligned}
&\tilde{B}_f^{-1} = \mathcal{M}^{-1} + \mathcal{M}^{-1}R_f\mathcal{M}^{-1} + \mathcal{M}^{-1}R_f\mathcal{M}^{-1}R_f\mathcal{M}^{-1} + \cdots ,\\
&R_f = \mathcal{M}/\mathfrak{n} - \tilde{B}_f ,\\
&R_{\exp(u)} = -\pi\, ad_u\mathcal{M} + \pi\mathcal{M}\, ad_u - \tfrac{1}{2}\pi(ad_u^2\mathcal{M} + \mathcal{M}\, ad_u^2) +\\
&+\pi\, ad_u\,\mathcal{M}\, ad_u + O(|u|^2) = -\tfrac{1}{2}\pi(ad_u^2\mathcal{M} + \mathcal{M}\, ad_u^2) +\\
&+\pi\, ad_u\mathcal{M}\, ad_u + O(|u|^2), \quad u \in \mathfrak{n}^\perp .
\end{aligned}\right\}(19)$$

We have used the equality $\pi\cdot ad_u/\mathfrak{n} = \pi ad_u^*/\mathfrak{n} = 0$ which is related to $\mathrm{Im}(ad_u/\mathfrak{n}) \subset \mathfrak{n}^\perp$.

3. Stationary rotations

Rotation is called stationary if it is a solution of Eq. (2) with constant angular velocity. If $(r, \gamma) > 0$, stationary rotation is called standing. Stationary rotation is called non-degenerate if the angular velocity matrix has a maximal rank. It follows from (1) (see also (6) and (7)) that the velocity Ω of stationary rotation satisfies the equation

$$[\Omega^2, M] + \lambda r\wedge\gamma = 0, \quad \dot{\gamma} + \Omega\gamma = 0 .$$

In particular, for rotations preserving the Poisson vector (γ = const) we have $\Omega\cdot\gamma = 0$.

Proposition 3. If $\gamma = r = e_n$, no stable stationary rotations can exist in an even-dimensional space.

Proof. Let W be angular velocity of a given stationary rotation. Since, according to the condition, the Poisson vector is preserved, we have $W\gamma = 0$ and $W^2M = MW^2$. Since $\mathrm{rk}\, W < n$, there exists $g \in G$ such that $W = g^{-1}\mathcal{E}\cdot g$, where $\mathcal{E} = \sum_{k=1}^{2l-1} \alpha_k \mathcal{E}_{2k-1,2k}$. Hence, $W^2 = g^{-1}\mathcal{E}^2 g$ and, therefore, W is a diagonal matrix (we have taken into account that M is a simple diagonal matrix). Let us consider the motion $f = g^{-1}\exp(u\,\mathcal{E}_{2l-1,2l})\exp(t\mathcal{E})g$ with angular velocity $\Omega = u' g^{-1}\mathcal{E}_{2l-1,2l}\, g + W$. Using (2) and taking into consideration (18), we obtain the condition for u:

$$(m_{2\ell-1} + m_{2\ell})\, u'' - \lambda \sin u = 0 \qquad (20)$$

(from (18) it follows $(\exp u\,\mathcal{E}_{2\ell-1,2\ell})e_{2\ell} = e_{2\ell}\cos u + e_{2\ell-1}\sin u$). Analysing the motion of a body under condition (20), we may conclude that this stationary rotation corresponds to zero initial condition: $u(0) = u'(0) = 0$; if $u(0) = 0$ but $u'(0) \neq 0$, we obtain a motion which differs noticeably from the given stationary rotation.

In the above consideration we used commutation with rotation in the plane $e_{2\ell-1}$, $e_{2\ell}$. Clearly, an analogous situation also arises in an odd-dimensional space in the case of degeneration $\mathrm{rk}W < n-1$. Thus, the following statement is valid.

Proposition 4. If $\gamma = r = e_n$, any degenerate stationary rotation is unstable.

Remark. If $f = \exp(tW)$ is a stationary rotation, then

$$MW + WM = P_{\exp(tW)}(W), \quad \forall t. \qquad (21)$$

Indeed, from the definition of P_f we have $P_{\exp(tW)} = \pi \mathrm{Ad}_{\exp(tW)}(\mathcal{M}(W))$, but since $P_f(\Omega)$ is an integral, we can assume $t=0$ to obtain (21).

Theorem 5. Any stationary rotation in the gravity field preserves the Poisson vector.

Proof. Let W be angular velocity of stationary rotation. Then, $[W^2,M] + \lambda\, r\wedge\gamma = 0$ and, therefore, $r\wedge\dot\gamma = 0$ or $\dot\gamma = \alpha(t)r$. We thus obtain $\dot\alpha r + W\alpha r = \ddot\gamma + W\dot\gamma = 0$. Since $(Wr,r) = 0$, we have $\alpha(t)|r|^2 = 0$ and $\dot\alpha(t) = 0$, whence $\alpha(t) = \alpha(0) = \mathrm{const}$ and $\gamma = \mathrm{const} + \alpha(0)rt$. But since $|\gamma| = \mathrm{const}$, $\alpha(0) = 0$. Consequently, $\gamma = \mathrm{const}$. The theorem is proved.

Cartan's subalgebra is called canonical if after a certain renumbering of the vectors $e_1, \dots, e_{2\ell}$ $(n = 2\ell+1)$ it can be represented as $\mathcal{H} = \mathrm{Lin}\{\mathcal{E}_{1,2}, \mathcal{E}_{3,4}, \dots, \mathcal{E}_{2\ell-1,2\ell}\}$. The subgroup H corresponding to the subalgebra $\mathcal{H}$ is also called canonical.

It is known that a commutative subgroup in G, which is a torus, corresponds to each commutative subalgebra in $\mathfrak{M}$. If all one-dimensional subgroups of a torus are integral curves of Eq. (1), the torus is called stationary. The following statement shows that a maximal stationary torus is a canonical (Cartan) subgroup.

Proposition 6. Let $W \in \mathrm{Reg}\,\mathfrak{M}$ be angular velocity of stationary rotation for $\gamma = r$ and let P be the corresponding value of the relative kinetic moment. Then, W and P belong to a common canonical Cartan subalgebra.

The proof is obvious and follows from the relations $[P,W] = 0$ and $W^2M = MW^2$, which are consequences of (1) and (21).

In the next section we shall formulate stability conditions for stationary rotations. In the narrow sense of the word, stability means that small perturbations of the kinetic moment give rise to a new (perturbed) stationary torus which is close to the initial one. In the broad sense of the word, stability means that small perturbations of the initial conditions for a given stationary rotation give rise to a new motion (not necessarily stationary) which is sufficiently close to the given torus. This type of stability is called T-stability.

3. Sleeping tops

In this section we consider T-stability of stationary rotations under the condition $\gamma = r = e_n$. These rotations are called sleeping tops. The study is based on the principle that the reduced effective potential is minimal. We shall proceed from function (17).

Proposition 7. The function k is even in the variable u:

$$k(-u,v,p) = k(u,v,p).$$

Proof. The following representation is valid for k:

$$k(u,v,p) = \frac{1}{2}\langle \widehat{B}^{-1}_{\exp(v)\exp(u)}(p), p\rangle \qquad (22)$$

which can easily be derived from the definition of k. Let

$$T = \mathrm{diag}\{-1, \dots, -1, 1\}.$$

Then, $\exp(v)\exp(-u) = \exp(Ad_T(v))\exp(Ad_T(u)) = T\exp(v)\exp(u)T^{-1}$, which is what was required ($\tilde{B}_{TfT^{-1}} = Ad_T \tilde{B}_f Ad_T^*$, $\overline{Ad}_T = I$).

From Proposition 7 we have

$$k(u,v,P) = k_{0,0}(P) + k_{2,0}(u,P) + k_{0,1}(v,P) + k_{0,2}(v,P) + o(|u|^2+|v|^2), \quad (23)$$

where $K_{r,s}(u,v,P)$ is a homogeneous form of orders r and s in the variables u and v.

Proposition 8. The following representations are valid:

$$k_{0,0}(P) = \frac{1}{2}\langle \mathcal{M}^{-1}(P), P\rangle, \quad k_{0,1}(v,P) = \langle [\mathcal{M}^{-1}(P), P], v\rangle, \quad (24)$$

$$k_{2,0}(u,P) = \frac{1}{2}\left(\langle ad_P(u), ad_{\mathcal{M}^{-1}(P)}(u)\rangle - \langle \mathcal{M}(ad_{\mathcal{M}^{-1}(P)}(u)), ad_{\mathcal{M}^{-1}(P)}(u)\rangle\right), \quad (25)$$

$$k_{0,2}(v,P) = \frac{1}{2}\left(\langle \mathcal{M}^{-1} ad_P(v), ad_P(v)\rangle - \langle ad_P(v), ad_{\mathcal{M}^{-1}(P)}(v)\rangle\right). \quad (26)$$

Proof. We obtain from (19)

$$\tilde{B}^{-1}_{\exp(v)} = \exp(ad_v^*)\mathcal{M}^{-1}\exp(ad_v) = \mathcal{M}^{-1} + ad_v\mathcal{M}^{-1} - \mathcal{M}^{-1}ad_v + \frac{1}{2}(ad_v^2\mathcal{M}^{-1} + \mathcal{M}^{-1}ad_v^2) - ad_v\mathcal{M}^{-1}ad_v + o(|v|^2),$$

$$\tilde{B}^{-1}_{\exp(u)} = \mathcal{M}^{-1} - \frac{1}{2}\pi(ad_u^2\mathcal{M}^{-1} + \mathcal{M}^{-1}ad_u^2) + \pi\mathcal{M}^{-1}ad_u\mathcal{M}ad_u\mathcal{M}^{-1} + o(|u|^2)$$

Hence,

$$\langle \tilde{B}^{-1}_{\exp(v)}(P), P\rangle = \langle \mathcal{M}^{-1}(P), P\rangle - 2\langle ad_v(P), \mathcal{M}^{-1}(P)\rangle +$$

$$+\langle ad_V^2(p), \mathcal{M}^{-1}(p)\rangle + \langle \mathcal{M}^{-1} ad_V(p), ad_V(p)\rangle + o(|V|^2) =$$

$$=\langle \mathcal{M}^{-1}(p), p\rangle + \langle [\mathcal{M}^{-1}(p), p], V\rangle - \langle ad_p(V), ad_{\mathcal{M}^{-1}(p)}(V)\rangle +$$

$$+\langle \mathcal{M}^{-1} ad_p(V), ad_p(V)\rangle + o(|V|^2),$$

$$\langle \tilde{B}^{-1}_{\exp(u)}(p), p\rangle = \langle \mathcal{M}^{-1}(p), p\rangle + \langle ad_{\mathcal{M}^{-1}(p)}(u), ad_p(u)\rangle -$$

$$-\langle \mathcal{M} ad_{\mathcal{M}^{-1}(p)}(u), ad_{\mathcal{M}^{-1}(p)}(u)\rangle + o(|u|^2).$$

which is what was required.

Remark. It follows from Proposition 8 that $\frac{\partial k}{\partial v}(0,0,P) = [W,P]$, where $W = \mathcal{M}^{-1}(P)$. Thus, for stationary rotation $\exp(tW)$ ($\gamma = r = e_n$) we obtain by virtue of Proposition (6): $\frac{\partial k}{\partial v}(0,0,P) = 0$.

Theorem 9. Let $r=e_n$, $n=2l+1$, and let the following relations be valid for $\forall\{j,k \mid 1 \le j < k \le l\}$:

$$(m_{2j-1} - m_q)(m_{2j} - m_{\hat{q}})(\omega_k^2 - \omega_j^2)(P_k^2 - P_j^2) > 0, \quad (27)$$

$$(P_j^2 - P_k^2)(P_j\omega_j(m_p - m_q) + P_k\omega_k(m_{\hat{p}} - m_{\hat{q}})) > 0, \quad (28)$$

$$p,\hat{p} \in \{2j-1, 2j\},\ p \ne \hat{p},\quad q,\hat{q} \in \{2k-1, 2k\},\ q \ne \hat{q},$$

$$\omega_j^2 \min\{(m_{2j-1} - m_n), (m_{2j} - m_n)\} > l,\ \forall j \in \{1,\dots,l\}. \quad (29)$$

Then $f = \exp(tW)$, where $W = \sum \omega_j \mathcal{E}_{2j-1,2j}$ is T-stable stationary rotation ($P = \sum P_j \mathcal{E}_{2j-1,2j} = \mathcal{M}(W)$).

The proof is a direct consequence of the following propositions.

Proposition 10. If conditions (27) and (28) are satisfied, the form $k_{0,2}$ is positive definite.

Proof. Let L_{jk} denote the linear span of the elements $\{\mathcal{E}_{2j,2k}, \mathcal{E}_{2j-1,2k}, \mathcal{E}_{2j,2k-1}, \mathcal{E}_{2j-1,2k-1}\}$. Then $\mathfrak{n}$ is decomposed into a direct sum of four-dimensional subspaces L_{jk}:

$$\mathfrak{n} = \sum_{1 \le j < k \le l} L_{j,k} \tag{30}$$

The subspace L_{jk} is invariant relative to the operators ad_p and $\mathcal{M}$. This assertion follows from apparent relations

$$\left.\begin{aligned}
ad_p(\mathcal{E}_{2j,2k}) &= -p_j \mathcal{E}_{2j-1,2k} - p_k \mathcal{E}_{2j,2k-1} ,\\
ad_p(\mathcal{E}_{2j-1,2k}) &= p_j \mathcal{E}_{2j,2k} - p_k \mathcal{E}_{2j-1,2k-1} ,\\
ad_p(\mathcal{E}_{2j,2k-1}) &= -p_j \mathcal{E}_{2j-1,2k-1} + p_k \mathcal{E}_{2j,2k} ,\\
ad_p(\mathcal{E}_{2j-1,2k-1}) &= p_j \mathcal{E}_{2j,2k-1} + p_k \mathcal{E}_{2j-1,2k} ,\\
\mathcal{M}(\mathcal{E}_{k,s}) &= (m_k + m_s)\mathcal{E}_{k,s} , \quad \forall k,s .
\end{aligned}\right\} \tag{31}$$

The following representation is derived from (30):

$$k_{0,2}(v,p) = \sum_{j,k} k_{0,2}(v_{j,k}, p) , \tag{32}$$

where v_{jk} is the component of v corresponding to decomposition (30). From (31) we obtain

$$\langle \mathcal{M}^{-1}(ad_p(v_{j,k})), ad_p(v_{j,k}) \rangle =$$

$$= \frac{\tilde{v}^2_{2j,2k}}{m_{2j}+m_{2k}} + \frac{\tilde{v}^2_{2j-1,2k}}{m_{2j-1}+m_{2k}} + \frac{\tilde{v}^2_{2j,2k-1}}{m_{2j}+m_{2k-1}} + \frac{\tilde{v}^2_{2j-1,2k-1}}{m_{2j-1}+m_{2k-1}} , \tag{33}$$

where $\tilde{v}_{r,k}$ is the coordinate of $ad_p(v)$ with respect to $\mathcal{E}_{r,k}$. Let $v_{r,k}$ and $\tilde{\tilde{v}}_{r,k}$ denote the coordinates of v and $ad_{\mathcal{M}^{-1}(p)}(v)$ with respect to $\mathcal{E}_{r,k}$. Then it follows from (31)

$$\begin{pmatrix} \tilde{\tilde{V}}_{2j,2k} \\ \tilde{\tilde{V}}_{2j-1,2k} \\ \tilde{\tilde{V}}_{2j,2k-1} \\ \tilde{\tilde{V}}_{2j-1,2k-1} \end{pmatrix} = \begin{pmatrix} 0 & \omega_j & \omega_k & 0 \\ -\omega_j & 0 & 0 & \omega_k \\ -\omega_k & 0 & 0 & \omega_j \\ 0 & -\omega_k & -\omega_j & 0 \end{pmatrix} \cdot \begin{pmatrix} V_{2j,2k} \\ V_{2j-1,2k} \\ V_{2j,2k-1} \\ V_{2j-1,2k-1} \end{pmatrix}$$

Similarly,

$$\begin{pmatrix} \tilde{V}_{2j,2k} \\ \tilde{V}_{2j-1,2k} \\ \tilde{V}_{2j,2k-1} \\ \tilde{V}_{2j-1,2k-1} \end{pmatrix} = \begin{pmatrix} 0 & p_j & p_k & 0 \\ -p_j & 0 & 0 & p_k \\ -p_k & 0 & 0 & p_j \\ 0 & -p_k & -p_j & 0 \end{pmatrix} \begin{pmatrix} V_{2j,2k} \\ V_{2j-1,2k} \\ V_{2j,2k-1} \\ V_{2j-1,2k-1} \end{pmatrix}$$

Hence,

$$(\tilde{\tilde{V}}_{2j,2k}, \tilde{\tilde{V}}_{2j-1,2k}, \tilde{\tilde{V}}_{2j,2k-1}, \tilde{\tilde{V}}_{2j-1,2k-1})^T = A_{j,k}\, (\tilde{V}_{2j,2k}, \tilde{V}_{2j-1,2k}, \tilde{V}_{2j,2k-1}, \tilde{V}_{2j-1,2k-1})^T$$

where

$$A_{j,k} = \frac{1}{(p_j^2 - p_k^2)} \begin{pmatrix} \omega_j p_j - \omega_k p_k & 0 & 0 & \omega_j p_k - \omega_k p_j \\ 0 & \omega_j p_j - \omega_k p_k & \omega_j p_k - \omega_k p_j & 0 \\ 0 & \omega_j p_k - \omega_k p_j & \omega_j p_j - \omega_k p_k & 0 \\ \omega_j p_k - \omega_k p_j & 0 & 0 & \omega_j p_k - \omega_k p_j \end{pmatrix}$$

Here $P_k = (m_{2k} + m_{2k-1})\omega_k$.

Consequently, the quadratic form $2k_{0,2}(V_{j,k}, P)$ in the variables $\tilde{V}_{2j,2k}, \tilde{V}_{2j-1,2k}, \tilde{V}_{2j,2k-1}, \tilde{V}_{2j-1,2k-1}$ has the following matrix (see (33)):

$$diag\left(\frac{1}{m_{2j}+m_{2k}}, \frac{1}{m_{2j-1}+m_{2k}}, \frac{1}{m_{2j}+m_{2k-1}}, \frac{1}{m_{2j-1}+m_{2k-1}}\right) - A_{j,k}$$

Applying the conventional Silvester criterion, which shows whether the matrix spectrum is positive, we obtain the conditions under which the form is positive definite:

$$\frac{1}{m_{2j}+m_{2k}} - \Delta_{j,k} > 0, \qquad \frac{1}{m_{2j-1}+m_{2k}} - \Delta_{j,k} > 0,$$

$$\left(\frac{1}{m_{2j-1}+m_{2k}} - \Delta_{j,k}\right)\left(\frac{1}{m_{2j}+m_{2k-1}} - \Delta_{j,k}\right) > \frac{\omega_j^2\omega_k^2}{(P_j^2-P_k^2)^2}(m_{2k}+m_{2k-1}-m_{2j}-m_{2j-1})^2,$$

$$\left(\frac{1}{m_{2j}+m_{2k}} - \Delta_{j,k}\right)\left(\frac{1}{m_{2j-1}+m_{2k-1}} - \Delta_{j,k}\right) > \frac{\omega_j^2\omega_k^2}{(P_j^2-P_k^2)^2}(m_{2k}+m_{2k-1}-m_{2j}-m_{2j-1})^2$$

$$\left(\Delta_{j,k} = \frac{\omega_j P_j - \omega_k P_k}{P_j^2 - P_k^2}\right).$$

After simple manipulations these inequalities are reduced to the form of (27).

<u>Proposition 11</u>. The following representation is valid for the form $2k_{2,0}$:

$$2k_{2,0} = \sum_{j=1}^{\ell} \omega_j^2 [(m_{2j} - m_n) u_{2j}^2 + (m_{2j-1} - m_n) u_{2j-1}^2]$$

Proof. Let $u = \sum_{j=1}^{\ell} (u_{2j-1}\mathcal{E}_{2j-1,n} + u_{2j}\mathcal{E}_{2j,n})$,
$p = \sum_{k=1}^{\ell} p_k \mathcal{E}_{2k-1,2k}$. Then

$$ad_p(u) = -\sum_{j=1}^{\ell} (p_j u_{2j}\mathcal{E}_{2j-1,n} - p_j u_{2j-1}\mathcal{E}_{2j,n}),$$

$$ad_{\mathcal{M}^{-1}(p)}(u) = -\sum_{j=1}^{\ell} (\omega_j u_{2j}\mathcal{E}_{2j-1,n} - \omega_j u_{2j-1}\mathcal{E}_{2j,n}).$$

and, therefore,

$$2k_{2,0}(u,p) = \sum_{j=1}^{\ell} \omega_j^2 [(m_{2j-1} + m_{2j} - m_{2j-1} - m_n) u_{2j}^2 +$$

$$+ (m_{2j-1} + m_{2j} - m_{2j} - m_n) u_{2j-1}^2].$$

This is what was required.

Proposition 12. The potential energy depends only on the variable u, and the following representation is valid for $r=e_n$:

$$U = \lambda(1 - \frac{|u|^2}{2}) + o(|u|^2)$$

Proof. By virtue of (18) we have

$$\gamma = \exp(-u)\exp(-v)e_n = \exp(-u)e_n =$$

$$= -ue_n + (1 - \frac{|u|^2}{2})e_n + o(|u|^2)$$

Hence, $U = \lambda(1 - \frac{|u|^2}{2})(e_n, r) + o(|u|^2)$

and for $r=e_n$ we obtain $U = \lambda(1 - \frac{|u|^2}{2}) + o(|u|^2)$, which is what was required.

To derive inequalities (29), it remains to point out that the quadratic (in u) part of the effective potential can be represented as

$$\frac{1}{2}\sum_{j=1}^{\ell} [(\omega_j^2(m_{2j} - m_n) - \lambda)u_{2j}^2 + (\omega_j^2(m_{2j-1} - m_n) - \lambda)u_{2j-1}^2].$$

REFERENCES

1. Routh E.J., Dynamics of a system of rigid bodies.- N.Y.,1905, P.2.
2. Chetaev N.G., Stability of motion.- Moscow, 1965 (in Russian).
3. Rumyantsev V.V., Stability of permanent rotations of a heavy rigid ball.- PMM, v. 1, No. 1, 1956 (in Russian).
4. Arnol'd V.I., Hamiltonian character of the Euler equations describing the dynamics of a rigid body and an ideal liquid .- Usp. Matem. Nauk, 1969, v. 24, No. 3 (in Russian).
5. Smale S., Topology and mechanics.- Invent. Math., 1970, v.10, No.6; v.11, No.1.
6. Katok S.B., Supplement to the translation of Smale's article "Topology and mechanics".- Usp. Matem. Nauk, 1972, v.27, No.2 (in Russian).
7. Arnol'd V.I., Mathematical methods in classical mechanics.- Moscow, 1974 (in Russian).
8. Kuz'min P.A., Small vibrations and stability of motion.- Moscow, 1973 (in Russian).
9. Tatarinov Ya.V.,On the study of phase topology of compact configurations with symmetry.- Vestn. MGU. Ser. Matem. Mekh., 1973, No.5 (in Russian).
10. Tatarinov Ya.V., Patterns of classical integrals in the problem on the rotation of a rigid body about a fixed point. Vestn. MGU. Ser. Matem. Mekh., 1974, No.6 (in Russian).
11. Mishchenko A.S., Integrals of geodesic flows on Lie groups.- Funkts. Analiz i ego Prilozh., 1970, v.4, No.3 (in Russian).
12. Manakov S.V., Comments on integration of Euler's equations for the dynamics of an n-dimensional rigid body.- Funkts. Analiz i ego Prilozh., 1976, v.10, No.10 (in Russian).
13. Mishchenko A,S. and Fomenko A.T., Euler's equations on finite-dimensional Lie groups.- Izv. Akad Nauk SSSR, 1978, v.42, No.2 (in Russian).
14. Vishik S.V. and Dolzhansky F.V., Analogues of the Euler-Poisson equations and magnetic hydrodynamics equations related to Lie groups.- Dokl. Akad. Nauk SSSR, 1978, v.238, No.5 (in Russian).
15. Kozlov V.V., Methods of qualitative analysis in solid state dynamics.- Moscow, 1980 (in Russian).
16. Fomenko A.T., Differential geometry and topology.- Moscow, 1983 (in Russian).
17. Kovalev A.M. and Savchenko A.Ya., Stability of stationary rotations of a rigid body about the principal axis.- PMM, 1975, v.39, No.4 (in Russian).
18. Sergeev V.S., On stability of permanent rotations of a heavy rigid body about a fixed axis.- PMM, 1976, v.40, No.3 (in Russian).
19. Chudnenko A.N., On stability of stationary rotations of a rigid body about the principal axis.- PMM, 1980, v.44, No.2 (in Russian).
20. Andreev D.V., On stability of permanent rotations of an asymmetric heavy rigid body.- PMM, 1983, v.47, No.3 (in Russian).
21. Sapronov Yu.I., Destruction of spherical symmetry in non-linear variational problems.- In: Analysis on manifolds and differential equations. Voronezh, 1986 (in Russian).

LAPLACE-RADON INTEGRAL OPERATORS AND SINGULARITIES OF SOLUTIONS OF DIFFERENTIAL EQUATIONS ON COMPLEX MANIFOLDS

B.Yu. Sternin
Moscow Institute of
Electronic Building
Moscow, 109208, URSS

V.E. Shatalov
Moscow Institute of
Civil Aviation Engineers
Moscow, 135838, USSR

The paper deals with constructing of asymptotics by smoothness of the solutions of partial differential equations in spaces of multiple-valued analytical functions. Construction of these asymptotics is based on "Laplace-Radon integral operators" theory introduced by authors. By means of these operators the asymptotic expansions of solutions are constructed both in noncharacteristical case and in neighbourhoods of characteristical manifolds of ramification.

In this paper the authors aimed at outworking a method for effective construction of asymptotic expansions by smoothness of differential equations in the space of multiple-valued analytical functions.

The classical works of J. Leray seemed to be the first in this field (see [1] , [2] , [3]). Nowadays the subject is being widely developed in this country as well as abroad (see, for example, [8] - [13] , [15] , [18] - [28] , [30] , [31] etc.).

The method of asymptotic expansions construction given in this paper is in fact the Maslov's canonic operator method. It is based on the presentation of asymptotic expansions in the form of Feunmann integral* of Laplace-Radon type with a ramifying contour. In this case Landau manifold of the integral exactly is a set of singularities of the solution.

We would like to add the following. To an experienced mathematician the basic idea of our constructions may resemble Maslov's canonical operator (the Fourier integral operators theory) [4] -

*We mean Feunmann diagram technique integrals.

[7], [14] . Naturally, the resemblance is not accidental. The lately developed in the theory of differential equations schemes of global asymptotics constructions may be of use in our case. But one should mind that our theory is based on Laplace-Radon transformation [10], [12] unlike the Fourier integral operators theory based on Fourier transformation. We found it possible to give the name of Laplace-Radon to the transformation introduced in [10], [12] because (as one can see from the text below) in our transformation we integrate over a relative homology class (as in Laplace transformation) laying on a set of hyperplanes (as in Radon transformation). An additional property of our transformation is the ramification of integration contours.

Briefly about the content of the paper. § 1 contains the definitions of main functional spaces which are filtered (if speaking of asymptotic expansions).

§ 2, being the central part of the paper, contains a detailed development of the Maslov canonical operators apparatus and, in particular, the main composition theorems. It also contains the notion of a pseudodifferential operator - not a trivial one in the complex analytical theory - and the composition formula for such operators.

The final § 3 contains some applications of Maslov's canonical operator apparatus. An asymptotic is constructed of the solution of a differential equation in case when the right side of the equation has a certain singularity on a noncharacteristic surface as well as in case when the entire surface lies in a characteristic set of the differential operator.

1. Functional spaces

Let $\mathbb{C}^n_x$ be a complex arithmetical n-dimensional space and $\mathbb{C}^n_{x_1} \oplus \mathbb{C}_{np_n}$ - phase symplectic space with coordinates $(x, p) = (x^1, \ldots, x^n, p_1, \ldots, p_n)$ and a structural form $dx \wedge dp = dx^1 \wedge dp_1 + \ldots + dx^n \wedge dp_n$.

Let L be a Lagrangian manifold with a determining action function $\xi = \xi(x)$. It means that $\xi = \xi(x)$ is a generating function of a nonsingular chart of L, that is a chart with coordinates x. Note that $\xi(x)$ is an analytical function with ramification on a certain analytical set of codimension 1.

In the space $\mathbb{C}^{n+1}_{x,\xi}$ with coordinates $(\xi, x^1, \ldots, x^n)$ we define a set X by formula

$$X = \{(\xi, x) | \ \xi - \xi(x) = 0\}.$$

The spaces of functions $f = f(x, \xi)$ below will be constructed only for arguments sufficiently near X in a certain neighbourhood of each point $(x, \xi) \in X$.

Let us start with the nonsingular situation. In this case in some neighbourhood U of a nonsingular point (x, ξ) the set X can be determined by the equation:

$$\xi = \xi(x)$$

where $\xi(x)$ is a holomorphic function. Denote by Aq(U, X) the space of analytical functions* for which in any neighbourhood U' compactly embedded in U ($U' \Subset U$) the estimate

$$|f(x, \xi)| \leq c \, | \, \xi - \xi(x)| ^q$$

holds.

Evidently, Aq(U, X) is a Frechet space with a countable set of seminorms of the form

$$\|f\|_g = \sup_{\tilde{U}} \frac{|f(x, \xi)|}{|\xi - \xi(x)|^q}, \tilde{U} \Subset U .$$

Now let x be a singular point, (x^I, p_I) - coordinates in a corresponding chart of the Lagrangian manifoli. On this chart the action

$$\xi_I(x^I, p_{\bar{I}}) = \xi(x^I, x^{\bar{I}}(x^I, p_{\bar{I}})) - x^{\bar{I}}(x^I, p_{\bar{I}})p_{\bar{I}}$$

is defined.

The function $\xi_I(x^I, p_{\bar{I}})$ is holomorphic. The set defined by the equation

$$\xi - \xi_I(x^I, p_{\bar{I}}) = 0$$

is denoted by X_I.

The space $Aq(U_I, X_I)$ is defined as a space of functions $f_I(x^I, p_{\bar{I}}, \xi)$ such that in any neighbourhood $U_I' \Subset U_I$

$$|f(x^I, p_I, \xi)| \leq c \, | \, \xi - \xi_I(x^I, p_{\bar{I}})| ^q$$

holds.

When the neighbourhood U contains a singular point x the set Aq(U, X) is defined as a space of functions of the form

---

* Here and below by "analytical function" we mean a multiplevalued analytical function.

$$F^{X}_{p_{\bar I}\to x^{\bar I}}\; f(x^{I}, p_{\bar I}, \xi).$$

Here $F^{X}_{p_{\bar I}\to x^{\bar I}}$ is a Laplace-Radon transformation. Mark that the transformation is defined correctly in case under consideration because $\mathrm{Hess}_{x^I, p_{\bar I}}\ \xi_I(x^I, p_{\bar I}) \neq 0$ at least in one point of the chart U_I (in fact on a dense set of it).

Evidently the spaces Aq form a decreasing filtration

$$Ao(U, X) \supset \ldots \supset Aq(U, X) \supset A_{q+1}(U, X) \quad \ldots$$

Henceforth two lemmas will be of use.

Lemma 1.1

The derivation $\partial/\partial\xi$ defines an operator

$$\frac{\partial}{\partial\xi} : Aq(U, X) \to A_{q-1}(U, X), \; q > 0,$$

and the integration $\int_{\xi(x)}^{\xi} f(x, \xi')\,d\xi'$ - an operator

$$\hat{I}: Aq-1(U, X) \longrightarrow Aq(U, X), \; q > 0.$$

Proof. It is clear that $\xi(x)$ may be assumed zero without loss of generality. In this case the coordinates x are parameters and, in fact, in the affirmation of the lemma we speak of an ordinary differentional operator. Now, due to the integral Cauchy formula

$$\frac{\partial f}{\partial \xi} = \frac{1}{2\pi i}\int_{\gamma} \frac{f(\zeta)\,d\zeta}{(\zeta-\xi)^2},$$

where $\gamma = S_{r/2}(\xi)$ is a circle with center in ξ and of radius $r/2$ ($r = |\xi|$).

Let us estimate the expression above:

$$\left|\frac{\partial f}{\partial \xi}\right| \leq \frac{1}{2\pi}\int_{\gamma}\frac{|f(\zeta)|\cdot|d\zeta|}{|\zeta-\xi|^2} \leq \frac{C}{2\pi r^2}\int_{\gamma}\|f\|_{q,\tilde U}\,|\zeta|^q\,|d\zeta| \leq \frac{\|f\|_{q,\tilde U}}{2\pi r^2}\left(\frac{3}{2}r\right)^q \times$$

$$\times 2\pi r = C\|f\|_{q,\tilde U}\cdot r^{q-1}.$$

The estimate for the operator $\hat I$ follows:

$$|\hat I f(x,\xi)| \leq \int_{\xi(x)}^{\xi} |f(x,\xi')|\cdot|d\xi'| \leq C\,|\xi - \xi(x)|^{q+1}.$$

This completes the proof of the lemma.

Lemma 1.2. Let $f = f(x, \xi)$ be an analytical function with all derivatives with respect to ξ up to order N being bounded:

$$\left|\frac{\partial^j f}{\partial \xi^j}\right| \leq C_j, \; j \leq N \quad in \; U' \Subset U.$$

Then the function f can be presented in the form

$$f = f_1 + f_2$$

where $f_1 \in A_N(U, X)$ and f_2 is holomorphic.

Proof. We denote $\partial^N f/\partial \xi^N$ by $f_N(x, \xi)$. Then $f_N \in A_o(U,X)$, and the functions

$$f_j(x, \xi) = \int_{\xi(x)}^{\xi} f_{j+1}(x, \xi')d\xi' = \hat{I} f_{j+1}, \; j = N-1, \ldots, 0$$

belong to the space A_{N-j} (U, X).

Let us consider the difference

$$f(x, \xi) - f_o(x, \xi).$$

Evidently

$$\frac{\partial^N}{\partial \xi^N}\left[f(x, \xi) - f_o(x, \xi)\right] = 0,$$

and consequently

$$f(x, \xi) - f_o(x, \xi)$$

is a polynomial with respect to ξ of degree N-1. In particular, the function $f(x, \xi) - f_o(x, \xi)$ is holomorphic. Now, the proof of the Lemma follows from the identity

$$f = f_o(x, \xi) + \left[f(x, \xi) - f_o(x, \xi)\right].$$

Here the first term $f_o(x, \xi) \in A_N$ and the second is holomorphic due to the proof above. This completes the proof.

2. Laplace-Radon integral operators

2.1. Auxiliary material. Let $\mathbb{C}^n \oplus \mathbb{C}_n$ be a 2n-dimensional complex arithmetical space with coordinates $(x^1, \ldots, x^n, p_1, \ldots, p_n)$ and a structural form $\omega = dx^1 \wedge dp_1 + \ldots + dx^n \wedge dp_n$.

We denote by $(y^1, \ldots, y^n, q_1, \ldots, q_n)$ the coordinates in another copy of the space $\mathbb{C}^n \oplus \mathbb{C}_n$ with a structural form $\omega' = dy \wedge dq$. Let (x_o, p^o) be a point in $\mathbb{C}^n + \mathbb{C}_n$ and W a neighbourhood of it. Similarly by (y_o, q^o) we denote a point of the second copy of the space and by W' a neighbourhood of it. At last, let

$$g: W' \longrightarrow W$$

be a canonical transformation of these neighbourhoods, i.e. a holomorphic map such that

$$g^* \omega = \omega' .$$

Let then $\xi = \xi(x)$ be a function analytical in a neighbourhood U of the point x_0, $\partial\xi/\partial x(x_0) = p_0$ *) and

$$L = \{(x,p) \mid p = \frac{\partial \xi(x)}{\partial x}\}$$

a Lagrangian manifold in $dx \wedge dp$-structure with a generating function $\xi = \xi(x)$.

Similarly with the help of a function $\eta = \eta(y)$ analytical in a neighbourhood U^1 of the point y_0 we define the Lagrangian manifold

$$L' = \{(y,q) \mid q = \frac{\partial \eta(y)}{\partial y}\}$$

in $dy \wedge dq$ - structure with a generating function $\eta = \eta(y)$.

Suggestion 2.1

$$g(L') = L$$

Now let the function $\Phi(x, y, \theta)$ of the variables $(x^1, \ldots, x^n, y^1, \ldots, y^n, \theta_1, \ldots, \theta_m)$ be analytical in some neighbourhood V of (x_0, y_0, θ_0). Let us consider an analytical set

$$C_\Phi = \{(x,y,\theta) \mid \frac{\partial \Phi}{\partial \theta}(x,y,\theta) = 0\}.$$

Suggestion 2.2. In the point (x_0, y_0, θ_0)

$$\operatorname{rank} \left\| \frac{\partial^2 \Phi}{\partial\theta \partial x} \quad \frac{\partial^2 \Phi}{\partial\theta \partial y} \quad \frac{\partial^2 \Phi}{\partial\theta \partial\theta} \right\| = m$$

holds.

The function Φ canonically determines a 2n-dimensional Lagrangian manifold L in 4n-dimensional complex space with coordinates (x, y, p, q) and the form $\omega - \omega'$. That is we consider the mapping

$$\alpha : \mathbb{C}^{2n} \oplus \mathbb{C}_m \to \mathbb{C}^{2n} \oplus \mathbb{C}_{2n}$$

---

* Here and below by $\frac{\partial}{\partial x}$ we denote for simplicily the whole set of partial derivatives: $\frac{\partial}{\partial x} = \left(\frac{\partial}{\partial x^1}, \ldots, \frac{\partial}{\partial x^n}\right)$.

determined by the formula

$$(x, y, \theta) \mapsto (x, y, p, q) = (x, y, \frac{\partial\Phi}{\partial x}, -\frac{\partial\Phi}{\partial y}),$$

and $\alpha(C\ \ ) = \mathcal{L}$. The mapping α will be called classifying.

Suggestion 2.3. The manifold $\mathcal{L}$ is a graph of the canonical transformation g :

$$\mathcal{L} = \text{graph } g$$

This suggestions settles a connection between Φ and g. To emphasize this fact we shall say Φ to be a determining function of g.

Lemma 2.1. Under suggestions above the following system of equations

$$\begin{cases} \dfrac{\partial\eta(y)}{\partial y} + \dfrac{\partial\Phi(x,y,\theta)}{\partial y} = 0, \\ \dfrac{\partial\Phi(x,y,\theta)}{\partial\theta} = 0 \end{cases} \tag{2.1}$$

is one-to-one solvable with respect to (y, θ) in some neighbourhood of (x_0, y_0, θ_0) and the corresponding functional determinant does not vanish.

Proof. We supplement the system (2.1) up to system:

$$\begin{cases} q(y) + \dfrac{\partial\Phi(x,y,\theta)}{\partial y} = 0; \\ \dfrac{\partial\Phi(x,y,\theta)}{\partial\theta} = 0; \\ p = \dfrac{\partial\Phi(x,y,\theta)}{\partial x}; \end{cases} \tag{2.2}$$

where $q(y) = \frac{\partial\eta(y)}{\partial y}$ is the equation of the manifold L'.

Due to Suggestion (2.3) one can solve the system (2.2) with respect to variables (x, p, θ). Due to Suggestion (2.1) the obtained system

$$\begin{cases} x = x(q(y), y); \\ p = p(q(y), y) \end{cases} \tag{2.3}$$

gives us a parametric representation of the Lagrangian manifold L. The Lagrangian manifold L being regular, y can be expressed by x using (2.2). Substituting $y = y(x)$ in the expression $\theta = \theta(q(y), y)$ we finally express y and θ by x. The affirmation about the determinant can be proved by similar arguments concerning the linearization of the system (2.1) instead of the system itself. This completes the proof.

Lemma 2.2. The equality

$$\xi(x) = \eta(y(x)) + \Phi(x, y(x), \theta(x)) + C \qquad (2.4)$$

holds, $(y(x), \theta(x))$ being the solution of the system (2.1).

Proof. Let us calculate the derivatives of the right-hand side of the equality (2.4)

$$\frac{\partial}{\partial x}\{\eta(y(x)) + \Phi(x, y(x), \theta(x))\} = \frac{\partial\eta}{\partial y}(y(x))\frac{\partial y(x)}{\partial x} + \frac{\partial\Phi}{\partial x}(x, y(x), \theta(x)) +$$

$$+ \frac{\partial\Phi}{\partial y}(x, y(x), \theta(x))\frac{\partial y(x)}{\partial x} + \frac{\partial\Phi}{\partial\theta}(x, y(x), \theta(x))\frac{\partial\theta(x)}{\partial x}.$$

Due to (2.1) we have:

$$\frac{\partial\eta}{\partial y}(y(x)) + \frac{\partial\Phi}{\partial y}(x, y(x), \theta(x)) \equiv 0,$$

$$\frac{\partial\Phi}{\partial\theta}(x, y(x), \theta(x)) \equiv 0;$$

and consequently

$$\frac{\partial}{\partial x}\{\eta(y(x)) + \Phi(x, y(x), \theta(x))\} = \frac{\partial\Phi}{\partial x}(x, y(x), \theta(x)).$$

But Suggestions 2.1 and 2.3 give us

$$\frac{\partial\Phi}{\partial x}(x, y(x), \theta(x)) = \frac{\partial\xi(x)}{\partial x},$$

and hence

$$\frac{\partial}{\partial x}\{\eta(y(x)) + \Phi(x, y(x), \theta(x))\} = \frac{\partial\xi(x)}{\partial x}$$

The last formula gives (2.4).

Suggestion 2.4. The constant C in (2.4) is equal to 0.

Now let us consider in the space $\mathbb{C}\oplus\mathbb{C}^n \oplus \mathbb{C}_{m+k}$ with the coordinates $(\eta, y^1, \dots, y^n, \theta_1, \dots, \theta_m, \tau_1, \dots, \tau_k)$ the set of manifolds

$$M_{x,\xi} = \{(\eta, y, \theta, \tau)\,|\, \eta = \xi - \Phi(x, y, \theta) + \tau^2\} \qquad (2.5)$$

dependent on parameters (x, ξ) (here $\tau^2 = \sum_{j=1}^{k} \tau_j^2$) and the manifold

$$Y = \{(\eta, y, \theta, \tau) \mid \eta = \eta(y)\} \; ^* \qquad (2.6)$$

The manifold $M_{x,\xi}$ is tangent to the manifold Y if

$$\begin{cases} \dfrac{\partial\eta}{\partial y} + \dfrac{\partial\Phi(x,y,\theta)}{\partial y} = 0, \quad \dfrac{\partial\Phi(x,y,\theta)}{\partial\theta} = 0, \quad \tau = 0, \\ \eta(y) = \xi - \Phi(x,y,\theta) + \tau^2. \end{cases}$$

Due to Lemma 2.2 and Suggestion 2.4 the manifold $M_{x,\xi}$ is tangent to Y when $\xi = \xi(x)$, i.e. when $(x, \xi) \in X$, where

$$X = \{(x, \xi) \mid \xi = \xi(x)\}. \qquad (2.7)$$

Lemma 2.1 shows that there is only one point of tangency in this case, and the tangency in this point is of simple quadratic type. Evidently, if $(x, \xi) = (x_0, \xi_0)$ (where $\xi_0 = \xi(x_0)$) the tangency point is (y_0, η_0) (where $\eta_0 = \eta(y_0)$).

Hence, if all the neighbourhoods defined above are sifficiently small

$$H_{n+m+k}(M_{x,\xi}, M_{x,\xi} \cap Y) = \mathbb{Z}. \qquad (2.8)$$

Here H_{n+m+k} is a relative compact homology group of dimension n+m+k. Let (x, ξ) be sufficiently close to (x_0, ξ_0). Let us choose the vanishing cycle (see [1]) as a generator of the group (2.8), and denote by $h(x, \xi)$ the homology class of the cycle with an orientation continuously depending on ξ.

2.2. Definition of Laplace-Radon integral operators. Continuity

Let $a(x, y, \theta)$ be a holomorphic function in some neighbourhood of the point (x_0, ξ_0, θ_0). For each integer $r \in \mathbb{Z}$ we define a Laplace-Radon integral operator $\hat{\Phi}_r[a]$ supposing for each function $f(y, \eta) \in A_{q+r}(U', Y)$

---

* The manifold $Y = \{(\eta, y) \mid \eta = \eta(y)\}$ in the space $\mathbb{C}^{n+1}$ with the coordinates (y, η) we shall also denote by Y.

$$\hat{\Phi}_\tau[a]f = \pi^{-\frac{m+k+n}{2}}\left(\frac{i}{2\pi}\right)^n\left(\frac{\partial}{\partial\xi}\right)^{\frac{m+k+n}{2}}\int\limits_{h(x,\xi)} a(x,y,\theta)\left[\left(\frac{\partial}{\partial\eta}\right)^\tau f\right](y,$$

$$\xi-\Phi(x,y,\theta)+\tau^2)\,dy\wedge d\theta\wedge d\tau\;^{*} \tag{2.9}$$

Here $h(x,\xi)$ is a homology class defined in section 1, the operator $\left(\frac{\partial}{\partial\eta}\right)^\tau$ is defined as $-r$-th power of the operator $\hat{I}$ $\left(\hat{I}f = \int\limits_{\eta(x)}^{\eta} f(y,\alpha)d\alpha\right)$ for negative r.

Theorem 2.1. The operator $\hat{\Phi}_r[a]$ is a continuous map in spaces $\hat{\Phi}_r[a] : A_{q+r}(U', Y) \to A_q(U, X)$.

Proof. We shall mark first of all that the class $h(x,\xi)$, as shown in section 1, is ramifying on the manifold X and consequently the function $\Phi_r[a]f$ can be singular (ramifies) only on X.

Assuming that $\xi \neq \xi(x)$ and using Morse lemma (applicable due to Lemma 2.1) there exists a change of variables

$$y = y(z), \quad \theta = \theta(z) \tag{2.10}$$

such that

$$\eta(y) + \Phi(x, y, \theta) - \tau^2 = \xi(x) + \sum_{j=1}^{n+m+k} z_j^2 = \xi(x) + z^2$$

The equation of intersection $M_{x,\xi} \cap Y$ in these coordinates can be written in the form

$$\xi(x) - \xi = z^2 .$$

Hence, the equation of vanishing cycle in coordinates z has the form

$$z = u\sqrt{\xi(x) - \xi}\ ,$$

where u belongs to a real $n + m + k$ -dimensional unit sphere $|u| \leq 1$. Consequently, a representative (contour) $\gamma(x,\xi)$ of the class $h(x,\xi)$ can be chosen in such a way that

$$\text{mes}\ \gamma(x,\xi) \leq c\ |\xi(x) - \xi|^{\frac{m+n+k}{2}} \tag{2.11}$$

where mes is the Lebesgue measure**.

---

* The integer k is selected so that $m + n + k$ is even (see Lemma 2.3).

** We estimate the Lebesgue measure in z-coordinates, but the analogous estimate holds also in previous variables (probably with another constant) due to smoothness of change variables. This is also true for the estimate of distance below.

Let us estimate now the distance between the point $(y, \eta) = (y, \xi - \Phi(x, y, \theta) + \tau^2)$ and the manifold y along the coordinate η, i.e.

$$|\xi - \Phi(x, y, \theta) + \tau^2 - \eta(y)|. \tag{2.12}$$

Mark that the expression (2.12) on the contour $\gamma(x, \xi)$ in z-coordinates is equal to

$$|\xi - \xi(x) - z^2| = |\xi - \xi(x) - u^2(\xi - \xi(x))|,$$

not exceeding $|\xi - \xi(x)|$. Thus, the distance between $\gamma(x, \xi)$ and Y along η is estimated in the following way:

$$|\eta - \eta(y)| \Big|_{\gamma(x,\xi)} \leq c\,|\xi - \xi(x)|.$$

Due to Lemma 1.1 it obviously suffices to prove the theorem for $r = 0$.

Since

$$\Big|\int_{h(x,\xi)} a(x,y,\theta) f(y, \xi - \Phi(x,y,\theta) + \tau^2)\, dy \wedge d\theta \wedge d\tau\Big| \leq C \int_{\gamma(x,\xi)} |f(y, \xi - \Phi(x,y,\theta) + \tau^2)| \cdot |dy \wedge d\theta \wedge d\tau| \leq C \cdot C_1 \cdot |\xi - \xi(x)|^{\frac{n+m+k}{2}} \sup_{\gamma(x,\xi)} |f(y, \xi - \Phi(x,y,\theta) + \tau^2)| \leq C\,|\xi - \xi(x)|^{q + \frac{n+m+k}{2}},$$

hence

$$\int_{h(x,\xi)} a(x,y,\theta) f(y, \xi - \Phi(x,y,\theta) + \tau^2)\, dy \wedge d\theta \wedge d\tau \in \mathcal{A}_{q + \frac{n+m+k}{2}}(U, X). \tag{2.13}$$

This completes the proof.

2.3. Canonical form of Laplace-Radon integral operators. Pseudodifferential operators

Lemma 2.3 (stabilization lemma). The representation 2.9 doesn't depend on the number k of variables τ (under condition that $m + n + k$ is even).

Proof. Let us show that

$$\left(\frac{i}{2\pi}\right)^n \pi^{\frac{n-m-k}{2}} \left(\frac{\partial}{\partial \xi}\right)^{\frac{n+m+k}{2}} \int_{h(x,\xi)} a(x,y,\theta) \left[\left(\frac{\partial}{\partial \eta}\right)^r f\right](y, \xi - \Phi(x,y,\theta) +$$

$$+\sum_{j=1}^{k}\tau_j^2)dy\wedge d\theta\wedge d\tau=\left(\frac{i}{2\pi}\right)^n\pi^{\frac{n-m-k+2}{2}}\left(\frac{\partial}{\partial\xi}\right)^{\frac{n+m+k-2}{2}}\int\limits_{h_1(x,\xi)} a(x,$$

$$y,\theta)\left[\left(\frac{\partial}{\partial\eta}\right)^{\tau}f\right](y,\xi-\Phi(x,y,\theta)+\sum_{j=1}^{k-2}\tau_j^2)dy\wedge d\theta\wedge d\tau' \tag{2.14}$$

where $\tau'=(\tau_1,\ldots,\tau_{k-2})$ and the class $h_1(x,\xi)$ is defined analogously to the class $h(x,\xi)$ with the number of variables k being diminished by 2. We shall perform a change of variables in the integral of the right-hand side of the formula (2.14)

$$\begin{cases} v=\tau_{k-1}+i\tau_k, \\ w=\tau_{k-1}-i\tau_k \end{cases}$$

so that

$$dv\wedge dw=(d\tau_{k-1}+id\tau_k)\wedge(d\tau_{k-1}-id\tau_k)=$$

$$=-2id\tau_{k-1}\wedge d\tau_k .$$

Under the change of variables above the integral is transformed to the form

$$\left(-\frac{1}{2i}\right)\pi^{-\frac{m+k-n}{2}}\left(\frac{i}{2\pi}\right)^n\left(\frac{\partial}{\partial\xi}\right)^{\frac{n+m+k-2}{2}}\int\limits_{h(x,\xi)} a(x,y,\theta)\frac{\partial}{\partial\xi}g(y,\xi-$$

$$-\Phi(x,y,\theta)+\sum_{j=1}^{k-2}\tau_j^2+vw)dy\wedge d\theta\wedge d\tau'\wedge dv\wedge dw \tag{2.15}$$

where

$$g=\left(\frac{\partial}{\partial\eta}\right)^{\tau}f.$$

Consider the submanifold $v = 0$ in a space with coordinates (y,θ,τ',v,w). Homology classes $h(x,\xi)$ and $h_1(x,\xi)$ belong respectively to:

$$h(x,\xi)\in H_{n+m+k}(\mathbb{C}^{n+m+k}\ Z),$$

$$h_1(x,\xi)\in H_{n+m+k-2}(\mathbb{C}^{n+m+k-2},\ Z_1)$$

where

$$Z=\left\{\xi-\Phi(x,y,\theta)+\sum_{j=1}^{k-2}\tau_j^2+vw=\eta(y)\right\},$$

$$Z_1 = \left\{ \xi - \Phi(x,y,\theta) + \sum_{j=1}^{k-2} \tau_j^2 = \eta(y) \right\}.$$

Let

$$i : Z_1 \hookrightarrow \mathbb{C}^{n+m+k-1}_{y,\theta,\tau',w} = \mathbb{C}^{n+m+k}_{y,\theta,\tau',v,w} \cap \{v = 0\}$$

be an embedding at $w = 0$. The set $Z \cap \{v = 0\}$ contracts on $i(Z_1)$. Hence the mapping

$$i : H_j(\mathbb{C}^{n+m+k-2}_{y,\theta,\tau'}, Z_1) \to H_j(\mathbb{C}^{n+m+k-1}_{y,\theta,\tau',w}, Z \cap \{v=0\})$$

is an isomorphism.

Let $\gamma(x,\xi)$ be a representative of the class $h(x,\xi)$, defined in the proof of theorem 2.1 (obviously, the variables τ_{k-1} and τ_k can be included into the set of Morse variables Z). Then the intersection $\gamma(x,\xi) \cap \{v = 0\}$ is defined by equality

$$(u_{k-1} + iu_k)\sqrt{\xi - \xi(x)} = 0 ,$$

which yields $u_{k-1} = u_k = 0$. Hence, if $\xi \neq \xi(x)$ the intersection is transversal. Besides, this intersection belongs to $\{v = 0\} \cap \{w = 0\}$. If $\overline{\omega}$ is the Leray homomorphism ([1], [2]) the considerations above give us

$$h_1(x,\xi) = i_*^{-1}\, \overline{\omega}\, h(x,\xi) \tag{2.16}$$

We consider now the integrand in (2.15)

$$\alpha = a(x,y,\theta)\frac{\partial g}{\partial \xi}\Big(y, \xi - \Phi(x,y,\theta) + \sum_{j=1}^{k-2}\tau_j^2 + vw\Big)dy\wedge d\theta\wedge d\tau'\wedge dv\wedge dw.$$

Evidently,

$$\alpha = d\left\{a(x,y,\theta)\,\frac{g(y,\xi - \Phi(x,y,\theta) + \sum_{j=1}^{k-2}\tau_j^2 + vw)}{v}\,dy\wedge d\theta\wedge d\tau'\wedge dv\right\} = d\beta.$$

Next, the residue of the form β at the submanifold $\{v = 0\}$ is equal to

$$\operatorname*{res}_{\{v=0\}} \beta = a(x,y,\theta)\,g\Big(y, \xi - \Phi(x,y,\theta) + \sum_{j=1}^{k-2}\tau_j^2\Big)dy\wedge d\theta\wedge d\tau'$$

in the space $\mathbb{C}^{n+m+k-1}_{y,\theta,\tau,w}$, i.e.

$$i^*(\operatorname{res}\beta) = \alpha_1$$

where

$$\alpha_1 = a(x,y,\theta)\,g\Big(y, \xi - \Phi(x,y,\theta) + \sum_{j=1}^{k-2}\tau_j^2\Big)\,dy\wedge d\theta\wedge d\tau'$$

is an integrand of the right-hand side of the equality (2.14).

Hence, ([1] , [2])

$$\alpha = -\frac{1}{2\pi i}\,\bar{\omega}^*\,(i^*)^{-1}\alpha_1 .$$

Now due to duality we have

$$\pi^{-\frac{m+k-n}{2}}\left(\frac{i}{2\pi}\right)^n\left(-\frac{1}{2i}\right)\left(\frac{\partial}{\partial\xi}\right)^{\frac{n+m+k-2}{2}}\int\limits_{h(x,\xi)}\alpha = \pi^{-\frac{m+k-n}{2}}\left(\frac{i}{2\pi}\right)^n\pi\,\times$$

$$\times\left(\frac{\partial}{\partial\xi}\right)^{\frac{n+m+k-2}{2}}\int\limits_{h(x,\xi)}\bar{\omega}^*(i^*)^{-1}\alpha_1 = \pi^{-\frac{m+k-2-n}{2}}\left(\frac{i}{2\pi}\right)^n\left(\frac{\partial}{\partial\xi}\right)^{\frac{n+m+k-2}{2}}\int\limits_{h_1(x,\xi)}\alpha_1 .$$

The last expression coincides with the right-hand side of (2.14). It completes the proof of Lemma 2.3.

<u>Remark 2.1.</u> The lemma is also valid in case functions a and f depend also on one of the variables v or w (but not on the other).

Let us denote the gradient ideal, that is the ideal generated by derivatives of the determining function Φ with respect to θ , by $\mathcal{J}$:

$$\mathcal{J} = \{\Phi_{\theta_1}, \ldots, \Phi_{\theta_m}\}.$$

<u>Lemma 2.4.</u> If the function a (x, y, θ) belongs to $\mathcal{J}$, the operator $\hat{\Phi}_r[a]$ is an operator of order r-1 in the scale A_q.

Proof. Let us transform the integral (2.9), assuming that

$$a(x,y,\theta) = \sum_{j=1}^{m} b_j(x,y,\theta)\,\frac{\partial\Phi}{\partial\theta_j}(x,y,\theta).$$

We have (a numerical multiplier being denoted by c)

$$\hat{\Phi}_r[a]f = c\left(\frac{\partial}{\partial\xi}\right)^{\frac{n+m+k}{2}}\int\limits_{h(x,\xi)} a(x,y,\theta)\frac{\partial}{\partial\xi}\left[\left(\frac{\partial}{\partial\eta}\right)^{r-1}f\right](y,\xi -$$

$$-\Phi(x,y,\theta)+\tau^2)\,dy\wedge d\theta\wedge d\tau = c\left(\frac{\partial}{\partial\xi}\right)^{\frac{n+m+k}{2}}\int\limits_{h(x,\xi)}\sum_{j=1}^{m} b_j(x,y,\theta)\,\times$$

$$\times\,\frac{\partial\Phi}{\partial\theta_j}(x,y,\theta)\frac{\partial}{\partial\xi}\left[\left(\frac{\partial}{\partial\eta}\right)^{r-1}f\right](y,\xi-\Phi(x,y,\theta)+\tau^2)\,dy\wedge d\theta\wedge d\tau =$$

$$= c\left(\frac{\partial}{\partial\xi}\right)^{\frac{n+m+k}{2}} \sum_{j=1}^{m} \int_{h(x,\xi)} b_j(x,y,\theta)\frac{\partial}{\partial\theta_j}\left[\left(\frac{\partial}{\partial\eta}\right)^{r-1} f\right](y,\xi-\Phi(x,y,\theta)+\tau^2)\times$$

$$\times\, dy\wedge d\theta\wedge d\tau = c\left(\frac{\partial}{\partial\xi}\right)^{\frac{n+m+k}{2}} \int_{h(x,\xi)} \left[\sum_{j=1}^{m}\frac{\partial b_j(x,y,\theta)}{\partial\theta_j}\right]\left[\left(\frac{\partial}{\partial\eta}\right)^{r-1} f\right](y,\xi-$$

$$-\Phi(x,y,\theta)+\tau^2)\,dy\wedge d\theta\wedge d\tau.$$

The operator obtained is an operator of order r-1 due to Theorem 2.1. This completes the proof of the lemma.

The lemmas 2.3 and 2.4 permit us to reduce the operator $\hat{\Phi}_r[a]$ to the canonical form up to operators of order r-1 (one can find details in [8]). Namely, if $(x^I, p_{\bar{I}}, q)$ are canonical coordinates on graph g and $|\bar{I}|$ is even, the formula

$$\hat{\Phi}_r[a]f = \pi^{-|\bar{I}|/2}\left(\frac{i}{2\pi}\right)^n\left(\frac{\partial}{\partial\xi}\right)^{n+|\bar{I}|/2}\int_{h(x,\xi)} a(x^I,p_{\bar{I}},q)\left[\left(\frac{\partial}{\partial\eta}\right)^r f\right](y,$$

$$\xi-S_I(x^I,p_{\bar{I}},q)-x^{\bar{I}}p_{\bar{I}}+yq)\,dy\wedge dp_{\bar{I}}\wedge dq$$

holds, S_I being the generating function of the canonical transformation g with respect to canonical coordinates $(x^I, p_{\bar{I}}, q)$.

Inversely, if $|\bar{I}|$ is odd, we have

$$\hat{\Phi}_r[a]f = \pi^{-(|\bar{I}|+1)/2}\left(\frac{i}{2\pi}\right)^n\left(\frac{\partial}{\partial\xi}\right)^{n+(|\bar{I}|+1)/2}\int_{h(x,\xi)} a(x^I,p_{\bar{I}},q)\left[\left(\frac{\partial}{\partial\eta}\right)^r f\right](y,$$

$$\xi-S_I(x^I,p_{\bar{I}},q)-x^{\bar{I}}p_{\bar{I}}+yq+\tau^2)\,dy\wedge dp_{\bar{I}}\wedge dq\wedge d\tau,$$

where $\tau\in\mathbb{C}^1$.

In order to define the operator $\hat{\Phi}_r[a]$ with the help of some function on the Lagrangian manifold L = graph g we must consider a non-degenerated holomorphic (2n, 0)-form μ on $\mathcal{L}$ and a non-degenerated holomorphic (2n, 0)-form v on $\mathbb{C}^{2n}_{x,y}$ (one can use $v = dx\wedge dy$ if changes of variables in the base space are of no interest).

More precisely, we define a function*

$$F(\Phi,\mu,v) = \frac{\alpha^*\mu\wedge d\Phi_{\theta_1}\wedge\ldots\wedge d\Phi_{\theta_m}}{v\wedge d\theta_1\wedge\ldots\wedge d\theta_m} \qquad (2.17)$$

---

*Throughout what follows we shall for simplicity omit mappings of the type α *.

extending all the objects under consideration from the set C to some of its neighbourhoods, where α is the classifying mapping (see point 1).

If now $\Phi_1(x, y, \theta_1, \ldots, \theta_{m+1}) = \Phi(x, y, \theta_1, \ldots, \theta_m) + \theta_{m+1}^2$, we evidently have

$$F(\Phi_1, \mu, v) = 2F(\Phi, \mu, v), \tag{2.18}$$

and if $\Phi_1(x, y, \theta') = \Phi(x, y, \theta(x, y, \theta'))$, we have

$$F(\Phi_1, \mu, v) = \left[\det \frac{\partial\theta(x,y,\theta')}{\partial\theta'}\right]^2 F(\Phi, \mu, v) \tag{2.19}$$

Indeed, in this case

$$F(\Phi_1, \mu, v) = \frac{\mu\wedge d\Phi_{1\theta_1'}\wedge\ldots\wedge d\Phi_{1\theta_m'}}{v\wedge d\theta_1'\wedge\ldots\wedge d\theta_m'}\cdot\frac{v\wedge d\theta_1\wedge\ldots\wedge d\theta_m}{\mu\wedge d\Phi_{\theta_1}\wedge\ldots\wedge d\Phi_{\theta_m}}\times$$

$$\times F(\Phi,\mu,v) = \frac{\mu\wedge d\Phi_{1\theta_1'}\wedge\ldots\wedge d\Phi_{1\theta_m'}}{\mu\wedge d\Phi_{\theta_1}\wedge\ldots\wedge d\Phi_{\theta_m}}\cdot\frac{v\wedge d\theta_1\wedge\ldots\wedge d\theta_m}{v\wedge d\theta_1'\wedge\ldots\wedge d\theta_m'}\times$$

$$\times F(\Phi,\mu,v) = \left[\det \frac{\partial\theta(x,y,\theta')}{\partial\theta'}\right]^2 F(\Phi,\mu,v)$$

on C_Φ, since $d\Phi_{1\theta_k'} = d\left(\partial\theta_j/\partial\theta_k'\,\Phi_{\theta_j}\right) = \partial\theta_j/\partial\theta_k'\, d\Phi_{\theta_j}$ on C_Φ.

Using formulae (2.18), (2.19) and the usual classifying lemma (see [10]), one can find that the operator

$$\hat{\Phi}_r[\varphi]f = \frac{i^n}{(2\pi)^{(n+m+k)/2}}\left[\frac{\partial}{\partial\xi}\right]^{\frac{n+m+k}{2}}\int_{h(x,\xi)}\varphi(x,y,\theta)\sqrt{F(\Phi,\mu,v)}\times$$

$$\times\left[\left(\frac{\partial}{\partial\eta}\right)^r f\right]\left(y, \xi-\Phi(x,y,\theta)+\sum_{j=1}^{k}\tau_j^2\right)dy\wedge d\theta\wedge d\tau$$

depends only on the manifold $\mathcal{L}$ = graph g and on the function φ on $\mathcal{L}$ up to operators of order r-1 (v, μ being fixed).

As changes of variables in base space are not of interest for us we stand $v = dx\wedge dy$. For unification we choose measure μ equal to the raising onto $\mathcal{L}$ = graph g of n-th exterior power of symplectic form ω' on the space $\mathbb{C}^n\oplus\mathbb{C}_n$ (with coordinates (y, q)). One can also assume the function φ on $\mathcal{L}$ to be a function on the same symplectic space. Under this convention we shall denote the operator (2.20) by

$$\hat{\Phi}_r = \hat{\Phi}_r[g, \varphi]$$

the (analytical) function φ being called the symbol of the operator $\hat{\Phi}_r$.

Example 2.1. (pseudodifferential operators). Let us now concretize our constructions, assuming that

1) $g = \mathrm{id}$
2) $F \equiv 1$
3) $\varphi(x,y,\theta) = A(x,\theta)$
4) $k = 0$

In this case we have

$$\Phi(x, y, \theta) = \theta\cdot(x - y)$$

and we obtain the operator

$$A_m\left(x, -\frac{\partial}{\partial x}\right)f = \frac{i^n}{(2\pi)^{\frac{n+m}{2}}}\left(\frac{\partial}{\partial \xi}\right)^{\frac{n+m}{2}} \int_{h(x,\xi)} A(x,p) \times$$
$$\times \left(\frac{\partial}{\partial \eta}\right)^{\frac{n+m}{2}} f(y, \xi - p(x-y))\, dp \wedge dy \tag{2.21}$$

where we imposed a standard notation $\theta = p$. The last operator will be named a pseudodifferential operator (p.d.o) of order $m \in \mathbb{Z}$.

Proposition 2.1. The operator (2.21) is a continuous mapping

$$A_m(x, -\partial/\partial x) : A_q(U, X) \to A_{q-m}(U, X),$$

for $q > -1$, $q-m > -1$.

Proof follows from Theorem 2.1.

Proposition 2.2. In the case when $A(x, p)$ is a homogeneous polynomial in p :

$$A(x, p) = \sum_{|\alpha| = m} a_\alpha(x) p^\alpha$$

the corresponding operator is a homogeneous differential operator of order m :

$$A_m(x, -\partial/\partial x) = \sum_{|\alpha| = m} a_\alpha(x)(-\partial/\partial x)^\alpha,$$

in particular, $A(x, p) = 1$, yields $A_0(x, -\frac{\partial}{\partial x}) = \hat{1}$ ($\hat{1}$ being an identity operator).

Proof of the proposition follows from Lemma 2.3.

2.4. Composition of Laplace-Radon differential operators

Theorem 2.2. Let

$$g_1 : U_{(z_0, r_0)} \longrightarrow U_{(y_0, q_0)}, \quad g_2 : U_{(y_0, q_0)} \longrightarrow U_{(x_0, p_0)}$$

be canonical transformations and $\hat{\Phi}_s[g_1, \varphi_1]$, $\hat{\Phi}_r[g_2, \varphi_2]$ be Laplace-Radon operators defined in corresponding neighbourhoods. The equality

$$\hat{\Phi}_r[g_2, \varphi_2] \circ \hat{\Phi}_s[g_1, \varphi_1] = \pm \hat{\Phi}_{r+s}\left[g_2 \circ g_1, (g_1^*, \varphi_2) \cdot \varphi_1\right] \qquad (2.22)$$

holds up to operators of order $r + s - 1$.

The sign in (2.21) depends on the choice of arguments of the functions $F(\Phi, \mu, v)$ for all the three operators in formula (2.22).

Proof. Let us calculate the composition

$$\hat{\Phi}_r[g_2, \varphi_2] \circ \hat{\Phi}_s[g_1, \varphi_1] f = \frac{i^n}{(2\pi)^{(n+m_2+k_2)/2}} \left(\frac{\partial}{\partial \xi}\right)^{(n+m_2+k_2)/2} \times$$

$$\times \int_{h_2(x,\xi)} \varphi_2(x, y, \theta) \sqrt{F(\Phi, \mu, v)} \left[\left(\frac{\partial}{\partial \eta}\right)^r \hat{\Phi}_s[g_1, \varphi_1] f\right] \Big(y, \xi - \qquad (2.23)$$

$$- \Phi(x, y, \theta) + \sum_{j=1}^{k_2} \tau_j^2\Big) dy \wedge d\theta \wedge d\tau,$$

where

$$\hat{\Phi}_s[g_1, \varphi_1] f = \frac{i^n}{(2\pi)^{(n+m_1+k_1)/2}} \left(\frac{\partial}{\partial \eta}\right)^{(n+m_1+k_1)/2} \int_{h_1(x,\xi)} \varphi_1(y, z, \theta') \times \qquad (2.24)$$

$$\times \sqrt{F(\Phi_1, \mu_1, v_1)} \left[\left(\frac{\partial}{\partial \zeta}\right)^s f\right]\Big(z, \eta - \Phi_1(y, z, \theta') + \sum_{j=1}^{k_1} \nu_j^2\Big) dz \wedge d\theta' \wedge d\nu.$$

Substituting (2.24) in (2.23) we obtain

$$\hat{\Phi}_r[g_2, \varphi_2] \circ \hat{\Phi}_s[g_1, \varphi_1] f = \frac{i^{2n}}{(2\pi)^{(2n+m_1+m_2+k_1+k_2)/2}} \left(\frac{\partial}{\partial \xi}\right)^{(n+m_2+k_2)/2} \times$$

$$\times \int\limits_{h_2(x,\xi)} \varphi_2(x,y,\theta)\sqrt{F(\Phi,\mu,v)}\left(\frac{\partial}{\partial\eta}\right)^{r}\left[\left(\frac{\partial}{\partial\eta}\right)^{(n+m_1+k_1)/2}\int\limits_{h_1(y,\eta)} \varphi_1(y,z,\theta')\sqrt{F(\Phi_1,\mu_1,v_1)}\,\times\right.$$

$$\times \left.\left[\left(\frac{\partial}{\partial\zeta}\right)^{s} f\right]\left(z,\eta-\Phi_1(y,z,\theta')+\sum_{j=1}^{k_1} v_j^2\right)dz\wedge d\theta'\wedge dv\right]\Bigg|_{\eta=\Phi(x,y,\theta)+\sum_{j=1}^{k_2}\tau_j^2}\times$$

$$\times\, dy\wedge d\theta\wedge d\tau.$$

For $r \geqslant 0$ the operator $(\partial/\partial\eta)^r$ can be allowed within the interior integral sign and transformed into $(\partial/\partial\zeta)^r$. For $r < 0$ k_1 should be chosen in such a way that $n + m_1 + k_1 > 2r$, and then operators $(\partial/\partial\eta)^r$ and $(\partial/\partial\eta)^{(n+m_1+k_1)/2}$ should be united into the operator $(\partial/\partial\eta)^{(n+m_1+k_1+2r)/2}$. After that the operator $(\partial/\partial\zeta)^s$ is substituted by $(\partial/\partial\zeta)^{-r}(\partial/\partial\zeta)^{s+r}$ and operator $(\partial/\partial\zeta)^{-r}$ is factored outside the interior integral sign as positive powered operator in the form $(\partial/\partial\eta)^{-r}$. Then the operator $(\partial/\partial\eta)^{(n+m_1+k_1)/2}$ is factored outside the exterior integral sign in the form $(\partial/\partial\xi)^{(n+m_1+k_1)/2}$.

Using the described transformations we obtain

$$\hat{\Phi}_r[g_2,\varphi_2]\circ\hat{\Phi}_s[g_1,\varphi_1]f = \frac{i^{2n}}{(2\pi)^{(2n+m_1+m_2+k_1+k_2)/2}}\left(\frac{\partial}{\partial\xi}\right)^{(2n+m_1+m_2+k_1+k_2)/2}\times$$

$$\times\int\limits_{h_2(x,\xi)}\ \int\limits_{h_1(y,\xi-\Phi_1(x,y,\theta)+\sum_{j=1}^{k_2}\tau_j^2)} \varphi_2(x,y,\theta)\sqrt{F(\Phi,\mu,v)}\,(x,y,\theta)\,\varphi_2(x,y,\theta)\times \tag{2.25}$$

$$\times\varphi_1(y,z,\theta')\sqrt{F(\Phi_1,\mu,v)}\,(y,z,\theta')\left[\left(\frac{\partial}{\partial\zeta}\right)^{r+s} f\right]\left(z,\xi-\Phi(x,y,\theta)-\Phi_1(y,z,\theta')+\right.$$

$$\left.+\sum_{j=1}^{k_2}\tau_j^2+\sum_{j=1}^{k_1}v_j^2\right)dz\wedge d\theta'\wedge dv\wedge dy\wedge d\theta\wedge d\tau.$$

The phase function of the integral (2.25) has the form

$$\tilde{\Phi}(x,y,z,\theta,\theta') = \Phi(x,y,\theta)+\Phi_1(y,z,\theta'). \tag{2.26}$$

The equations of the set $C_{\tilde{\Phi}}$ (taking into account that y belongs to the class of variables "of the type Θ") are

$$\begin{cases} \dfrac{\partial \Phi}{\partial \theta}(x,y,\theta)=0, \\ \dfrac{\partial \Phi_1}{\partial \theta'}(y,z,\theta')=0, \\ \dfrac{\partial \Phi}{\partial y}(x,y,\theta)+\dfrac{\partial \Phi_1}{\partial y}(y,z,\theta')=0. \end{cases}$$

It is convenient to rewrite the system of equations as a set of two systems

$$\begin{cases} \dfrac{\partial \Phi}{\partial \theta}(x,y,\theta)=0, \\ q=-\dfrac{\partial \Phi}{\partial y}(x,y,\theta), \end{cases} \qquad \begin{cases} \dfrac{\partial \Phi_1}{\partial \theta'}(x',z,\theta')=0, \\ p'=\dfrac{\partial \Phi_1}{\partial x'}(x',z,\theta'). \end{cases}$$

connected by equalities $p' = q$ and $x' = y$.

The last formulae show that the function (2.26) is the determining function of the composition $g_2 \circ g_1$. The product

$$\varphi_2(x,y,\theta)\cdot\varphi_1(y,z,\theta')\Big|_{C_{\tilde\Phi}}$$

corresponds to the product $\varphi_1 \cdot g_1^* \varphi_2$.

Since

$$\mu = dy\wedge dp,\ v = dx\wedge dy,\ \mu_1 = dz\wedge dq',\ v_1 = dy\wedge dz,$$

we have

$$F(\Phi,\mu,v) = \frac{\mu\wedge d\Phi_{\theta_1}\wedge\ldots\wedge d\Phi_{\theta_{m_2}}}{v\wedge d\theta_1\wedge\ldots\wedge d\theta_{m_2}} = \frac{dy\wedge dq\wedge d\Phi_{\theta_1}\wedge\ldots\wedge d\Phi_{\theta_{m_2}}}{dx\wedge dy\wedge d\theta_1\wedge\ldots\wedge d\theta_{m_2}} =$$

$$= (-1)^n \frac{dy\wedge d\Phi_{y^1}\wedge\ldots\wedge d\Phi_{y^n}\wedge d\Phi_{\theta_1}\wedge\ldots\wedge d\Phi_{\theta_{m_2}}}{dx\wedge dy\wedge d\theta_1\wedge\ldots\wedge d\theta_{m_2}};$$

$$F(\Phi_1,\mu_1,v_1) = \frac{dz\wedge d(-\Phi_{1z^1})\wedge\ldots\wedge d(-\Phi_{1z^n})\wedge d\Phi_{1\theta_1'}\wedge\ldots\wedge d\Phi_{1\theta'_{m_1}}}{dy\wedge dz\wedge d\theta_1'\wedge\ldots\wedge d\theta'_{m_1}}$$

Hence, we have

$$F[\Phi,\mu,v]\cdot F[\Phi_1,\mu_1,v_1]=$$

$$=(-1)^n \frac{dz\wedge d(-\Phi_{1z})\wedge d\Phi_y\wedge d\Phi_\theta\wedge d\Phi_{1\theta'}}{dx\wedge dz\wedge dy\wedge d\theta\wedge d\theta'}.$$

Taking the equality $\tilde{\Phi}=\Phi+\Phi_1$ into account, we have

$$F[\Phi,\mu,v]\cdot F[\Phi_1,\mu_1,v_1]=(-1)^n F[\tilde{\Phi},\tilde{\mu},\tilde{v}]$$

where $\tilde{v} = dx\wedge dz$.

Substituting the obtained formulae in the integral (2.25) we have

$$\hat{\Phi}_r[g_2,\varphi_2]\circ\hat{\Phi}_s[g_1,\varphi_1]f=\frac{\pm i^n}{(2\pi)^{(\tilde{m}+\tilde{k}+n)/2}}\left(\frac{\partial}{\partial\xi}\right)^{(\tilde{m}+\tilde{k}+n)/2}\times$$

$$\times\int\limits_{h_1\times h_2(x,\xi)}\tilde{\varphi}(x,z,\tilde{\theta})\sqrt{F[\tilde{\Phi},\tilde{\mu},\tilde{v}]}\left[\left(\frac{\partial}{\partial\zeta}\right)^{r+s}f\right]\Big(z,\xi-\tilde{\Phi}(x,z,\tilde{\theta})+\sum_{j=1}^{\tilde{k}}\tilde{\tau}_j^2\Big)\times$$

$$\times\, dz\wedge d\tilde{\theta}\wedge d\tilde{\tau}$$

up to operators of order r+s-1. Here $\tilde{\Phi}$ is defined by the formula (2.26), $\tilde{\theta}=(y,\theta,\theta')$, $\tilde{\tau}=(\tau,\nu)$, $\tilde{m}=n+m_1+m_2$, $\tilde{k}=k_1+k_2$. And now, proceeding from the relation (2.8) it is not hard to see that $(h_1\times h_2)(x,\xi)=h(x,\xi)$.

Corollary 2.1. Let $P(x,-\partial/\partial x)$ be a differential operator of order r. For any p.d.o. $A_m(x,-\partial/\partial x)$ the formulae

$$P(x,-\partial/\partial x)\circ A_m(x,-\partial/\partial x)=(P\cdot A)_{m+r}(x,-\partial/\partial x)+Q_1,$$

$$A_m(x,-\partial/\partial x)\circ P(x,-\partial/\partial x)=(A\cdot P)_{m+r}(x,-\partial/\partial x)+Q_2$$

hold, where Q_j, j = 1, 2, are p.d.o. of order m+r-1.

The more precize information on the composition of two p.d.o contains in the following theorem.

Theorem 2.3. For any p.d.o. $A_r(x,-\partial/\partial x)$, $B_m(x,-\partial/\partial x)$ the formula

$$A_r(x,-\partial/\partial x)\circ B_m(x,-\partial/\partial x)=\sum_{j=0}^{N}P_{m+r-j}(x,-\partial/\partial x)$$

holds up to operators of order r+m-N-1, where

$$P_{m+r-j}(x,p)=\sum_{|\alpha|=j}\frac{(-1)^{|\alpha|}}{\alpha!}\,\frac{\partial^{|\alpha|}A(x,p)}{\partial p^{\alpha}}\,\frac{\partial^{|\alpha|}B(x,p)}{\partial x^{\alpha}}.$$

<u>Proof</u>. We have

$$A_r\left(x,-\frac{\partial}{\partial x}\right)f=\left(\frac{i}{2\pi}\right)^n\left(\frac{\partial}{\partial\xi}\right)^n\int\limits_{h(x,\xi)}A(x,p)\left[\left(\frac{\partial}{\partial\eta}\right)^r f\right](y,\xi-p(x-y))\,dp\wedge dy,$$

$$B_m\left(x,-\frac{\partial}{\partial x}\right)f=\left(\frac{i}{2\pi}\right)^n\left(\frac{\partial}{\partial\xi}\right)^n\int\limits_{h(x,\xi)}B(x,p)\left[\left(\frac{\partial}{\partial\eta}\right)^m f\right](y,\xi-p(x-y))\,dp\wedge dy.$$

Let us calculate the composition of the operators above. Analogously to the theorem 2.2, we have

$$A_r\left(x,-\frac{\partial}{\partial x}\right)\circ B_m\left(x,-\frac{\partial}{\partial x}\right)f=\left(\frac{i}{2\pi}\right)^{2n}\left(\frac{\partial}{\partial\xi}\right)^{2n}\int\limits_{H(x,\xi)}A(x,p)B(y,q)\times$$

$$\times\left[\left(\frac{\partial}{\partial\eta}\right)^{r+m}f\right](z,\xi-p(x-y)-q(y-z))\,dp\wedge dy\wedge dq\wedge dz.$$

Introducing the variables

u = y - x, v = p - q, q, z

we transform the last integral

$$A_r\left(x,-\frac{\partial}{\partial x}\right)\circ B_m\left(x,-\frac{\partial}{\partial x}\right)f=\left(\frac{i}{2\pi}\right)^{2n}\left(\frac{\partial}{\partial\xi}\right)^{2n}\int\limits_{H(x,\xi)}A(x,q+v)B(x+u,$$

$$q)\left[\left(\frac{\partial}{\partial\eta}\right)^{r+m}f\right](z,\xi-q(x-z)+uv)\,du\wedge dv\wedge dq\wedge dz. \tag{2.27}$$

Let us expand the function B(x + u, q) in a series in u.

$$B(x+u,q)=\sum_{|\alpha|\le N}\frac{1}{\alpha!}\,\frac{\partial^{|\alpha|}B}{\partial x^{\alpha}}(x,q)u^{\alpha}+\sum_{|\alpha|=N+1}u^{\alpha}B_{\alpha}(x,q,u). \tag{2.28}$$

Integrating by parts with respect to v shows that the term of the right-hand side of (2.27) corresponding to the last sum in (2.28) is an operator of order r+m-N-1. Using the same method for the group of terms corresponding to the first sum of the right-hand side of (2.28), we obtain

$$A_r\left(x,-\frac{\partial}{\partial x}\right)\circ B_m\left(x,-\frac{\partial}{\partial x}\right)f=\sum_{|\alpha|\leq N}\left(\frac{i}{2\pi}\right)^{2n}\left[\frac{\partial}{\partial \xi}\right]^{2n}\times$$

$$\times\int_{H(x,\xi)}(-1)^{|\alpha|}\frac{\partial^{|\alpha|}A}{\partial q^{\alpha}}(x,q+v)\frac{\partial^{|\alpha|}B}{\partial x^{\alpha}}(x,q)\frac{1}{\alpha!}\left[\left(\frac{\partial}{\partial \eta}\right)^{r+m-|\alpha|}f\right](z,$$

$$\xi-q(x-z)+uv)\,du\wedge dv\wedge dq\wedge dz$$

Using Lemma 2.3 and Remark 2.1 and taking into account that the amplitude doesn't depend on u, we transform the last sum to the form

$$\sum_{|\alpha|\leq N}\left(\frac{i}{2\pi}\right)^{n}\left[\frac{\partial}{\partial \xi}\right]^{n}\int_{h(x,\xi)}\frac{(-1)^{|\alpha|}}{\alpha!}\,\frac{\partial^{|\alpha|}A(x,q)}{\partial q^{\alpha}}\,\frac{\partial^{|\alpha|}B(x,q)}{\partial x^{\alpha}}\times$$

$$\times\left[\left(\frac{\partial}{\partial \eta}\right)^{r+m-|\alpha|}f\right](z,\xi-q(x-z))\,dq\wedge dz.$$

This completes the proof of the theorem.

3. Asymptotics of solutions of differential equations

Let

$$P\left(x,-\frac{\partial}{\partial x}\right)=\sum_{|\alpha|\leq r}a_{\alpha}(x)\left(-\frac{\partial}{\partial x}\right)^{\alpha} \tag{3.1}$$

be a differential expression of order r with holomorphic coefficients.

We consider a solution u(x,) of the equation

$$P(x,-\partial/\partial x)u(x,\xi)=f(x,\xi) \tag{3.2}$$

in a neighbourhood of a point $(x_o,\xi_o)\in\mathbb{C}^{n+1}_{x,}$.

To begin with we shall consider the non-characteristical case. Suppose that the function $u(x,\xi)$ has singularities on a surface X:

$$X=\{(\xi,x)|\ \ \xi=\xi(x)\ \ , \tag{3.3}$$

the function $\xi(x)$ being holomorphic on U and that $u(x,\xi)$ has N bounded derivatives with respect to ξ :

$$\left|\frac{\partial^j u}{\partial \xi^j}\right| \leq C_j, \ j \leq N \quad \text{in} \quad \tilde{U} \Subset U \tag{3.4}$$

Theorem 3.1. If the surface X is non-characteristical with respect to the operator $P(x, -\partial/\partial x)$:

$$P_r(x, \partial\xi/\partial x) \neq 0 \tag{3.5}$$

the asymptotics of the function $u(x, \xi)$ has the form

$$u(x, \xi) = \left(\frac{1}{P_r}\right)_{-r}(x, -\partial/\partial x)f \pmod{A_{N+1}(U, X)} \tag{3.6}$$

Proof. Due to Lemma 1.1 the function $u(x, \xi)$ has the form

$$u(x, \xi) = u_1 + u_2 \tag{3.7}$$

where $u_1 \in A_N(U,X)$, u_2 is holomorphic. Substituting (3.7) in the equation (2.2) we obtain the equation on the term u_1 of the function u:

$$P(x, -\partial/\partial x)u_1 = f(x, \xi) - P(x, -\partial/\partial x)u = f_1(x, \xi)$$

Due to Theorem 2.1 the right-hand side $f_1(x, \xi)$ of this equation belongs to the space $A_{N-r}(U, X)$. Using now the composition Theorem 2.2, we obtain

$$\left(\frac{1}{P_r}\right)_{-r}(x, -\partial/\partial x)P(x, -\partial/\partial x)u_1 = u_1 + Qu_1,$$

where ord Q = -1. The latter relation gives us $(P(x, -\partial/\partial x)u_1 = f_1)$

$$u_1 \equiv \left[\frac{1}{P_r}\right]_{-r}(x, -\frac{\partial}{\partial x})f_1 \pmod{A_{N+1}(U, X)}$$

This completes the proof.

Now let us examine the characteristical case. Let

$$X = \{\xi = \xi(x)\}$$

be a characteristic manifold with respect to the expression (3.1), i.e. $P_r(x, \partial\xi(x)/\partial x) = 0$, and

$$V(P) = \frac{\partial P_r}{\partial p}\frac{\partial}{\partial x} - \frac{\partial P_r}{\partial x}\frac{\partial}{\partial p} \neq 0$$

on the set

$$\text{char}\,\hat{P} = \{(x,p) \mid P_r(x,p) = 0\}.$$

Let g be a canonical transformation which reduces the principal symbol $P_r(x, p)$ of the operator $P(x, -\partial/\partial x)$ to the symbol p_1.

Hence, due to composition theorem 2.2 we have

$$\hat{\Phi}_o[g,1]\circ P(x,-\frac{\partial}{\partial x})\circ\hat{\Phi}_o[g^{-1},1] = -\frac{\partial}{\partial x^1}\circ\left(\frac{\partial}{\partial\xi}\right)^{r-1} + \hat{Q} \qquad (3.8)$$

where $\hat{Q}$ has the order r-1.

The formula (3.8) reduces the problem

$$P(x, -\partial/\partial x)u = f,$$

$u \in A_q(U, x)$, $f \in A_{q-r}(U, X)$ to the following problem

$$-\left(\frac{\partial}{\partial\xi}\right)^{r-1}\frac{\partial\tilde{u}}{\partial x^1} + \hat{Q}\tilde{u} = -\hat{\Phi}_o[g^{-1},1]f, \qquad (3.9)$$

where $u = \hat{\Phi}_o[g^{-1}, 1]\,u$, and $\hat{Q}$ has the order r-1.

Using now the theorem 2.3 we can obtain such an operator $U(x, -\partial/\partial x)$ that the problem (3.9) is equivalent to the problem*)

$$\frac{\partial\tilde{v}}{\partial x^1} + \hat{Q}_1 v = -\hat{U}\,\hat{\Phi}_o[g, 1]\,f,$$

$\hat{Q}_1$ being the operator of order - N (for details see [16]).

(3.9) yields

$$\tilde{v} = -\int\hat{U}\,\hat{\Phi}_o[g, 1]\,f\,dx^1$$

and hence

$$u(x,\xi) \equiv -\hat{\Phi}_o[g^{-1},1]\left(\frac{\partial}{\partial\xi}\right)^{1-r}\hat{U}^{-1}\int\hat{U}\,\hat{\Phi}_o[g,1]\,f\,dx^1$$

$$(\mathrm{mod}\ A_{q+N}(U, X)).$$

References

1. Leray J. Le calcul différentiel et integral sur une variété analitique complexe (Problème de Cauchy, III), Bull. Soc. Math. de France, 1959, vol. 87, fasc. II.
2. Leray J. Un prolongement de la transformation de Laplace qui transforme la solution unitaire d'un opérateur hyperbolique en sa solution élémentaire (Problème de Cauchy, IV). Bull. Soc. Math. de France, 1962, vol. 90, fasc. I.

*) We denote $v = \left(\frac{\partial}{\partial\xi}\right)^{r-1}\tilde{u}$, $\tilde{v} = \hat{U}v$.

3. Leray J., Garding L. et al. Problème de Cauchy, Paris, 1964.
4. Maslov V.P., Teoriya vozmushchenii i asimptoticheskie metody (Theory of Perturbations and Asymptotic Methods) Moscow State University, Moscow, 1965 (in Russian) (French translation: Théorie des perturbations et méthodes asymptotiques, Dunod, Ganthier Villars, Paris, 1972).
5. Maslov V.P., Operatornye metody (Operator Methods) Nauka, Moscow, 1973 (in Russian). MR 56 = 3647.
6. Mistchenko A.S., Sternin B.Yu., Shatalov V.E. Lagranjevy mnogoobraziya i metod kanonitcheskogo operatora (Lagrangian Manifolds and Canonical Operator Method), Nauka, Moscow, 1978 (in Russian).
7. Nazaikinskii V.E., Oshmyan V.G., Sternin B.Yu., Shatalov V.E., Fourier integral operators and the canonical operator, Uspehi Mat. Nauk 36:2 (1981), 81-140 = Russian Math. Surveys, 36 : 2 (1981), 93-161.
8. Sternin B.Yu., Shatalov V.E., Analitic Lagrangian Manifolds and Feunmann integrals, Uspehi Mat. Nauk, 1979, vol. 34, issue 6.
9. Sternin B.Yu., Shatalov V.E., Legendre Uniformization of Multiple-Valued Analytic Functions, Math. Sb., 1980, vol. 113, issue 2.
10. Sternin B.Yu., Shatalov V.E., Characteristic Cauchy Problem On a Complex-Analytical Manifold, Voronez State Univ., 1982 (English translation: Lect. Notes Math., 1984, N 1108).
11. Sternin B.Yu., Shatalov V.E., On an integral transformation of complex analytic functions, Doklady Akad. Nauk SSSR, 280 (1985) N 3 = Soviet Math. Dokl. Vol. 31 (1985) No. 1.
12. Sternin B.Yu., Shatalov V.E. On a integral transformation of complex analytic functions, Izvestija Akad. Nauk SSSR, v. 5 (1986).
13. Pham F., Introduction à l'étude topologique des singularités de Landau, Paris, 1967.
14. Hörmander L., Fourier Integral Operators, I., Acta Math. 127 (1971), 79-183. MR 52 = 9299.
15. Shatalov V.E., Asymptotical Solutions of Complex Analytic Characteristical Cauchy Problem, Doklady Akad. Nauk SSSR, 273 (1983), N 2.
16. Sternin B.Yu., Metod kanonicheskogo operatora Maslova. (Maslov Canonical Operator Method) Kompleksnii analiz i mnogoobrazija,

Kijev, Naukova Dumka, 1978 (in Russian).

17. Lychagin V.V., Sternin B.Yu. O mikrolokalnoi strukture psevdodifferenzialnich operatorov (On a microlocal structure of pseudodifferential operators), Matem. Sb. 1985, vol.12 (in Russian).

18. Hamada Y., Leray J., Wagshal C. Systèmes d'equations aux dérivées patielles a charactéristiques multiples: problème de Cauchy ramifié; hyperbolicité partielle. - J. Math. Pure et Appl., 1976, 55, N 3.

19. Hamada Y., Nakamura G. On the singularities of the solution of the Cauchy problem for the operators with non multiple characteristics. - Ann. Sci norm. super. Pisa CL. Sci, 1977, N 4.

20. Ishii T. On a representation of the solution of the Cauchy problem with singular initial data. - Proc. Jap. Acad., 1980, A-56, N 2.

21. Ishimura R. On the existence of holomorphic solutions of a singular partial differential equations. - Mem. Fac. Sci. Kyushu Univ., 1978, A-32, N 2.

22. Kobayashi T. On the singularities of the solutions to the Cauchy problem with singular data in the complex domain. - Math. Ann. 1984, Bd 269, N 2, S, pp. 217 - 234.

23. Leray J. Probleme de Cauchy. - Bul. Soc. Math de France, I, 1957, N 85; II, 1958, N 86; III, 1958, N 87; IV, 1962, N 90.

24. Nakamura G. The singularities of solutions of the Cauchy problems for systems whose characteristic roots are non-uniform multiple. - Publ. Res. Inst. Math. Sci., 1977, 13, N 1.

25. Nakamura G., Sasai T. The singularities of solutions of the Cauchy problem for second order equations in case the initial manifold includes characteristic points. - Tohoku Math. J., 1976, 28, N 4.

26. Noutchequeme N. Application de la théorie de Leray au problème de Cauchy sur deux hypersurfaces charactéristiques sécantes. - C.r. Acad. Sci., 1979, AB288, N 20.

27. Persson J. On the Cauchy problem in C^n with singular data. - Matematiche, 1975 (1976), 30, N 2.

28. Pham F. Caustiques, phase stationnaire et microfonctions. - Acta Math. Vietnamica, 1977, 2, N 2.

29. Thom R. La stabilité topologique des applications polynomiales. - L'enseignement mathématique, 1962, N 8.

30. Wagschal C. Problème de Cauchy a caractéristiques multiples dans les classes Gevrey. - Lect. Notes Math., 1978, N 660.

31. Wagschal C. Problème de Cauchy ramifié à caractéristiques multiples holomorphes de multiplicité variable. J. Math. Pures et Appl., 1983, 62, N 1, 10 B 224.

Translated from Russian by M.G.Shatalova

ON THE NUMBER OF SOLUTIONS FOR CERTAIN BOUNDARY-VALUE PROBLEMS

V.G. Zvyagin

Department of Mathematics,
Voronezh State University,
394693, Voronezh, USSR

In the present paper we consider boundary-value problems for non-linear equations of the elliptic type. In Ref. [1] methods have been proposed for reducing these problems to operator equations with operators satisfying condition α), a theory has been constructed for the degree of mappings satisfying condition α), and examples have been presented which show how the degree of mapping can be applied to prove the solvability of certain boundary-value problems. In this paper we describe another operator approach to these problems: namely, they are treated as equations with non-linear Fredholm operators, and the theory of non-linear Fredholm mappings is used to study these equations [2] . As to boundary-value problems, a priori estimates are assumed to be available (except for the cases where they are known) because of primary interest to us is to study the number of solutions and, where it is necessary, all the conditions related to solvability are assumed to be satisfied.

1. Auxiliary concepts and facts

Let X and Y be Banach spaces. The mapping $f:X \to Y$ is called a Fredholm mapping of index n if f is C^1-smooth and at each point $u \in X$ the Frechet derivative Df(u) is a linear Fredholm operator of index n, i.e. dim Ker Df(u) $< \infty$, dim Coker Df(u) $< \infty$, and n = ind Df(u) = dim Ker Df(u) - dim Coker Df(u).

Note that since the space X is connected, the index ind Df(u) of the linear operator Df(u) does not depend on the point u X, so that this common value may be called the index of the non-linear mapping f. The set of C^r-smooth Fredholm mappings from X into Y is denoted by $\Phi_n C^r(X,Y)$.

Furthermore, the mapping $f:X \to Y$ is called σ-proper if $f^{-1}(K)$ can be represented as no more than a countable number of compact sets for any compact K in Y, and $f:X \to Y$ is called proper if for any compact $K \subseteq Y$ the inverse image $f^{-1}(K)$ is compact in X. Clearly, a proper mapping is a σ-proper mapping.

Let $f:X \to Y$ be a C^1-smooth mapping. A point $u \in X$ is called a regular point of the mapping f if the Frechet derivative at this point, $Df(u)$, is a surjective mapping. Otherwise, the point u is called critical or singular. A point $y \in Y$ is called a regular value for the mapping f if the complete inverse image of this point is either empty or consists only of regular points. Otherwise, y is called a singular value. The set of regular values for the mapping f is denoted by R_f. For the proper $\Phi_n C^1$-mapping $f:X \to Y$ $(n \geq 0)$, the inverse image $f^{-1}(y)$ of a regular value is either an empty set or a compact n-dimensional submanifold in X. In particular, if $n = 0$, the inverse image $f^{-1}(y)$ is either empty or consists of finitely many points. The set of regular values is an interesting topic closely related to this fact. Smale's theorem on the set of regular values for a Fredholm mapping [3] is a well-known theorem in the field. We shall need a modification of Smale's theorem proved by Quinn even for a more general case (than it is required in this paper).

<u>Theorem 1 (Quinn [4])</u>. If the $\Phi_n C^r$-mapping $f:X \to Y$ is σ-proper and $r > \max\{0,n\}$, then R_f is the residual set.

We recall that a subset of a topological space is called residual if it can be represented as a countable intersection of open, everywhere dense sets. According to the Baire theorem on categories, the residual subset of a complete metric space is dense.

We shall apply Theorem 1 if f is a proper mapping. In this case the set of singular values for the mapping f is closed and, therefore, the set of regular values for f is open and, if the conditions of Theorem 1 are satisfied, is dense in Y.

2. Fredholm and proper mappings in elliptic-type boundary-value problems

Let Ω be a bounded domain in R^n with infinitely differentiable boundary $\partial\Omega$, and let $m, m_1, \ldots, m_m$ be non-negative integer numbers, $m \geq 1$, $m_j < 2m$. Let also $M(q)$ denote the number of distinct multiindices $\alpha = (\alpha_1, \ldots, \alpha_n)$ with coordinates α_i and with length $|\alpha| = \alpha_1 + \ldots + \alpha_n$ not greater than q. Suppose the functions $F: \bar{\Omega} \times R^{M(2m)} \to R^1$ and $G_j: \bar{\Omega} \times R^{M(m_j)} \to R^1$, $j = 1, \ldots, m$, have continuous derivatives with respect to all the arguments up to orders $\ell + 1$ and $\ell + 2m - m_j + 1$, respectively ($\ell \geq [\frac{n}{2}] + 1$). It is convenient to represent the functions $F(x, \xi)$ and $G_j(x, \eta)$, $\xi = (\xi_\alpha ; |\alpha| \leq 2m)$ and $\eta = (\eta_\beta ; |\beta| \leq m_j)$, in the form $F(x, \xi) = F(x, \xi_0, \xi_1, \ldots, \xi_{2m})$ and $G_j(x, \eta) = G_j(x, \eta_0, \eta_1, \ldots, \eta_{m_j})$, where $\xi_k = (\xi_\alpha ; |\alpha| = k)$ and $\eta_k = (\eta_\alpha ; |\alpha| = k)$.

Below we shall use the following notation. We put for an arbitrary multiindex α:

$$D^\alpha u = \left(\frac{\partial}{\partial x_1}\right)^{\alpha_1} \ldots \left(\frac{\partial}{\partial x_n}\right)^{\alpha_n} u , \qquad D^k u = \{ D^\alpha u ; |\alpha| = k \},$$

$$F_\alpha(x, \xi) = \frac{\partial F(x, \xi)}{\partial \xi_\alpha} , \qquad G_{j\beta}(x, \eta) = \frac{\partial G_j(x, \eta)}{\partial \eta_\beta} .$$

Let us consider the boundary-value problem

$$F(x, u, \ldots, D^{2m} u) = h(x) , \quad x \in \Omega \tag{1}$$

$$G_j(x, u, \ldots, D^{m_j} u) = g_j(x) , \quad x \in \partial\Omega . \tag{2}$$

and define the non-linear mapping $f: W_2^{\ell+2m}(\Omega) \to W_2^{\ell}(\Omega) \times \prod_{j=1}^{m} W_2^{\ell+2m-m_j-\frac{1}{2}}(\partial\Omega)$ by the relation

$$f(u) = \left(F(x, u, \ldots, D^{2m} u), G_1(x, u, \ldots, D^{m_1} u), \ldots, G_m(x, u, \ldots, D^{m_m} u)\right) \tag{3}$$

The problems concerning the solvability and the number of solutions to boundary-value problem (1)-(2) in the eigenspace $W_2^{\ell+2m}(\Omega)$ are reduced to studying the mapping f. We now analyse some properties of this mapping.

Note, first of all, that f is a C^1-smooth mapping, and its Frechet differential Df(u) is calculated by the formula

$$Df(u) = (L(u)v,\ B_1(u)v,\ \ldots,\ B_m(u)v), \tag{4}$$

where

$$L(u)v = \sum_{|\alpha| \le 2m} F_\alpha(x, u, \ldots, D^{2m}u)\, D^\alpha v, \tag{5}$$

$$B_j(u)v = \sum_{|\beta| \le m_j} G_{j,\beta}(x, u, \ldots, D^{m_j}u)\, D^\beta v, \tag{6}$$

$$j = 1, \ldots, m.$$

To deduce non-trivial facts, the following additional conditions should be imposed on the function F_α and $G_{j,\beta}$.

Condition 1 (ellipticity condition). There exists a positive continuous function $C:\ R^{M(2m)} \to R^1$ such that for all $x \in \bar{\Omega}$, $\xi = \{\xi_\alpha;\ |\alpha| \le 2m\} \in R^{M(2m)}$, $\eta = (\eta_1, \ldots, \eta_n) \in R^n$

$$\sum_{|\alpha| = 2m} F_\alpha(x, \xi)\, \eta^\alpha \ge C(\xi)|\eta|^{2m},$$

where $\eta^\alpha = \eta_1^{\alpha_1} \ldots \eta_n^{\alpha_n}$.

Condition 2. For each function $u \in W_2^{\ell+2m}(\Omega)$ the operator

$$Df(u):\ W_2^{\ell+2m}(\Omega) \to W_2^{\ell}(\Omega) \times \prod_{j=1}^{m} W_2^{\ell+2m-m_j-\frac{1}{2}}(\partial\Omega)$$

is a Fredholm operator of index 0.

Remark. That the linear operator Df(u) is a Fredholm one follows from Condition 1 and from Ya.B. Lopatinsky's condition for L(u) and $B_j(u)$ at each point $x \in \partial\Omega$. Furthermore, Condition 2 demands that ind Df(u) = 0.

Thus, the mapping f defined by relation (3) is a $\Phi_0 C^1$-mapping. This follows from Condition 2 and from C^1-smoothness of the mapping f.

Lemma 1. Let Conditions 1 and 2 be satisfied. Then the restriction $f|_M$ of the mapping f onto any closed bounded subset M of the space $W_2^{\ell+2m}(\Omega)$ is a proper mapping.

Proof. Let K be an arbitrary compact in the space $W_2^{\ell}(\Omega) \times \prod_{j=1}^{m} W_2^{\ell+2m-m_j-\frac{1}{2}}(\partial\Omega)$ and M, an arbitrary bounded closed subset of the space $W_2^{\ell+2m}(\Omega)$. We now demonstrate that $f^{-1}(K)\cap M$ is a compact. Let $\{u_n, n\in N\} \subseteq f^{-1}(K)\cap M$ be an arbitrary sequence. We may assume, without loss of generality, that the sequence $\{u_n; n\in N\}$ weakly converges in $W_2^{\ell+2m}(\Omega)$ to a certain element u_0, $\{u_n; n\in N\}$ is fundamental with respect to the norm of the space $\mathcal{L}_2(\Omega)$, and that the sequence $\{f(u_n); n\in N\}$ is fundamental in the space $W_2^{\ell}(\Omega)\times\prod_{j=1}^{m} W_2^{\ell+2m-m_j-\frac{1}{2}}(\partial\Omega)$. In this case,

$$\begin{aligned}
&\| f(u_k) - f(u_s)\| = \\
&\|(F(x,u_k,\dots,D^{2m}u_k) - F(x,u_s,\dots,D^{2m}u_s), \\
&G_1(x,u_k,\dots,D^{m_1}u_k) - G_1(x,u_s,\dots,D^{m_1}u_s), \\
&\dots, G_m(x,u_k,\dots,D^{m_m}u_k) - G_m(x,u_s,\dots,D^{m_m}u_s))\| = \\
&= \|(L(tu_k+(1-t)u_s)(u_k-u_s), B_1(tu_k+(1-t)u_s)(u_k-u_s), \\
&\dots, B_m(tu_k+(1-t)u_s)(u_k-u_s))\|,
\end{aligned}$$

whence

$$\begin{aligned}
&\| f(u_k) - f(u_s)\| \geq \\
&\geq \| L(u_0)(u_k-u_s), B_1(u_0)(u_k-u_s), \dots, \qquad (7)\\
&B_m(u_0)(u_k-u_s))\| - \|\omega(u_k,u_s)\|,
\end{aligned}$$

where

$$L(tu_k+(1-t)u_s)(u_k-u_s) =$$

$$= \sum_{|\alpha|\le 2m} \int_0^1 F_\alpha(x, tu_k+(1-t)u_s, \ldots, tD^{2m}u_k+(1-t)D^{2m}u_s)\,dt\, D^\alpha(u_k-u_s)$$

$$B_j(tu_k+(1-t)u_s)(u_k-u_s) =$$

$$= \sum_{|\beta|\le m_j} \int_0^1 G_{j,\beta}(x, tu_k+(1-t)u_s, \ldots, tD^{m_j}u_k+(1-t)D^{m_j}u_s)\,dt\, D^\beta(u_k-u_s),$$

$L(u_o)(u_k-u_s)$ and $B_j(u_o)(u_k-u_s)$ are defined by the relations (5) and (6),

$$\omega(u_k,u_s) = ([L(tu_k+(1-t)u_s) - L(u_o)](u_k-u_s),$$
$$[B_1(tu_k+(1-t)u_s) - B_1(u_o)](u_k-u_s), \ldots,$$
$$[B_m(tu_k+(1-t)u_s) - B_m(u_o)](u_k-u_s)).$$

We have then

$$\|\omega(u_k,u_s)\|^2 = \Big\{\|[L(tu_k+(1-t)u_s) - L(u_o)](u_k-u_s)\|^2_{\ell,\Omega} + \sum_{j=1}^{m} \|[B_j(tu_k+(1-t)u_s) - B_j(u_o)](u_k-u_s)\|^2_{\ell+2m-m_j-\frac{1}{2},\partial\Omega}\Big\} \tag{8}$$

and

$$\|\omega(u_k, u_s)\| \to 0 \quad \text{for} \quad k,s \to \infty \tag{9}$$

Relation (9) can be verified by reasoning similar to that used in the proof of Theorem 6 of Ref. [1]. We now outline this reasoning. For the first term in (8) the following inequality is valid:

$$\|[L(tu_k+(1-t)u_s) - L(u_o)](u_k-u_s)\|_\ell \le$$

$$\le C_1 \sum_{k_1+\ldots+k_q\le \ell} \|v_{k,s}-u_o\|_{W_{p_1}^{2m+k_1}(\Omega)} \times \|v_{k,s}-u_o\|_{W_{p_2}^{2m+k_2}(\Omega)} \times$$

$$\prod_{i=s}^{q} \max_{0\le\tau\le 1} \| \tau V_{k,s} + (1-\tau) u_0 \|_{W_{p_i}^{2m+k_i}(\Omega)} ,$$

where $V_{k,s}(x) = t u_k(x) + (1-t) u_s(x)$, $p_i = 2\ell/k_i$, C_1 is a constant dependent on F, Ω, ℓ, and M; $\| V_{k,s} - u_0 \|_{W_{p_i}^{2m+k_i}} \to 0$ for $k_i = 0$, according to the embedding theorem, and $0 < k_i < \ell$, according to the Nierenberg-Gaglardo inequality. While verifying that the norms of the following sums really tend to zero, one should use the well-known estimate $\|V\|_{k-\frac{1}{2},\partial\Omega} \le C_2 \|V\|_{k,\Omega}$ which is valid for the derivative of the function $v \in W_2^k(\Omega)$, $k \ge 1$.

Using the a priori estimate for linear elliptic operators

$$\| u_k - u_s \|^2_{\ell+2m} \le C_3 \Big\{ \| L(u_0)(u_k - u_s) \|^2_{\ell,\Omega} +$$

$$+ \sum_{j=1}^{m} \| B_j(u_0)(u_k - u_s) \|^2_{\ell+2m-m_j-\frac{1}{2},\partial\Omega} + \| u_k - u_s \|^2_{L_2(\Omega)} \Big\}$$

and also relations (7) and (9), we may conclude that the sequence $\{ u_k , k \in N \}$ is fundamental in $W_2^{\ell+2m}(\Omega)$, which implies that the set $f^{-1}(K) \cap M$ is compact. The lemma is proved.

Theorem 2. Let, for Problem (1)-(2), Conditions (1) and (2) be satisfied. Let also the following a priori estimate be valid for Problem (1)-(2):

$$\| u \|_{\ell+2m} \le C \Big(\| h \|_{\ell,\Omega} + \sum_{j=1}^{m} \| g_j \|_{\ell+2m-m_j-\frac{1}{2},\partial\Omega} \Big), \quad (10)$$

where $C: R_+ \to R_+$ is a function which is bounded on bounded subsets of R_+ (R_+ is the set of non-negative real numbers). Then f, defined by relation (3), is a proper mapping.

The proof of Theorem 2 follows from Lemma 1 and from the following lemma.

Lemma 2. Let $f: X \;\; Y$ be a mapping such that its restriction $f|_B : B \to Y$ onto any closed ball (with centre at zero) in the space X is a proper mapping. Suppose that the following estimate is valid:

$$\|u\|_X \le C(\|f(u)\|_Y), \tag{11}$$

where the function $C : R_+ \to R_+$ is bounded on bounded subsets of R_+. Then f is a proper mapping on the entire space X.

Proof. Let $K \subseteq Y$ be a compact. Then K is a bounded subset of the space Y. It follows from (11) that $f^{-1}(K)$ is also a bounded set. Suppose B is a closed ball in the space X, which contains $f^{-1}(K)$. Since the restriction $f|_B$ is a proper mapping and $f^{-1}(K) = (f|_B)^{-1}(K)$, then $f^{-1}(K)$ is a compact in X. The lemma is proved.

3. Finiteness of the set of solutions to a boundary-value problem

Theorem 3. Let, for Problem (1)-(2), Conditions (1) and (2) be satisfied and, furthermore, a priori estimate (10) be valid. Then there exists an open everywhere dense set $\mathcal{O} \subseteq W_2^{\ell}(\Omega) \times \prod_{j=1}^{m} W_2^{\ell+2m-m_j-\frac{1}{2}}(\partial\Omega)$ such that for each collection $(h, g_1, \ldots, g_m) \in \mathcal{O}$ the set of solutions to Problem (1)-(2) is finite.

Proof. As was noted above, the mapping f defined by relation (3) is a $\Phi_0 C^1$-mapping. According to Theorem 2, f is a proper mapping. Hence, it obeys Theorem 1. Let $\mathcal{O} = R_f$, the set of regular values for the mapping f. Then, for each $y = (h, g_1, \ldots, g_m) \in \mathcal{O}$ the set of solutions to the equation $f(u) = y$ and, therefore, to Problem (1)-(2) is finite. The theorem is proved.

Remark. It follows from the theory of regular values for $\Phi_n C^r$-mappings that the number of solutions to the equation $f(u) = y$ and, therefore, to Problem (1)-(2) is constant on each connected component of the set $\mathcal{O}$; furthermore, on each such component any solution of Problem (1)-(2) is a C^1-smooth function of the collection $(h, g_1, \ldots, g_m)$.

Let us now consider the Dirichlet problem

$$F(x, u, \ldots, D^{2m}u) = h(x), \quad x \in \Omega \qquad (12)$$

$$D^{\alpha}u(x) = 0, \quad x \in \partial\Omega, \quad |\alpha| \leq m-1 \qquad (13)$$

Introduce the notation $X = W_2^{2m+\ell}(\Omega) \cap \overset{\circ}{W}{}_2^{m}(\Omega)$ and $Y = W_2^{\ell}(\Omega)$, where $\ell \geq [\frac{n}{2}] + 1$. The function $F(x, \xi)$ is assumed to be $C^{\ell+1}$-smooth with respect to all the arguments.

A question naturally arises: whether the number of solutions to Problem (12)-(13) is finite, depending on the right-hand side of Eq. (12). Theorem 3 does not answer this question because the boundary conditions for Problem (12)-(13) are fixed.

Theorem 4. Suppose that Condition (1) is satisfied for the function $F(x, \xi)$ and that the following a priori estimate is valid for solutions to Problem (12)-(13):

$$\|u\|_X \leq C(\|h\|_Y),$$

where $C: R_+ \to R_+$ is a function bounded on bounded subsets of R_+. Then there exists an open everywhere dense set $\mathcal{O} \subseteq Y$ such that for each $h \in \mathcal{O}$ the set of solutions to Problem (12)-(13) in the space X is finite.

The proof of Theorem 4 is similar to that of Theorem 3. Namely, we define the mapping $f: X \to Y$ by the relation $f(u) = F(x, u, \ldots, D^{2m}u)$ and verify that f is a proper Φ_0 C^1-mapping. Putting $\mathcal{O} = R_f$, we find, by virtue of Theorem 1, that $\mathcal{O}$ is an open everywhere dense set in Y. Then, as was noted above, $f^{-1}(h)$ is a finite set for any $h \in \mathcal{O}$.

Note that a similar statement for the set of solutions to the Navier-Stokes homogeneous stationary equation, which describes the motion of viscous incompressible liquid, has been proved in [5], and the inhomogeneous Navier-Stokes equation has been studied in [6]. The result obtained in those papers can be formulated in brief as follows. For a fixed right-hand side there exists an open everywhere dense set of boundary conditions for which the Navier-Stokes problem has finitely many solutions. For the boundary-value problem

$$\sum_{i,j=1}^{n} a_{ij}(x)\frac{\partial^2 u}{\partial x_i \partial x_j} + g(x,u,\operatorname{grad} u) = 0\ , \quad x \in \Omega$$

$$u(x) = \varphi(x)\ , \quad x \in \partial\Omega$$

this result has been obtained in [7].

Here we present a similar result for the general non-linear Dirichlet problem

$$F(x,u,Du,D^2u) = h(x)\ , \quad x \in \Omega \tag{14}$$

$$u(x) = \varphi(x)\ , \quad x \in \partial\Omega \tag{15}$$

Theorem 5. Let $h \in W_2^{\ell}(\Omega)$ be a fixed function. Suppose Condition (1) is satisfied and the following a priori estimate is valid for solutions to Problem (14)-(15):

$$\|u\|_{\ell+2} \leq C\left(\|h\|_{\ell} + \|\varphi\|_{\ell+\frac{3}{2},\partial\Omega}\right),$$

where $C : R_+ \to R_+$ is a function bounded on bounded subsets of R_+. Then there exists, in the space $W_2^{\ell+2}(\Omega)$, an open everywhere dense set $\mathcal{O}$ such that for each function $\varphi \in \mathcal{O}$ Problem (14)-(15) has only a finite number of solutions for $u(x) = \varphi(x)$, $x \in \partial\Omega$.

To prove Theorem 5, we use the following transversality theorem [7] (cf. Abraham's results [8]).

Let X,Y,Z be real-valued Banach spaces; $U \subseteq X$ and $V \subseteq Y$, open sets; and $f: U \times V \to Z$, a C^k-smooth mapping such that:
(1.1) for each $v \in V$, $f(\cdot, v): u \mapsto f(u,v)$ is a Φ_s C^k-mapping, $s < k$.

Let also the following conditions be satisfied:

(1.2) z_0 is a regular value of the mapping f, i.e. the operator

$$f'(u_0,v_0)(u,v) = f'_u(u_0,v_0)u + f'_v(u_0,v_0)v$$

acts "onto" for each point (u_0,v_0) such that $F(u_0,v_0) = z_0$;

(1.3) the set $\{ u \in U;\ f(u,v) = z_0$, where v belongs to a compact set in Y $\}$ is a relative compact in U.

Then the set $\{ v \in V;\ z_0$ is a regular value of $f(\cdot, v)\}$ is open and dense in V.

Under the conditions of Theorem 5 we have

$$X = U = W_2^{\ell+2} \cap \overset{\circ}{W}{}_2^1(\Omega), \quad Y = V = W_2^{\ell+2}(\Omega)$$

$$Z = W_2^{\ell}(\Omega), \quad f(u,v) = F(\cdot, u+v, D(u+v), D^2(u+v)),$$

$$u \in X, \ v \in Y.$$

4. Uniqueness of the solution to a boundary-value problem

The following statement is presented in [2].

Theorem 6. Let X,Y be topological spaces and $f: X \to Y$, a continuous mapping. Suppose X is a Hausdorff space and the following conditions are satisfied:

a) f is locally reversible at point $x \in f^{-1}(\mathcal{N})$, where $\mathcal{N}$ is a fixed connected and locally connected closed subset;

b) the mapping f is proper only with respect to $\mathcal{N}$, i.e. $f^{-1}(K)$ is compact if the compact K belongs to $\mathcal{N}$.

Then, if $f^{-1}(\mathcal{N}) \neq \emptyset$, the inverse image of each point $y \in \mathcal{N}$ is not empty, and the number of elements in this image is equal to a constant finite number (y runs the set $\mathcal{N}$).

Banach and Mazur [9] deduced from this theorem the following statement.

Theorem 7. Let X and Y be Banach spaces, and let the continuous mapping $f: X \quad Y$ be proper and locally reversible. Then f is completely reversible.

We now apply this theorem to prove solvability of boundary-value problem (1)-(2).

Theorem 8. Let Conditions (1) and (2) be satisfied, and let *a priori* estimate (10) be valid. Suppose that for any function $u \in W_2$ the linear boundary-value problem

$$L(u)v = 0 \quad , \quad x \in \Omega$$

$$B_j(u)v = 0 \quad , \quad x \in \partial\Omega \quad , j=1,\dots,m,$$

where $L(u)$ and $B_j(u)$ are defined by relations (5) and (6), has only zero solution in the space $W_2^{\ell+2m}(\Omega)$.
Then, boundary-value problem (1)-(2) has a unique solution for any right-hand side $h \in W_2^{\ell}(\Omega)$ and for any boundary conditions

$$(g_1, \dots, g_m) \in \prod_{j=1}^{m} W_2^{\ell+2m-m_j-\frac{1}{2}}(\partial\Omega)$$

Proof. According to Theorem 2, the mapping defined by relation (3) is a proper mapping. Furthermore, f is a C^1-smooth mapping, and its Frechet derivative $Df(u)$ has zero kernel at any point $u \in W_2^{\ell+2m}(\Omega)$ and is an isomorphism of the corresponding spaces because ind $Df(u)=0$. According to the inverse function theorem, f is a local homeomorphism at each point $u \in W_2^{\ell+2m}(\Omega)$. It follows then from Theorem 7 that f is a global homeomorphism. The theorem is proved.

Remark. Clearly, if the conditions of Theorem 8 are satisfied, each solution of Problem (1)-(2) is a C^1-smooth function of the right-hand side and of the boundary conditions.

An analogue of Theorem 8 for the Dirichlet problem for a quasilinear elliptic equation has been proved in paper [10], which also reports particular examples of boundary-value problems where all the conditions of Theorem 8 are satisfied.

The following theorem is presented in [11].

Theorem 9. Let X and Y be Banach spaces, and let I denote the segment [0,1]. Suppose that $\Phi : X \times I \to Y$ is a continuous mapping whose derivative $\Phi_x(x, \tau)$ is continuous with respect to (x, τ), and that the following conditions are satisfied:

1) for any solution x of the equation $\Phi(x,\tau) = 0$ corresponding to an arbitrary $\tau \in I$ the operator $\Phi_x(x,\tau)$ has a bounded inverse operator $\Phi_x^{-1}(x,\tau):Y \to X$;
2) the set of all solutions of the equation $\Phi(x,\tau) = 0$, which correspond to all $\tau \in I$, is compact in X;
3) for a certain fixed $\tau_0 \in I$ there exists a unique solution x_0 of the equation $\Phi(x,\tau_0) = 0$.
Then, for each $\tau \in I$ the equation $\Phi(x,\tau) = 0$ has a unique solution in X.

The proof of Theorem 9 is immediately reduced to Theorem 6 by introducing the auxiliary mapping $f: X \times I \to Y \times I$ defined by the correspondence $(x,\tau) \to (\Phi(x,\tau),\tau)$.

We now apply Theorem 9 to prove unique solvability of the Dirichlet problem for a general elliptic equation in a narrow band. The existence theorem for this problem has been proved in [1] using degree theory (described therein). Below we shall use the notation and some facts of Ref. [1].

Let $\{\Omega_s,\ 0 < s \le 1\}$ stand for the family of domains in R^n with infinitely differentiable boundary such that:
a) $\Omega_{s_1} \subset \Omega_{s_2}$ for $s_1 < s_2$;
b) there exist open coverings $\{\mathcal{U}_i,\ i = 1, \ldots, \mathfrak{I}\}$ of the set Ω_1 and diffeomorphisms $\varphi_i : \mathcal{U}_i \cap \Omega_1 \to R^n$ of class C^∞ for which

$$\varphi_i(\mathcal{U}_i \cap \overline{\Omega}_s) = T_s = \{x \in R^n;\ |x'| < 1,\ 0 \le x_n \le s\},$$

$$x = (x', x_n),\quad x' = (x_1, \ldots, x_{n-1}).$$

Let $\{\psi_i(x)\}$, $i = 1, \ldots, \mathfrak{I}$, be a partirion of unity subordinate to a covering. If m, ℓ, k, and p > 1 are integer non-negative numbers, we denote by $W_p^{2m,\ell,k}(\Omega_s)$ the closure of the set of functions which are infinitely differentiable in Ω_s with respect to the norm

$$\|u\|^p_{W_p^{2m,\ell,k}(\Omega)} = \sum_{i=1}^{\mathfrak{I}} \|\psi_i(\varphi_i^{-1}(y))\, u(\varphi_i^{-1}(y))\|_{W_p^{2m,\ell,k}(T_s)},$$

where

$$\| v(y) \|^p_{W_p^{2m,\ell,k}(T_s)} = \sum_{|\alpha| \le 2m} \int_{T_s} \{ \sum_{j=0}^{k} | D_n^j D^\alpha v(y) |^p + \sum_{|\beta| = \ell}' | D^\beta D^\alpha v(y) |^p \} dy.$$

The symbol $\sum'$ denotes summation over all multiindices the last coordinate of which is zero.

Let us fix $\ell > n+1$ and examine solvability of the non-linear Dirichlet problem

$$F(x, u, \ldots, D^{2m} u) = h(x), \quad x \in \Omega_s \tag{16}$$

$$D^\alpha u(x) = 0, \quad x \in \partial \Omega_s, \quad |\alpha| \le m-1 \tag{17}$$

in $W_2^{2m,\ell,1}(\Omega_s)$. It is assumed that the function $F(x, \xi)$ defined in $\bar{\Omega}_1 \times R^{M(2m)}$ has continuous derivatives of order up to ℓ, and that for $\eta \in R^n$ the condition of uniform ellipticity

$$\sum_{|\alpha| = 2m} F_\alpha(x, \xi) \eta^\alpha \ge \nu |\eta|^{2m}$$

is valid with a certain constant ν.

Suppose that $F(x,0) = 0$, the function $h(x)$ belongs to the space $W_2^{0,\ell,1}(\Omega_1) \cap C^{1,\alpha}(\bar{\Omega}_1)$, and

$$\| h \|_{W_2^{0,\ell,1}(\Omega_1)} + \| h \|_{C^{1,\alpha}(\bar{\Omega}_1)} \le R, \quad \| F(x,\xi) \|_{C^\ell(\bar{\Omega}_1 \times B_t)} \le g(t),$$

where $B_t = \{ \xi \in R^{M(2m)};\ |\xi| \le t \}$ and $g(t)$ is a non-decreasing positive function.

Theorem 10. There exists an s_0 such that for all $s \in (0, s_0]$ Problem (16)-(17) has a unique solution which belongs to $W_2^{2m,\ell,1}(\Omega_s) \cap \overset{\circ}{W}{}_2^m(\Omega_s)$.

Proof. Let us put $X_s = W_2^{2m,\ell,1}(\Omega_s) \cap \overset{\circ}{W}{}_2^m(\Omega_s)$, $Y_s = W_2^{0,\ell,1}(\Omega_s)$, and define the mapping $\Phi : X_s \times I \to Y_s$

by the relation

$$\Phi(u,\tau) = \tau F(x, u, \ldots, D^{2m} u) + (1-\tau) L_0(D)u - \tau h(x), \; x \in \Omega_{s},$$

where $L_0(D) = \sum_{|\alpha| = 2m} a_\alpha D^\alpha$ is a fixed elliptic operator of constant ellipticity ν .

It has been shown in [1] that there exists an a priori estimate for the equation $\Phi(u, \tau) = 0$. Using this fact and reasoning just as in Lemma 1 and Theorem 2, we can verify Condition (2) of Theorem 9. Applying the aforementioned a priori estimate for the linear equations ($\Phi_u(u, \tau)v = 0$, $u \in \Phi^{-1}(0)$) and taking into account that $\Phi_u(u, \tau)$ is a Fredholm operator of zero index, we obtain Condition (1) of Theorem 9. Using then the same a priori estimate for the equation $L_0(D)u = 0$, u X, and recalling that the operator $L_0(D)$ has zero index, we obtain Condition (3) for $\tau_0 = 0$. The theorem is proved.

REFERENCES

1. Skrypnik I.V., Topological methods for studying general non-linear elliptic boundary-value problems.- In: Geometry and topology in global non-linear problems, 1984, p.68-89 (in Russian).

2. Borisovich Yu.G., Zvyagin V.G., and Sapronov Yu.I., Non-linear Fredholm mappings and Lerey-Schauder theory.- Usp. Matem. Nauk, 1977, v.32, No.4 (in Russian).

3. Smale S., An infinite dimensional version of Sard's theorem.- Amer. J. Math., 1965, 87, p.861-866.

4. Quinn F., Transversal approximation on Banach manifolds.- Proc. Sympos. Pure Math. (Global analysis), 1970, 15, p.213-223.

5. Foias C. and Temam R., Structure of the set of stationary solutions of the Navier-Stokes equations.- Commun. Pure Appl.Math., 1977, v.30, p.149-164.

6. Saut J.C. and Temam R., Generic properties of Navier-Stokes equations: genericity with respect to the boundary value.- Indiana Univ. Math. J., 1980, v.29, No.3, 427-446.

7. Saut J.C. and Temam R., Generic properties of nonlinear boundary-value problems.- Commun. in Partial Diff. Equations, 1979, 4(3), p.293-319.

8. Abraham R. and Robbin J., Transversal mappings and flows.- Benjamin, New York, 1967.

9. Banach S. and Mazur S., Über mehrdeutige stetige Abbildungen.- Studia Math., 1934, No.5, p.174-178.

10. Pokhozhaev S.Ya. On non-linear operators with a weakly closed domain of values and on quasilinear elliptic equations.- Matem. Sbornik, 1969, 78, p.237-238 (in Russian).

11. Ladyzhenskaya O.A. and Ural'tseva N.N., Linear and quasilinear equations of elliptic type.- Moscow, Nauka, 1973 (in Russian).

CONTACT STRUCTURE, RELAXATION OSCILLATIONS AND SINGULAR POINTS OF IMPLICIT DIFFERENTIAL EQUATIONS*)

V.I.Arnol'd
Department of Mechanics and Mathematics
Moscow State University
119899, Moscow, USSR

Vertical (tangent to the fibres) "unperturbed" vector field, determining "fast motion" in the space of smooth fibre bundle and perturbed by arbitrary small field in the bundle space is studied in the theory of relaxation oscillations.

Generally speaking, zero manifold of a fast field is a smooth manifold of the same dimension as the bundle base but it is not necessarily a section; its projection on the base can have singularities. For example, if the fibres are one-dimensional, then the singularities of projection of the given manifold (called the slow manifold) on the bundle base for the generic systems are described by Whitney: they are folds, cusps and their multidimensional generalizations.

The perturbation defines a vector field on the slow manifold: the projection of the perturbed field on the tangential plane of the slow manifold along the fibres. On the projection folds this field has a polar singularity but its direction field extends to the smooth one at the generic points of the fold line on the slow manifold.

This paper describes up to diffeomorphisms, fibred over the base, the singularities of the direction field so defined on the slow manifold in the case when the base is two-dimensional while the fibres are one-dimensional (i.e. there are two slow and one fast variables).

Theorem. The direction field on a slow manifold in a generic system with two slow and one fast variables in the neighbourhood of each point of a fold is reduced by a fibred diffeomorphism either to the cubic family defined below or to the singular points of 4 types defined below: folded focus, folded saddle, folded node, dissected folded umbrella.

---

*) A report at a Geometrical Topology conference at Lwow, read the 1 October 1984

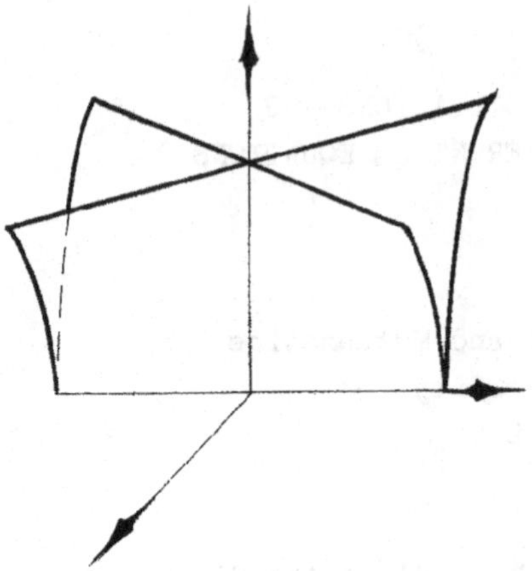

Fig.1.
Folded umbrella.

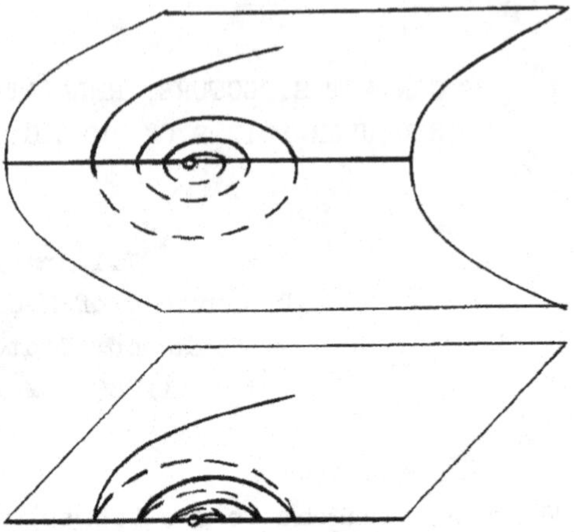

Fig.2.
Folded focus.

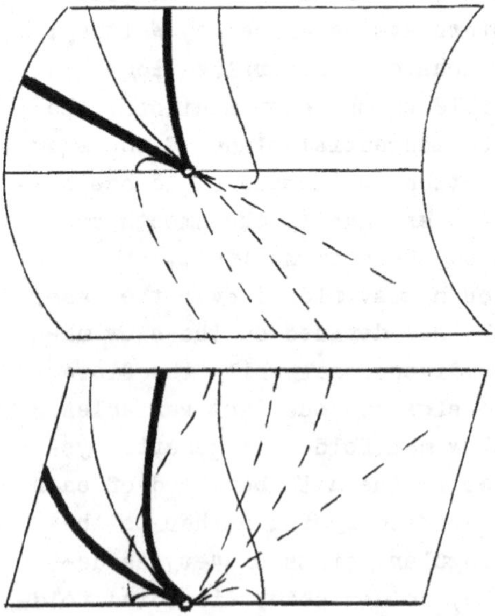

Fig.3.
Folded node.

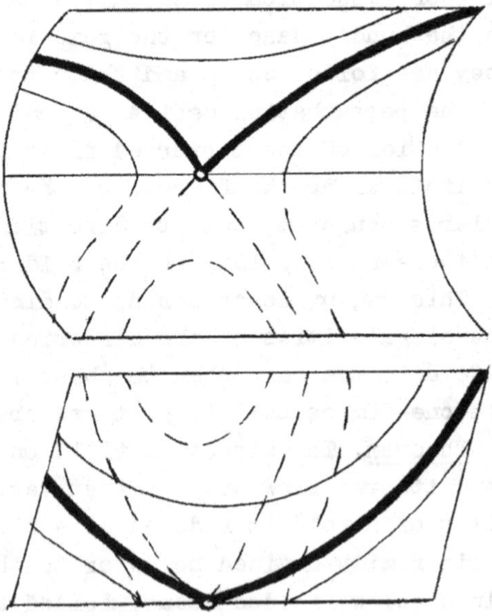

Fig.4.
Folded saddle.

The cubic family and the folded focus, saddle and node are also treated in some other problems, e.g. in the theory of asymptotic lines on the surfaces of three-dimensional space and in the theory of the equations unsolved with respect to derivatives. The folded umbrella $u^2 = v^3 w^2$ is considered in the problem on the covering of the caustic by the cuspidal edges of the moving fronts and in the problem on the by passing of an obstacle. However, unlike the folded singular points the emergence of the folded umbrella in our case does not follow from general considerations and is an unexpected consequence of intricate calculations.

As far as the folded focus, node and saddle are concerned, their emergence is explained by the contact structure implicitly present in all the three problems where they occur. In the relaxation problem this structure is the field of planes spanned by the vectors of the unperturbed fast field and the perturbed field in each point.

This is a field of planes in the three-dimensional space except a singular line where both fields are collinear (in the generic case). This line intersects the slow manifold at certain points, i.e. at the singular points of the slow motions field which we have constructed on the slow manifold having projected a perturbing field upon it. These points do not lie on the fold (in a generic system).

The obtained field of planes is not degenerate and gives a contact structure in the three-dimensional space except (in a generic system) a certain surface where the contact structure degenerates. This surface intersects the slow manifold along a curve and intersects the curve of folds on it at certain points (in a generic system). These points will give rise to the folded umbrellas. Meanwhile we shall not dwell on them but we shall consider other points where the contact structure does not degenerate.

The field of contact planes is vertical everywhere (i.e. it contains the direction of a fibre). Outside the fold it cuts out on the slow manifold exactly the direction field of the slow motions. At generic points of the fold the contact planes are transversal to the slow manifold and cut out vertical directions on it.

At certain points of the fold the contact plane can be tangent to the slow manifold. These are singular points of the direction field of the slow motions on the slow manifold. In a generic system these singular points are usual saddles, nodes and focuses and their normal forms are given by linear vector fields as was proved by Poincare for the node and focus and by Siegel for the saddle.

But these are normal forms on the slow manifold and the reduction to them proceeds by the diffeomorphism of the slow manifold neglecting the projection, whereas we are interested first of all in the projections of integral curves on the bundle base and in a formally more general problem of the reduction to a normal form by the group of fibred (transferring fibres to fibres) diffeomorphisms.

Now we shall see that the group can be further reduced without changing the answer with the requirement that the contact structure is preserved everywhere and not only on the slow manifold or it can be increased by considering all the diffeomorphisms on the slow manifold commuting with its involution (that involution which exchanges the places of any two points of the slow manifold lying at the same fibre).

For this purpose it is necessary to recall the theory of equations unsolved with respect to the derivative. Such an equation is given by a surface in a three-dimensional contact space of 1-jet of functions. The projection of the surface on the 0-jets space has as its singularities only folds and cusps (for a generic equation). On the surface contact planes cut out a direction field. This field is vertical at the fold points and isn't defined at singular points where the contact plane is tangent to the surface.

Theorem. The direction field of slow motions in the neighbourhood of a singular point of a generic system with one fast and two slow variables can be reduced by a fibred contact diffeomorphism to the direction field of one of the following equations $p^2 = x$ (cubic family), $(p + kx)^2 = y$ (folded singularities), unsolved with respect to the derivative $p = dy/dx$.

The classification of the folded singular points of implicit differential equations up to contact diffeomorphisms has been recently obtained by A.A.Davydov. In 1972 Thom began to study these singular points*) and in 1975 L.Dara continued Thom's work and put forward the conjecture that the normal forms mentioned above exhaust the topological types of the folded singularities. Davydov not only proved this conjecture but also showed that in the generic case normal forms exhaust the differentiable and analytical type of singularities of differentiable or analytical equations.

---

*) earlier studied by Cibrario, Pkhakadze, Shestakov, Sokolov whose works seem to be unknown to Thom.

The main idea of Davydov's proof is that not the equation is reduced to a normal form preserving the involution but the involution is reduced while the equation is preserved.

Consider on the plane a direction field with a singular point O and an involution the fixed points of which form a curve transversing O. The involution is called admissible if the field on the line of fixed points is antiinvariant under the involution. Davydov showed that all the admissible involutions for which the lines of fixed points are not divided by the directions of eigenvectors of the field (they lie in one component in the set of admissible involutions) are locally transferred one into another by diffeomorphisms at which each point remains on its integral curve.

The proof uses the homotopic method. First one can match the lines of fixed points. Then the involutions can be connected by homotopy with the same line of fixed points at all the values of the parameter. For the construction of vector field closing the infinitesimal commutative diagram we obtain the following homological equation

$$f v - \bar{f} \bar{v} = h$$

where v is the known vector field, specifying the direction field, f is the unknown function, a vinculum denotes the action of the intermediate involution included in homotopy and the right-hand side is the deformation velocity vector of this involution under homotopy; this vector is known and it satisfies the relation $\bar{h} = - h$. Since vectors v and $\bar{v}$ are collinear only on the line of fixed points (this excludes only nodes with equal and saddles with opposite eigenvalues), the right-hand side outside the line of fixed points can be expanded with respect to the basis: $h = fv + g\bar{v}$. From $\bar{h} = -h$ follows now $g = -\bar{f}$, Q.E.D. (the smoothness f at the line of fixed points is easily checked).

The investigation of the case when the contact structure degenerates leads in the same way to the investigation of the action of the involution upon the family of smooth curves which are tangent (on the line of fixed points) to the antiinvariant direction. In the generic point on the line of fixed points such curve is tangent to its image under the unvolution cubically (second-order tangency) but in certain points the order of tangency is raised up to the fourth.

The corresponding normal form on the level of formal series or infinitely differentiable functions (but not in the analytical case) is as follows:

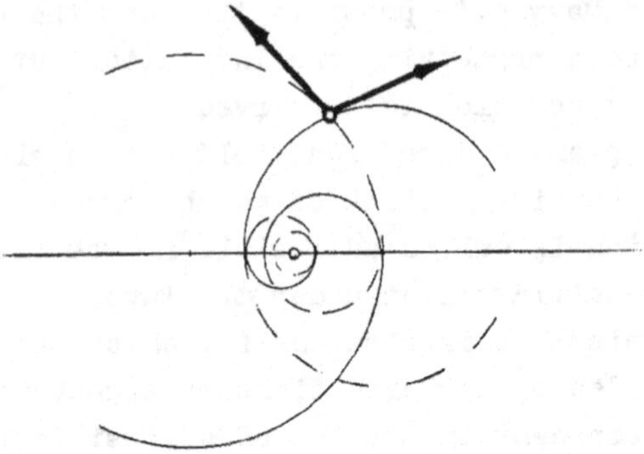

Fig.5. Vectors v and $\bar{v}$.

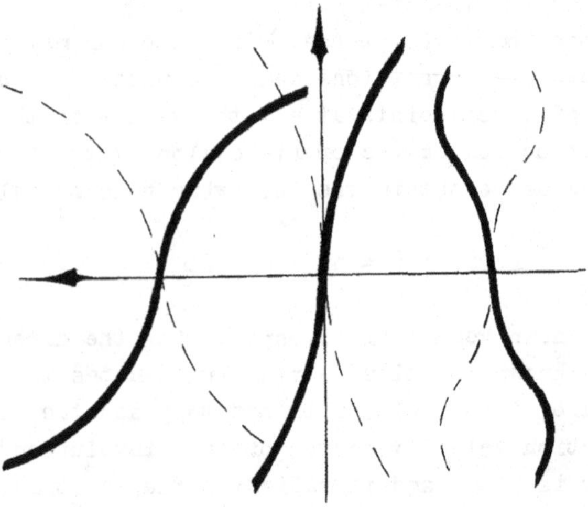

Fig.6. Trace of the degenerating contact structure and its reflection.

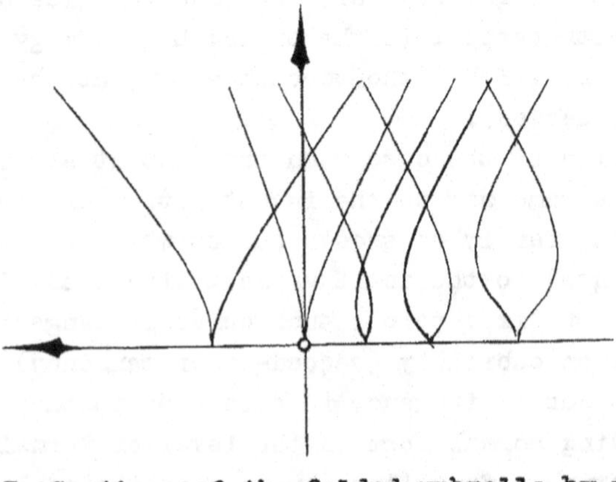

Fig.7. Sections of the folded umbrella by planes.

$$z + x^3 z + x^5 = c$$

The involution changes the sign of x. By raising each curve of the family to its height we obtain in a three-dimensional space ($y = x^2$, z, c) the surface split into flat sections.

Theorem. This surface is diffeomorphic to the folded umbrella $u^2 = v^3 w^2$ and the sections are diffeomorphic to the levels of a generic function on it, e.g. function $u + v + w$.

(The equivalence of all the families of level lines on the folded umbrella is easily proved using its double quasihomogeneity as well as for other surfaces with binomial equations, e.g. for a usual Whitney umbrella).

The detailed proof and generalisations of this theorem will be published in a forthcoming article by A.A.Davydov in Matemeticheskiĭ Sbornik.

TOPOLOGICAL INDEX ESTIMATES

N.M. Bliznyakov

Institute of Mathematics,
Voronezh State University,
394693, Voronezh, USSR

The paper deals with rotation estimates for a polynomial vector field on algebraic manifolds in $\mathbb{R}^n$; in particular, with index estimates for a singular point of a vector field. The problem in question was considered in [1,4,5,8,9,10,12,18,20,21,22] ; furthermore, the index estimate problem is related to Hilbert's 16-th problem (the corresponding references are presented in the review by V.I. Arnol'd and O.A. Oleinik [3]). The problem can be treated using various approaches. For a polynomial vector field, the most natural problem statement is the one in terms of the degrees of vector field components. This approach was used in [1,8,12,18,22] ; the final results have been obtained by V.I. Arnol'd [1] and A.G. Khovansky [18] . In the present paper, as the initial data we consider vector field characteristics which are more subtle than the degrees of its components (convex support spans, Newton diagrams, and the number of monomials of vector field components).

1. Theorems on the number of solutions of a system of equations

First, we **recall** some definitions, notation and facts used in this paper. The octant of an integer lattice $\mathbb{Z}^n$ in $\mathbb{R}^n$ for which the coordinates of all vectors $m = (m_1, \ldots, m_n)$ are non-negative is denoted by $\mathbb{Z}^n_+$ and the octant of the space $\mathbb{R}^n$ with positive vector coordinates, by $\mathbb{R}^n_+$. If $f = \sum_{m \in \mathbb{Z}^n} a_m x^m$ is a series ($x^m = x_1^m \ldots x_n^m$), the set $\operatorname{supp} f = \{m \in \mathbb{Z}^n : a_m \neq 0\}$ is called its support. For any set $M \subset \operatorname{supp} f$, we denote $f_M = \sum_{a_m \in M} a_m x^m$. The convex span of the union $\bigcup_m (m + \mathbb{Z}^n_+)$, $m \in \operatorname{supp} f$, is called a Newton polyhedron for the series f. The union of compact faces of a Newton polyhedron is called a Newton diagram. A Newton polyhedron (diagram) is called "suitable" if it intersects all the coordinate axes.

Let $p = (p_1, \ldots, p_n)$ be a non-zero integer vector and S, an integer polyhedron in $\mathbb{R}^n$. We put $m(p,S) = \min\{ \langle p,x\rangle , x \in S\}$, $S_p = \{x \in S: \langle p,x\rangle = m(p,S)\}$. The set S_p is a q-dimensional ($0 \leqslant q \leqslant n-1$) face of the polyhedron S, which is the intersection of S with the hyperplane along p and supported to S.

Let $F: F_1 = \ldots = F_n = 0$ be a system of equations in $\mathbb{C}^n$, where F_i, $1 \leqslant i \leqslant n$, are Laurent polynomials. Let Γ_i denote the convex span of the set supp F_i, $1 \leqslant i \leqslant n$. The system F is called non-degenerate if for any $p \in \mathbb{Z}^n \setminus 0$ the system $F_{1,\Gamma_{1,p}} = \ldots = F_{n,\Gamma_{n,p}} = 0$ does not have solutions in $(\mathbb{C} \setminus 0)^n$.

<u>Theorem 1</u> [6]. The number of isolated solutions of system F on the set $(\mathbb{C} \setminus 0)^n$ (with an allowance for multiple solutions) does not exceed

$$n!\,V(\Gamma_1,\ldots\Gamma_n) = (-1)^{n-1}\sum_{i=1}^{n} V(\Gamma_i) + (-1)^{n-2}\sum_{\substack{i,j=1\\ i<j}}^{n} V(\Gamma_i+\Gamma_j) + \ldots + V(\Gamma_1+\ldots+\Gamma_n)$$

and is equal to this number if F is non-degenerate.

Suppose $F: F_1 = \ldots = F_n = 0$ is a system of analytic equations in $\mathbb{C}^n$ ($F_1(0) = \ldots = F_n(0) = 0$), and the functions F_i have "suitable" Newton polyhedrons Γ_i, $1 \leqslant i \leqslant n$. The system F is called non-degenerate at zero if for any $p \in \mathbb{Z}^n_+$ the system $F_{1,\Gamma_{1,p}} = \ldots = F_{n,\Gamma_{n,p}} = 0$ does not have solutions in $(\mathbb{C} \setminus 0)^n$. The Newton number $\nu(\Gamma_1, \ldots, \Gamma_n)$ [2] for a system of polyhedrons $\Gamma_1, \ldots, \Gamma_n$ is said to be the quantity

$$(-1)^{n-1}\sum_{i=1}^{n} V(\mathbb{R}^n_+ \setminus \Gamma_i) + (-1)^{n-2}\sum_{\substack{i,j=1\\ i<j}}^{n} V(\mathbb{R}^n_+ \setminus (\Gamma_i+\Gamma_j)) + \ldots + V(\mathbb{R}^n_+ \setminus (\Gamma_1+\ldots+\Gamma_n)).$$

<u>Theorem 2</u> [2,16]. The multiplicity $\mu_0(F)$ of zero solution of a system F is not less than the Newton number $\nu(\Gamma_1, \ldots, \Gamma_n)$,

and is equal to this number if the system F is non-degenerate at zero.

Note that the nondegeneracy condition in Theorems 1 and 2 is "almost always" satisfied (except for the case of hyperplane in the space of systems F with given polyhedrons $\Gamma_1, \ldots, \Gamma_n$) (see, for example, [6,7,16]). Note also that the non-degeneracy condition in Theorems 1 and 2 should be verified only for a finite number of vectors $p \in \mathbb{Z}^n$; suffice it to choose one vector p for each face of the convex span of $\Gamma_1 + \ldots + \Gamma_n$, and faces of all dimensions i from 1 to $n-1$ must be verified.

Let $P: P_1 = \ldots = P_n = 0$ be a system of real-valued polynomial equations in $\mathbb{R}^n$, and let $m = \#(\bigcup_{i=1}^{n} \operatorname{supp} P_i)$ be the number of elements of the set $\bigcup_{i=1}^{n} \operatorname{supp} P_i$.

<u>Theorem 3</u> [19] . The number of regular real-valued solutions of the system P in the positive octant $\mathbb{R}^n_+$ is equal to or less than $(n+2)^m \, 2^{m(m+1)/2}$.

The following two statements are corollaries of Theorems 1 and 3. Let us consider a system of polynomial equations $P: P_1 = \ldots P_n = 0$ in $\mathbb{C}^n$, and let Γ_i denote the convex span of the set $(\operatorname{supp} P_i) \cup 0$ in $\mathbb{R}^n$, $1 \leqslant i \leqslant n$.

<u>Theorem 4</u>. The number of isolated solutions (with an allowance for multiplicity) of the system P in $\mathbb{C}^n$ does not exceed $n!\, V(\Gamma_1, \ldots, \Gamma_n)$.

<u>Proof</u>. We now demonstrate that for the mapping $\overline{P} = (P_1, \ldots, P_n)$: $: \mathbb{C}^n \longrightarrow \mathbb{C}^n$ there exists a small $\varepsilon = (\varepsilon_1, \ldots, \varepsilon_n)$ such that the mapping $\overline{P} - \varepsilon$ does not have zeros on the "coordinate cross" $\mathbb{C}^n \setminus (\mathbb{C} \setminus 0)^n$, and the number of isolated solutions for this mapping is not less than for $\overline{P}$. Let G_i denote the restriction of $\overline{P}$ onto the set $\mathbb{C}^{n-1}_i = \pi_i \mathbb{C}^n$ (π_i is the projection of "deleting the i-th coordinate"). According to Sard's theorem, the sets $G_i(\mathbb{C}_i^{n-1})$, $1 \leqslant i \leqslant n$, and the set L of non-regular values of the mapping $\overline{P}$ have non-zero Lebesgue measure in $\mathbb{C}^n$. Hence, the set $[\bigcup_{i=1}^{n} G_i(\mathbb{C}_i^{n-1})] \cup$ $\cup L$ - the union of the set of non-regular values of $\overline{P}$ and the image of the "coordinate cross" under the mapping $\overline{P}$ - has zero

Lebesgue measure in $\mathbb{C}^n$. Therefore, there exists an arbitrary small ε such that $G_i^{-1}(\varepsilon) = \emptyset$, $1 \leqslant i \leqslant n$, and $\varepsilon \,\bar{\in}\, L$, i.e. the mapping $\bar{P} - \varepsilon$ does not have zeros on the "coordinate cross", and all zeros of this mapping are regular. If the mapping $\bar{P}$ is replaced by the mapping $\bar{P} - \varepsilon$, each isolated μ-multiple zero of $\bar{P}$ splits into μ regular zeros. Thus, the number of isolated solutions of the system P is not less than the number of isolated solutions of the system $P - \varepsilon$. Applying Theorem 1 to the system $P_1 - \varepsilon_1 = \ldots = P_n - \varepsilon_n = 0$, we obtain the assertion in question.

Let now $P: P_1 = \ldots = P_n = 0$ be a system of real-valued polynomial equations in $\mathbb{R}^n$, and let $m = \#(\bigcup_{i=1}^{n} \operatorname{supp} P_i)$. We put $H(m) = 2^{n+[(m+1)(m+2)/2]}\,(n+2)^{m+1}$.

<u>Theorem 5</u>. The number of regular real-valued solutions of the system P in $\mathbb{R}^n$ does not exceed H(m).

<u>Proof</u>. According to Theorem 3, the number of regular real-valued solutions of the system P in each octant of the space $\mathbb{R}^n$ does not exceed $(n+2)^m\, 2^{m(m+1)/2}$. Therefore, the number of real-valued regular solutions of the system P in $(\mathbb{R} \setminus 0)^n$ does not exceed $X(m) = 2^n (n+2)^m\, 2^{m(m+1)/2}$. Proceeding then exactly as in Theorem 4 and perturbing the system P with constants, we obtain the system $P - \varepsilon$ all the **solutions** of which are regular in $\mathbb{R}^n$ and do not lie on the "coordinate cross" $\mathbb{R}^n \setminus (\mathbb{R} \setminus 0)^n$. In this case the number of real-valued regular solutions of the system $P - \varepsilon$ in $\mathbb{R}^n$ is not less than the number of real-valued regular solutions of the system P in $(\mathbb{R} \setminus 0)^n$. Since $\#(\bigcup_{i=I}^{n} \operatorname{supp}(P_i - \varepsilon_i)) \leqslant \#(\bigcup_{i=I}^{n} \operatorname{supp} P_i) + 1 = m+1$, the number of real-valued regular solutions of the system P in $\mathbb{R}^n$ does not exceed $X(m+1) = H(m)$. The theorem is proved.

It should be noted that the estimates for the number of regular real-valued solutions given by Theorems 4 and 5 do not, in general, majorate each other.

2. Cauchy indices. The degree of mapping of a manifold into sphere. Index calculation

Let $U \subset \mathbb{R}^n$ be an open set, $F: U \longrightarrow \mathbb{R}^n$ a smooth vector field, and $g:U \longrightarrow \mathbb{R}^1$ a smooth function. Let the set $F^{-1}(0)$ of singular points of the vector field be finite and let $F^{-1}(0) \cap g^{-1}(0) = \emptyset$.

The sum $\sum_{x \in F^{-1}(0)} \operatorname{ind}(F,x) \operatorname{sgn} g(x)$ is called the Cauchy index of the pair (F,g) on U and is denoted by $\operatorname{ind}_C(F,g,U)$ (here $\operatorname{ind}(F,x)$ is the index of the singular point x of the vector field F).

Remark 1. For $n = 1$ the Cauchy indices have been studied quite well. Let $U = (a,b)$ be an interval, and $F:U \longrightarrow \mathbb{R}^1$, $g:U \longrightarrow \mathbb{R}^1$ polynomials which do not have common real-valued roots in U. It is a simple matter to demonstrate that $\operatorname{ind}_C(F,g,U) = I_a^b(g/F)$, where $I_a^b(g/F)$ is a classical concept - the Cauchy index of a rational function $R = (g/F)$, which is defined as the difference between the number of discontinuities of the rational function $R(x)$ when it changes from $-\infty$ to $+\infty$ and the number of discontinuities when it changes from $+\infty$ to $-\infty$ for the argument x varying from a to b (see [14]). The procedure of calculating the Cauchy indices of rational functions has been developed quite well (see [14,17]).

Let us consider the smooth mapping $F = (F_1, \dots, F_n): M^{n-1} \longrightarrow \mathbb{R}^n \setminus 0$ of a smooth compact oriented manifold without boundary in $\mathbb{R}^n$. For any $x \in \mathbb{R}^n \setminus 0$ we put $\rho(x) = x/\|x\|$. Let $\pi_i: \mathbb{R}^n \longrightarrow \mathbb{R}^{n-1}$ be the projection of "deleting the i-th coordinate".

Theorem 6 [11]. Let the set $(\pi_i F)^{-1}(0)$ be finite, and let $\mathfrak{U} = \{(U_k, \varphi_k)\}$ be the set of charts from a given orienting atlas on M^{n-1}, which satisfies the conditions

1) $(\pi_i F)^{-1}(0) \subset \bigcup_{U_k \in \mathfrak{U}} U_k$ and
2) each point of the set $(\pi_i F)^{-1}(0)$ belongs to only one of the sets U_k.

Then the following relation is valid for the degree of the mapping $\rho F: M^{n-1} \longrightarrow S^{n-1}$:

$$\deg(\rho F) = \frac{(-1)^{i-1}}{2} \sum_{x \in (\pi_i F)^{-1}(0)} \operatorname{ind}(\pi_i F, x) \operatorname{sgn} F_i(x) = \qquad (1)$$

$$= \frac{(-I)^{i-I}}{2} \sum_{U_k \in \mathfrak{U}} \operatorname{ind}_C((\pi_i F \varphi_k^{-I}), (F \varphi_k^{-I})_i, \varphi_k(U_k)) ,$$

where $(F\varphi_k^{-1})_i$ denotes the i-th component of the mapping $F\varphi_k^{-1}$.

3. Estimates for the rotation of a vector field in $\mathbb{R}^n$

We now prove a theorem which makes it possible to apply Theorems 2, 4, and 5 for estimating the rotation of a vector field on algebraic manifolds and the index of a singular point of a vector field.

Let M^{n-1} be a smooth manifold consisting of one or several compact components of the smooth manifold $M_1 = V - \Sigma(V)$ of non-singular points of the algebraic hypersurface V in $\mathbb{R}^n$ defined by the equation $Q(x) = 0$. Let $P = (P_1, \dots, P_n)$ be a polynomial vector field in $\mathbb{R}^n$. Consider the system of polynomial equations: $T_i(x, \varepsilon): P_1(x) = \varepsilon_1, \dots, P_{i-1}(x) = \varepsilon_{i-1}, Q(x) = 0, P_{i+1}(x) = \varepsilon_{i+1}, \dots, P_n(x) = \varepsilon_n$ (here $\varepsilon = (\varepsilon_1, \dots, \varepsilon_{i-1}, 0, \varepsilon_{i+1}, \dots, \varepsilon_n)$).

<u>Lemma 1</u>. For any sufficiently small regular value ε of the mapping $\pi_i P: M^{n-1} \to \mathbb{R}^{n-1}$ the absolute value of the degree of the mapping $\rho P: M^{n-1} \to S^{n-1}$ does not exceed half the number of real-valued regular solutions of the system $T_i(x, \varepsilon)$.

<u>Proof</u>. We first demonstrate that for any smooth vector field F in $\mathbb{R}^n$ the degree of the mapping $\rho F: N^{n-1} \to S^{n-1}$ of a smooth compact oriented manifold N^{n-1} into S^{n-1} does not exceed half the number of inverse images $\{(\pi_i F)^{-1}(a)\}$ of the regular value $a \in \mathbb{R}^{n-1}$ of the mapping $\pi_i F: N^{n-1} \to \mathbb{R}^{n-1}$, which is rather close to the point 0. Indeed, let $0 \in \mathbb{R}^{n-1}$ be a regular value of the mapping $\pi_i F$ and let C_1, C_2 be inverse images of the point 0 on S^{n-1} under the projection π_i. The set $A = (\rho F)^{-1}(C_1) \cup (\rho F)^{-1}(C_2)$ is finite, since $A = (\pi_i F)^{-1}(0)$ and $(\pi_i F)^{-1}(0)$ is the inverse image of a regular value under the mapping of a compact manifold into a manifold of the same dimension. Hence, relation (1) is valid for the degree of the mapping $\rho F: N^{n-1} \to S^{n-1}$:

$$\deg(\rho F) = \frac{(-1)^{i-1}}{2} \sum_{x \in (\pi_i F)^{-1}(0)} \operatorname{ind}(\pi_i F, x) \operatorname{sgn} F_i(x). \qquad (2)$$

For $a = 0$, the statement in question follows from relation (2) because $|\operatorname{ind}(\pi_i F, x)| = 1$ for $x \in (\pi_i F)^{-1}(0)$. Let now $a \in \mathbb{R}^{n-1}$ be a regular value of the mapping $\pi_i F$, which is sufficiently close to the point 0. Then, 0 is a regular value of the mapping $\pi_i F - a = \pi_i(F - a)$ and, therefore, the absolute value of the degree of the mapping $\rho(F - a)$ does not exceed half the number of the inverse images $(\pi_i(F - a))^{-1}(0)$. For small a, however, the degrees of the mappings ρF and $\rho(F - a)$ are equal, and $(\pi_i(F - a))^{-1}(0) = (\pi_i F)^{-1}(a)$, which completes the proof for an arbitrary a. We now demonstrate that the set of regular points of the mapping $\pi_i P : M_1 \to \mathbb{R}^{n-1}$ coincides with the intersection of the set of regular points of the mapping $(P_1, \ldots, P_{i-1}, Q, P_{i+1}, \ldots, P_n) : \mathbb{R}^n \to \mathbb{R}^n$ and the set M_1. Indeed, since M_1 is a submanifold in $\mathbb{R}^n$, for any point $b \in M_1$ there exists a chart (U, φ) ($b \in U$, $\varphi \in \mathbb{R}^n \to U$) such that $\varphi^{-1}|_{U \cap M_1} = \mathbb{R}^{n-1}$. Note that $\frac{\partial(Q\varphi)}{\partial x_k} \equiv 0$, $1 \leq k \leq n-1$ (because $Q\varphi \equiv 0$ on $\mathbb{R}^{n-1}$) and $\left.\frac{\partial(Q\varphi)}{\partial x_n}\right|_{\varphi^{-1}(b)} \neq 0$ (since b is not a singular point of the manifold M_1, and $\varphi : \mathbb{R}^n \to \mathbb{R}^n$ is a diffeomorphism). Therefore, the condition

$$\det\left(\frac{\partial(\pi_i P\varphi)}{\partial(x_1, \ldots, x_{n-1})}\right)\bigg|_{\varphi^{-1}(b)} \neq 0$$

that the point b is regular for the mapping $\pi_i P : M_1 \to \mathbb{R}^{n-1}$ is equivalent to the condition

$$\det\left(\frac{\partial(P_1\varphi, \ldots, P_{i-1}\varphi, Q\varphi, P_{i+1}\varphi, \ldots, P_n\varphi)}{\partial(x_1, \ldots, x_n)}\right)\bigg|_{\varphi^{-1}(b)} \neq 0 .$$

Since φ is a diffeomorphism, the latter condition is equivalent to

$$\left(\frac{\partial(P_1,\dots,P_{i-1},Q,P_{i+1},\dots,P_n)}{\partial(x_1,\dots,x_n)}\right)\bigg|_b \neq 0 \quad .$$

Thus, the absolute value of the degree of the mapping $\rho P:M^{n-1}\longrightarrow S^{n-1}$ does not exceed half the number of inverse images $\{(\pi_i P)^{-1}(\varepsilon)\}$ ($\varepsilon = (\varepsilon_1, \dots, \varepsilon_{i-1}, \varepsilon_{i+1}, \dots, \varepsilon_n)$) of the regular value $\varepsilon \in \mathbb{R}^{n-1}$ (sufficiently close to the point 0) of the mapping $\pi_i P:M^{n-1}\longrightarrow \mathbb{R}^{n-1}$, and the number of inverse images $\{(\pi_i P)^{-1}(\varepsilon)\}$ does not exceed the number of real-valued regular solutions of the system $T_i(x, \varepsilon)$ (since the set of regular points of the mapping $\pi_i P:M_1 \longrightarrow \mathbb{R}^{n-1}$ coincides with the intersection of the set of regular points of the mapping $(P_1, \dots, P_{i-1}, Q, P_{i+1}, \dots, P_n)$: $\mathbb{R}^n \longrightarrow \mathbb{R}^n$ and the set M_1). The lemma is proved.

Let R_i denote the maximum number of regular real-valued solutions of systems of n polynomial equations in $\mathbb{R}^n$ having the same collection of supports as the system $T_i(x, \varepsilon)$ (for small ε), i.e.

$$\mathrm{supp}P_1 \cup 0, \dots, \mathrm{supp}P_{i-1} \cup 0, \mathrm{supp}Q, \mathrm{supp}P_{i+1} \cup 0, \dots, \mathrm{supp}P_n \cup 0.$$

Assuming that the number of real-valued regular solutions of the system $T_i(x, \varepsilon)$ is less than or equal to R_i and taking into account that Lemma 1 is valid for all $i = 1, \dots, n$, we obtain an inequality for $|\deg(\rho P)|$.

Theorem 7. The following estimate is valid for the absolute value of the degree of the mapping $\rho P:M^{n-1}\longrightarrow S^{n-1}$:

$$|\deg(\rho P)| \leqslant \frac{1}{2} \min_{1\leqslant i\leqslant n} R_i \quad .$$

Estimating the number R_i, $1 \leqslant i \leqslant n$, we shall obtain, by virtue of Theorem 7, estimates for $|\deg(\rho P)|$.

Let $\Gamma_1, \ldots, \Gamma_n, \Gamma$ stand for convex spans of the sets

$$\mathrm{supp}P_I \cup 0, \ldots, \mathrm{supp}P_n \cup 0, \mathrm{supp}Q \cup 0 \qquad (3)$$

respectively. Since R_i does not exceed the maximum number of regular solutions, in $\mathbb{C}^n$, for systems of polynomial equations with collection of supports (3), we obtain from Theorems 7 and 4

<u>Corollary 1</u>. The following inequality is valid:

$$|\deg(\rho F)| \leq \frac{n!}{2} \min_{I \leq i \leq n} \{V(\Gamma_I, \ldots, \Gamma_{i-I}, \Gamma, \Gamma_{i+I}, \ldots, \Gamma_n)\} . \qquad (4)$$

In particular, if the sphere $\sum_{i=I}^{n} (x_i - a_i)^2 - r^2 = 0$ is considered as the hypersurface, then $\Gamma = 2\Delta$ (Δ is the standard simplex in $\mathbb{R}^n$) and, therefore, for the absolute value of the degree of the mapping $\rho P : S_r^{n-1}(a) \longrightarrow S^{n-1}$ we have the inequality

$$|\deg(\rho P)| \leq \frac{n!}{2} \min_{I \leq i \leq n} \{V(\Gamma_I, \ldots, \Gamma_{i-I}, 2\Delta, \Gamma_{i+I}, \ldots, \Gamma_n)\} . \qquad (5)$$

Taking into account that the mixed volume is homogeneous, we obtain from (5)

<u>Corollary 2</u>. The following inequality is valid for the absolute value of the index of the singular point a, ind(P,a), and for the absolute value of the sum of indices of all singular points, ind(P, $\mathbb{R}^n$) of the vector field P:

$$|\mathrm{ind}(P,a)| , \ |\mathrm{ind}(P,\mathbb{R}^n)| \leq n! \min_{I \leq i \leq n} \{V(\Gamma_I, \ldots, \Gamma_{i-I}, \Delta, \Gamma_{i+I}, \ldots, \Gamma_n)\} . \qquad (6)$$

Proposition 1. For any algebraic hypersurface $Q(x) = 0$ we have the following estimate for the absolute value of the degree of the mapping $\rho P: M^{n-1} \longrightarrow S^{n-1}$:

$$|\deg(\rho P)| \leqslant \frac{n!}{2}\left(V(\Gamma_1,\dots,\Gamma_n)+\min_{1\leqslant i\leqslant n}\left\{V(\Gamma_1,\dots,\Gamma_{i-1},\Delta,\Gamma_{i+1},\dots,\Gamma_n)\right\}\right). \quad (7)$$

Proof. One may assume, without loss of generality, that all singular points of the vector field $P: \mathbb{R}^n \longrightarrow \mathbb{R}^n$ are regular. (Otherwise, instead of the field P we should consider the field $P - \varepsilon$, where ε is a small regular value of the mapping P. With this replacement of the field P by the field $P - \varepsilon$ the degree of mapping and the polyhedrons $\Gamma_1, \dots, \Gamma_n$ remain unchanged). In this case, $|\deg(\rho P)|$ does not exceed the maximum number, p or q, of singular points of the vector field P with positive or negative index, respectively. According to Theorem 4, we have

$$p+q \leqslant n!\,V(\Gamma_1,\dots,\Gamma_n). \quad (8)$$

Next, $p - q$ is equal to the sum of the indices of all singular points of the vector field P, and we obtain from Corollary 2

$$|p-q| \leqslant n! \min_{1\leqslant i\leqslant n}\left\{V(\Gamma_1,\dots,\Gamma_{i-1},\Delta,\Gamma_{i+1},\dots,\Gamma_n)\right\}. \quad (9)$$

The estimate in question directly follows from inequalities (8), (9).

Let us express estimates (4), (6), and (7) in terms of polynomial degrees. Note, first of all, that if $K_1, \dots, K_n$ are arbitrary natural numbers, then

$$V(K_1\Delta,\dots,K_n\Delta)=\left(\prod_{i=1}^{n}K_i\right)V(\Delta,\dots,\Delta)=\left(\prod_{i=1}^{n}K_i\right)V(\Delta)=\frac{1}{n!}\prod_{i=1}^{n}K_i. \quad (10)$$

Let now the degrees of polynomials $Q, P_1, \ldots, P_n$ be equal to m_0, $m_1, \ldots, m_n$, respectively. Then, obviously, the following inclusions are valid: $\Gamma \subset (m_0 \Delta)$, $\Gamma_1 \subset (m_1 \Delta)$, ..., $\Gamma_n (m_n \Delta)$. Taking into account relation (10) and monotonicity of mixed volumes, we obtain for estimates (4), (6), and (7):

$$|\deg(\rho P)| \leqslant \frac{m_0}{2} \min_{I \leqslant i \leqslant n} \left\{ \left(\prod_{k=I}^{n} m_k\right) / m_i \right\} \tag{11}$$

$$|\mathrm{ind}(P,a)| \;,\; |\mathrm{ind}(P,\mathbb{R}^n)| \leqslant \min_{I \leqslant i \leqslant n} \left\{ \left(\prod_{k=I}^{n} m_k\right) / m_i \right\} \tag{12}$$

$$|\deg(\ P)| \leqslant \frac{I}{2} \left(\prod_{k=I}^{n} m_k + \min_{I \leqslant i \leqslant n} \left\{ \left(\prod_{k=I}^{n} m_k\right) / m_i \right\} \right). \tag{13}$$

Let us now estimate the quantities R_i, $1 \leqslant i \leqslant n$, using Theorem 5. Introduce the notation $s_i = \# \left\{ \left(\bigcup_{\substack{k=I \\ i \neq k}}^{n} \mathrm{supp}\, P_k \right) \cup (\mathrm{supp}\, Q) \right\}$. From Theorems 7 and 5 we obtain the estimate similar to estimate (4).

<u>Corollary 1a</u>. The following inequality is valid:

$$|\deg(\rho P)| \leqslant \frac{I}{2} \min_{I \leqslant i \leqslant n} \{ H(s_i) \} . \tag{14}$$

In particular, if the sphere $Q(x) = \sum_{i=1}^{n} (x_i - a_i)^2 - r^2 = 0$ is considered as the hypersurface, inequality (14) leads to the index estimate which is similar to estimate (6).

<u>Corollary 2a</u>. The following inequality is valid for the absolute value of the index of the singular point a, $\mathrm{ind}(P,a)$, and for the absolute value of the sum of indices of all singular points of the vector field P, $\mathrm{ind}(P, \mathbb{R}^n)$:

$$|\mathrm{ind}(P,a)| \;,\; |\mathrm{ind}(P,\mathbb{R}^n)| \leq \frac{1}{2} \min_{1\leq i\leq n} \{H(l_i)\} \tag{15}$$

where l_i is the value of s_i for the hypersurface $Q(x) = 0$.

Denoting $l = \#(\bigcup_{k=1}^{n} \operatorname{supp} P_k)$, we arrive at the following analogue of Proposition 1:

<u>Proposition 1a</u>. For any algebraic hypersurface $Q(x) = 0$ the absolute value of the degree of the mapping $\rho P: M^{n-1} \longrightarrow S^{n-1}$ is estimated as follows:

$$|\deg(\rho P)| \leq \frac{1}{2}\big(H(l) + \min_{1\leq i\leq n} \{ H(l_i) \} \big). \tag{16}$$

In some cases more correct index estimates (Corollaries 2 and 2a) can be obtained using the concept of multiplicity. Let $F = (F_1, \ldots, F_n)$ be a vector field in $\mathbb{R}^n$, which has a singular point a. Let $\mu_a(L_i)$ denote the multiplicity of the solution a of the system

$$L_i \;:\; F_1 = \ldots = F_{i-1} = \sum_{k=1}^{n} (x_k - a_k)^2 = F_{i+1} = \ldots = F_n = 0.$$

<u>Lemma 2</u>. The following inequality holds for the absolute value of the index of the singular point a of the vector field F:

$$|\mathrm{ind}(F,a)| \leq \frac{1}{2} \min_{1\leq i\leq n} \{\mu_a(L_i)\} \;. \tag{17}$$

<u>Proof</u>. Since the index of the singular point a of the vector field F is equal to the degree of the mapping $\rho F: S_r^{n-1}(a) \longrightarrow S^{n-1}$ (for small r), we obtain by virtue of Theorem 7 that the absolute value of the index $\mathrm{ind}(F,a)$ does not exceed half the number of real-valued

regular solutions of the system

$$T_i: F_1=\varepsilon_1,\dots,F_{i-1}=\varepsilon_{i-1},\ \sum_{i=1}^{n}(x_i-a_i)^2-r^2=0, F_{i+1}=\varepsilon_{i+1},\dots,F_n=\varepsilon_n$$

where $\varepsilon=(\varepsilon_1,\dots,\varepsilon_{i-1},0,\varepsilon_{i+1},\dots,\varepsilon_n)$ is a regular (close to the point 0) value of the mapping $\pi_i F: S_r^{n-1}(a)\longrightarrow \mathbb{R}^{n-1}$. The number of real-valued regular solutions of the system T_i, for small ε and r, does not exceed the multiplicity $\mu_a(L_i)$ of the solution a of the system L_i, since the system T_i is obtained from L_i by perturbing the latter with small ε and r. To complete the proof, it remains to note that the above reasoning is valid for all $i = 1, \dots, n$.

Estimating the multiplicities $\mu_0(L_i)$, $1 \leq i \leq n$, we obtain by virtue of Lemma 2 an estimate for the absolute value of the index ind(F, 0).

Proposition 2. Let $F=(F_1, \dots, F_n)$ be an analytical vector field in $\mathbb{R}^n$ whose components have "suitable" Newton polyhedrons $\Gamma_1, \dots, \Gamma_n$, and for which 0 is an isolated singular point. If the systems

$$L_i: F_1=\dots=F_{i-1}=\sum_{k=1}^{n} x_k^2=F_{i+1}=\dots=F_n=0,\quad 1\leq i\leq n$$

are non-degenerate at zero, the following inequality is valid

$$|\mathrm{ind}(F,0)|\leq \min_{1\leq i\leq n}\{\nu(\Gamma_1,\dots,\Gamma_{i-1},\nabla,\Gamma_{i+1},\dots,\Gamma_n)\}$$

where $\nabla=(\overline{\mathbb{R}_+^n\setminus\Delta})$.

Proof. The proposition directly follows from Lemma 2, Theorem 2, and the relation $\nu(\Gamma_1,\dots,\Gamma_{i-1},2\nabla,\Gamma_{i+1},\dots,\Gamma_n) =$
$= 2\,\nu(\Gamma_1,\dots,\Gamma_{i-1},\nabla,\Gamma_{i+1},\dots,\Gamma_n)$.

·. Estimation of the index from Newton diagrams for the components of a vector field on a plane

.et $F = (F_1, F_2)$ be an analytical vector field in $\mathbb{R}^2$, $F(0) = 0$, and et F_1 and F_2 have "suitable" Newton polyhedrons, Γ_1 and Γ_2, res-ectively. A vector field F is called real-nondegenerate if for any $\in \mathbb{Z}_+^2$ the system $F_{1,\Gamma_{1,p}} = F_{2,\Gamma_{2,p}} = 0$ does not have solutions n $(\mathbb{R} \setminus 0)^2$.

.et $A = \|a_{ij}\|$ be an $m \times n$ matrix composed of zeros and unities, nd let $B(A) = \{M\}$ be the system of all sets of the form

$$M = \{a_{i_1 j_1}, a_{i_2 j_1}, a_{i_2 j_2}, a_{i_3 j_2}, a_{i_3 j_3}, \ldots, a_{i_{r-1} j_{r-1}}, a_{i_r j_{r-1}} :$$

$$1 = i_1 < i_2 < \ldots < i_r = m,\ 1 \leqslant j_1 < j_2 \ \ldots < j_{r-1} \leqslant n\}.$$

.et us put $d(A) = \max\limits_{M \in B(A)} \{\text{the sum of elements of the set } M\}$.

n other words, d(A) is defined as follows. For all possible ladders with vertices in positions corresponding to matrix elements) origi-ating in the first row and terminating in the last row of the mat-ix A

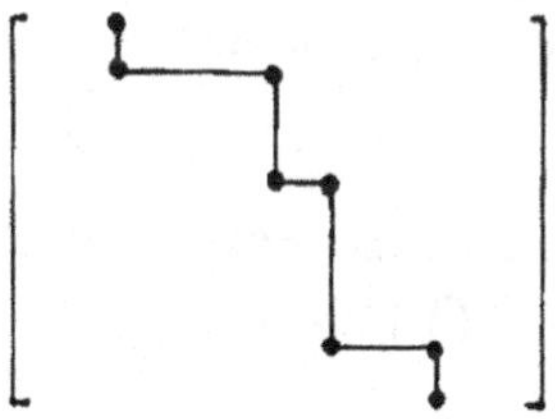

ve evaluate the sums of matrix elements located in ladder vertices. The maximum sum is precisely d(A).

Let $F = (F_1, F_2)$ be a vector field in $\mathbb{R}^2$, and let F_1 and F_2 have 'suitable" Newton diagrams, Γ_1 and Γ_2. Let $W(\Gamma_2)$ denote the set of all points of the support supp F_2 on the diagram Γ_2 which lie on straight lines supporting Γ_2 and parallel to edges of the diag-ram Γ_1. We shall number points of the sets supp $F_1 \cup \Gamma_1$ and $W(\Gamma_2)$, moving along the diagrams Γ_1 and Γ_2 from bottom to top.

Let Δ be a certain edge of the diagram Γ_1. Let $p(\Delta)$ denote the subset of points of the set $W(\Gamma_2)$ which lie on the straight line supporting Γ_2 and parallel to the edge Δ. Suppose $m^{(i)}, m^{(i+1)}, \ldots, m^{(i+\nu)}$ and $\mu^{(j)}, \mu^{(j+1)}, \ldots, \mu^{(j+\lambda)}$ are points of the sets $\Delta \cap \operatorname{supp} F_1$ and $p(\Delta)$, respectively (with the aforementioned numbering). We put

$$h_{kl} = \varkappa(m_1^{(k)}+\mu_1^{(l)},\ m_2^{(k)}+\mu_2^{(l)}),\quad i \leqslant k \leqslant i+\nu,\quad j \leqslant l \leqslant j+\lambda$$

where $m^{(k)} = (m_1^{(k)}, m_2^{(k)})$, $\mu^{(l)} = (\mu_1^{(l)}, \mu_2^{(l)})$,

$$\varkappa(pq) = \begin{cases} 1 & \text{for } pq \equiv 1 \pmod 2, \\ 0 & \text{for } pq \equiv 0 \pmod 2). \end{cases}$$

For each edge Δ of the diagram Γ_1 we find the numbers h_{kl} and consider the matrix $H = \|h_{kl}\|$, completing it with zeros at places where no elements are present.

<u>Theorem 8</u>. If $F = (F_1, F_2)$ is a vector field which is real-nondegenerate at zero and has "suitable" Newton diagrams Γ_1, Γ_2 for the components F_1 and F_2, then the following estimate is valid for the index of a singular point of the field F:

$$|\operatorname{ind}(F,0)| \leqslant d(H). \tag{18}$$

This estimate is accurate in the class of vector fields with given sets $\operatorname{supp} F_1 \cap \Gamma_1$ and $\operatorname{supp} F_2 \cap \Gamma_2$.

5. Remarks

Now a few words about the accuracy of the estimates just obtained. It follows from the results of Ref. [12] that estimate (12) is exact for polynomial vector fields with fixed components degrees $m_1, \ldots, m_n$ one of which, say m_k, is much greater than the other degrees, and $\sum_{i=1}^{n} m_i \not\equiv n \pmod 2$. (In this case the estimate is attained for homogeneous vector fields all the components of which are products of linear forms). The assertion is also valid if $\sum_{i=1}^{n} m_i \neq n \pmod 2$.

Indeed, in this case

$$\sum_{\substack{i=1 \\ i\neq k}}^{n} m_i + (m_k - 1) \equiv n \pmod 2$$

and, as was pointed out above, there exists a number l such that for $m_k - 1 \geqslant l$ we can find a homogeneous vector field $P = (P_1, \ldots, P_n)$ with component degrees $m_1, \ldots, m_{k-1}, m_k - 1, m_{k+1}, \ldots, m_n$ for which

$$\operatorname{ind}(P, o) = \min_{1 \leqslant j \leqslant n} \Big\{ \Big(\prod_{\substack{i=1 \\ i\neq k}}^{n} m_i\Big)(m_k - 1) / m_j \Big\} = \prod_{\substack{i=1 \\ i\neq k}}^{n} m_i .$$

In this case, zero is, apparently, an isolated singular point of the vector field $\widetilde{P} = (P_1, \ldots, P_{k-1}, P_k + x_1^{m_k}, P_{k+1}, \ldots, P_n)$ with component degrees $m_1, \ldots, m_{k-1}, m_k, m_{k+1}, \ldots, m_n$ $(m_k \geqslant l + 1)$, and $\operatorname{ind}(\widetilde{P}, 0) = \operatorname{ind}(P, 0) = \prod_{\substack{i=1 \\ k\neq i}}^{n} m_i$.

In the case of a homogeneous vector an exact index estimate $\operatorname{ind}(P, 0)$ for arbitrary collections of component degrees $(m_1, \ldots, m_n) = m$ has been obtained by V.I. Arnol'd [1] (see also [18]):

$$|\operatorname{ind}(P, 0)| \leqslant \Pi(m),$$

where $\Pi(m)$ is the number of integer points in the central section $\sum_{i=1}^{n} x_i = \frac{1}{2}(\sum_{i=1}^{n} m_i - n)$ of the parallelepiped $\Delta(m) = \{x \in \mathbb{R}^n : 0 \leqslant x_i \leqslant m_i - 1,\ 1 \leqslant i \leqslant n$.

The results of Ref. [18] imply, in particular, the following index estimate for an arbitrary (generally non-homogeneous) polynomial vector field P with component degrees $m_1, \ldots, m_n$:

$$|\mathrm{ind}(P,o)| \leqslant \begin{cases} O(m,2), \text{ if } \sum_{i=1}^{n} m_i \not\equiv n \pmod 2 \\ \min\limits_{1\leqslant i\leqslant n} \{ O((m_1,\dots,m_{i-1},m_i+1,m_{i+1},\dots,m_n),2)\}, \\ \quad \text{if } \sum_{i=1}^{n} m_i \equiv n \pmod 2 \end{cases} \tag{19}$$

where $O(m,2)$ is the number of integer points of the parallelepiped $\Delta(m)$ which satisfy the inequalities $\frac{1}{2}(\sum_{i=1}^{n} m_i - n - 2) \leqslant \sum_{i=1}^{n} x_i \leqslant \frac{1}{2}(\sum_{i=1}^{n} m_i - n)$. If $\sum_{i=1}^{n} m_i \not\equiv n \pmod 2$, estimate (19) is more accurate than estimate (12), and if $\sum_{i=1}^{n} m_i \equiv n \pmod 2$, neither of these two estimates majorates each other.

Furthermore, it can easily be seen from the expressions for estimates (11) and (13) that the latter estimate is more accurate than the former one for a hypersurface with the degree m_o much greater than the degrees $m_1, \dots, m_n$. To be more exact, direct calculation shows that estimate (13) is more accurate than estimate (11), provided $m_o > 1 + \max\limits_{1\leqslant i\leqslant n} \{m_i\}$. Similarly, if $\Gamma_1, \dots, \Gamma_n$ are convex bodies, estimate (7) is more accurate than estimate (4), provided the volume $V(\Gamma)$ is sufficiently large as compared with the volumes $V(\Gamma_1), \dots, V(\Gamma_n)$. Indeed, it follows from the Aleksandrov-Fenchel inequality (see [13]) that mixed volumes of convex bodies satisfy

$$V(\Gamma_1,\dots,\Gamma_{i-1},\Gamma,\Gamma_{i+1},\dots,\Gamma_n) \geqslant V(\Gamma)^{1/n} \Big(\prod_{\substack{k=1\\k\neq i}}^{n} V(\Gamma_k)\Big)^{1/n}. \tag{20}$$

It can be seen from inequality (20) that $V(\Gamma_1, \dots, \Gamma_{i-1}, \Gamma, \Gamma_{i+1}, \dots, \Gamma_n) \to \infty$ if $V(\Gamma) \to \infty$.

Note that problems similar to that for the rotation index of a polynomial vector field on algebraic manifolds have been considered by A.G. Khovansky [18] who obtained several different exact estimates in terms of vector field component degrees, and also in terms of the degrees of the polynomial defining an algebraic manifold.

As to the accuracy of estimates (14)-(16), we may say that they are eventually based on the estimate for the number of real-valued solutions of the system $P: P_1 = \ldots = P_n = 0$ of polynomial equations in the positive octant $\mathbb{R}^n_+$ (Theorem 3). As pointed out in [19], "... this estimate is, apparently, too high." According to the hypothesis of A.G. Kushnirenko, it can be reduced to $\prod_{i=I}^{n} (k_i - 1)$, where k_i is the number of monomials appearing in the polynomial P_i. If in the polynomials $P_1, \ldots, P_n$ we take into account not only the number of monomials but also the support geometry, even the estimate $\prod_{i=I}^{n} (k_i - 1)$ is sufficiently rough.

Note also that an estimate close to (17) was proposed in [21] : namely, $|\mathrm{ind}(F,0)| \leqslant (\mu_0(F))^{(n-1)/n}$. This estimate is almost always less accurate than (17) and, as was shown in [22] , is not exact for $n \geqslant 3$.

Index estimation from Newton diagrams was first carried out by V.I. Arnol'd [1] for gradient vector fields on a plane:

$$|\mathrm{ind}(\mathrm{grad}\ F,\ 0)| \leqslant \left\{ \begin{array}{l} \text{the number of interior integer points} \\ \text{on the Newton diagram of the} \\ \text{function } F \end{array} \right\} \quad (21)$$

Estimate (21) is exact in the class of gradient vector fields with a fixed Newton diagram for the function F. For arbitrary (generally non-gradient) vector fields on a plane, estimation of the singular point index from Newton diagrams of vector field components was carried out in [4,5,10] . These estimates, unlike estimate (18), do not allow for the specificity of component supports, and are not exact.

The author is greatly indebted to A.A. Talashev, translation editor.

REFERENCES

1. Arnol'd V.I., The index of a vector field singular point, Petrovský-Oleinik inequalities, and Hodge's mixed structures. Funkts. analiz i ego prilozh., 1978, v. 12, No. 1, p. 1-14 (in Russian).
2. Arnol'd V.I., Additional chapters in the theory of ordinary differential equations. Moscow, Nauka, 1978 (in Russian).
3. Arnol'd V.I. and Oleinik O.A., The topology of real-valued algebraic manifolds. Vestnik MGU. Ser.1, Matem. i Mekh., 1979, No. 6, p. 7-17 (in Russian).

4. Berezovskaya F.S., Complicated stationary point of a system on a plane: neighbourhood structure and index. Pushchino, 1978. Preprint - Research centre for biological studies and Research computing centre of the USSR Academy of Sciences (in Russian).

5. Berezovskaya F.S., The index of stationary point of a vector field on a plane. Funkts. analiz i ego prilozh., 1979, v. 13, No. 2, p. 77 (in Russian).

6. Bershteĭn D.N., The number of roots of a system of equations. Funkts. analiz i ego prilozh., 1975, v. 9, No. 3, p. 1-4 (in Russian).

7. Bershteĭn D.N., Kushnirenko A.G., and Khovansky A.G., Newton polyhedrons. Usp. Matem. Nauk, 1976, v. 31, No. 3, p. 201-202 (in Russian).

8. Bliznyakov N.M., On rotation estimates for vector fields on algebraic manifolds. Funkts. analiz i ego prilozh., 1979, v. 13, No. 2, p. 78 (in Russian).

9. Bliznyakov N.M., On topological index estimates for a singular point of a vector field. Deposited at VINITI, 1979, No. 589-79 (in Russian).

10. Bliznyakov N.M., Calculation and estimates of a vector field singular point on a plane. Deposited at VINITI, 1979, No. 3041-79 (in Russian).

11. Bliznyakov N.M., Cauchy indices and singular point index of a vector field. In: Application of topology in modern analysis. Voronezh State University, 1985, p. 3-21 (in Russian).

12. Bliznyakov N.M. and Mukhamadiev E.M., On the calculation of singular point index for a polylinear vector field. Trudy Matem. Fak. VGU, 1971, No. 4, p. 19-29 (in Russian).

13. Busemann H., Convex surfaces. Interscience, New York

14. Gantmakher F.R., Matrix theory. Moscow, Nauka, 1967 (in Russian).

15. Krasnosel'sky M.A., Vainiko G.M., Zabreiko P.P., Rutitsky Ya.B., and Stetsenko V.Ya., Approximate solution of operator equations. Moscow, Nauka, 1969 (in Russian).

16. Kushnirenko A.G., Newton polyhedron and the number of solutions for a system of k equations with k unknown variables. Usp. Matem. Nauk, 1975, v. 30, No. 2, p. 266-267 (in Russian).

17. Postnikov M.M., Stable polynomials. Moscow, Nauka, 1981 (in Russian

18. Khovansky A.G., The index of a polynomial vector field. Funkts. analiz i ego prilozh., 1979, v. 13, No. 1, p. 49-58 (in Russian).

19. Khovansky A.G., On a certain class of systems of transcendental equations. Doklady AN SSSR, 1980, v. 255, No.4, p.804-807 (in Russian).

20. Khovansky A.G. Newton polyhedrons and the index of a vector field. Usp. Matem. Nauk, 1981, v. 36, No. 4, p. 234 (in Russian).

21. Eisenbud D. and Levine H., An algebraic formula for the degree of a C^∞ map germ. Ann. Math., 1977, v. 106, No. 1, p. 19-44.

22. Granger M., Sur le degré local d'un germe d'application analytique réele. Comptes Rendus Acad. Sci. Paris. Sér. A et B, 1978, v. 287, No. 7, p. 531-534.

MODERN APPROACH TO THE THEORY OF TOPOLOGICAL CHARACTERISTICS OF NONLINEAR OPERATORS I.

Yu.G.Borisovich
Department of Mathematics
Voronezh State University
394693, Voronezh, USSR

Introduction.

In the present paper (and its 2nd part is presupposed) we shall ystematize the questions connected with the theory of topological haracteristics of non-compact operators generalizing the Leray-Schau- er degree theory and we shall deal, in particular, with the results btained recently and found in different publications, sometimes al- ost inaccessible. We mean the works by V.G.Zvyagin, V.T.Dmitrienko [1-4]), Yu.G.Borisovich, V.G.Zvyagin ([5 , 6]), N.M.Ratiner [7 , 12]), the reports ([8 - 12]) and review articles ([13 - 16]). New results are also given.

We shall consider operator equations

$$A\ (x) - g\ (x) = 0 \qquad (0)$$

ith Fredholm or monotomous (nonlinear) operator A perturbed by non- inear, non-compact but A-condensing with respect to (1) operator g. oreover, operator g is not assumed to be smooth and it can be mul- ivalued (in this case equation (1) is replaced by an inclusion). The ain aim of this paper is the construction and investigation of the opological characteristic of problem (1) on the domain Ω of the anach space L on which are defined operators A, g (single- or mul- ivalued), acting in the Banach space E. The construction is based on he following three topological principles developed to a great ex- ent in the 70ies and 80ies: compact contraction of condensing opera- ors (Yu.G.Borisovich, Yu.I.Sapronov [17 , 18 , 19]); bijectivi- y of homotopic classes of condensing and quite continuous vector ields (Yu.I.Sapronov [20]) and the principle of invariance for Fred- olm operators (Yu.I.Sapronov, Yu.G.Borisovich, V.G.Zvyagin [21 - 23]).

The three principles proved to be fundamental which is testified by the modern development of the degree theory (see e.g. review articles [8,9 , 13-16 , 22 , 24 , 25] and monographs [26 , 27].

Part I deals with a further development of the principles of compact contraction and bijective correspondence and with the establishment of the principle of the reduction of a topological characteristic in the case of A-condensing operator g to the case of a compact operator. In Part II we are going to describe the constructions of topological characteristics, their properties and the applications to the conditions of solvability of problem (1) and also to expand the bibliography.

The author is grateful to his disciples B.D.Gel'man, V.G.Zvyagin, V.V.Obukhovskiĭ and Yu.I.Sapronov for the useful discussion of the questions under consideration.

I. Principle of compact contraction and principle of bijective correspondence for non-compact mappings.

1. In paper [17] the authors posed the problem of the "investigation of non-compact operators by their "contraction" to compact subspaces". To put it more precise, if $F : \bar{\Omega} \to X$ is a mapping of the closed domain Ω of the topological space X into X, then is there a compact subspace $Y \subset X$ such that $F|_Y : \Omega|_Y \longrightarrow Y$? In fact, the subspace Y should satisfy special conditions which make it possible to investigate the mapping $F|_Y$ by topological methods" (p.43).

For example, if Y is Leray space [28] or a convex compact set in a local convex space (LCS) E, then on $\Omega|_Y$ a full Leray index or a relative rotation can be defined [30 , 29 , 14] .

In paper [17] the authors formulated an abstract principle of compact contraction and applied it to the condensing mappings developing their earlier work [18] . Yu.I.Sapronov in [19] investigated the structure of partial order of all compact convex sets, invariant for the continuous mapping F and constructed an algebraic index theory in a full structure of X , generalizing degree theories of condensing and limit compact mappings (see [18 , 31 , 32] and later papers [33 , 34] and also the results of [17]).

2. Let us state (with some refinements) the principle of compact contraction from [17] . It will be used below.

Let X be a topological space with an axiom of separability T_1, (X)-set of all its closed subsets($C(X)\ni\emptyset$), S be a real space with a one K of non-negative elements inducing the semi-ordering ($\leq$). enote by K' extensions $K' = K \cup \{\infty\}$ of the cone K by an ideal" element $+\infty$ and consider $x \leq +\infty$, if $x \in K$. The ystem $\mathfrak{M} = \{\Phi_\lambda\}$ of the closed subsets $\Phi_\lambda \in C(X)$ will be alled "admissible", if it is closed with respect to non-empty interections and majorizes any element $A \in C(x)$ by the inclusion (i.e. $\Phi_\lambda \supset A$ exists). Then consider the mapping $\chi : C(X) \to K'$. It s called discriminating if it satisfies the following conditions: . $\chi(A) \leq \chi(B)$ at $A \subset B$ (monotony); 2. $\chi(A \cup R) = \chi(A)$ if $\chi(R) = 0$; . $\chi(\cap \Phi_\lambda \mid \Phi_\lambda \supset A) = \chi(A)$ for every $A \subset C(X)$; 4. if x is a oint of X , then $\chi(x) = 0$.

A special role will be played by such sets $A \subset C(X)$ for which (A) = 0. The collection of such A will be denoted by Ker χ and ill be called the kernel of the discriminating mapping χ .

Now let $F : \bar{\Omega} \to X$ be single-valued or multi-valued mappings f the closure of an open set $\Omega \subset X$. Let $\Omega_\Phi = \Omega \cap \Phi$ be an pen set in Φ and let $\bar{\Omega}_\Phi$ be its closure (in a relative toology of Φ).

Definition 1.1. The mapping F will be called coordinated on Ω ith a discriminating χ, if 1. $\chi(FA) = 0$, when $\chi(A) = 0$, $A \subset \bar{\Omega}$. $\chi(FA) \not\geq \chi(A)$ ($\neq +\infty$), if $\chi(A) \neq 0$ ($= +\infty$).

Let us state the main property of the mapping coordinated with χ:

Theorem 1.1. If $F : \bar{\Omega} \to X$ is coordinated with χ , then for ny $R \in \operatorname{Ker}\chi$ there exists $\Phi^* \in \mathfrak{M}$ such that 1. $\Omega_{\Phi^*} = \Omega \cap \Phi^* \neq \emptyset$; 2. $\Phi^* \in \operatorname{Ker}\chi$; 3. $\Phi^* \supseteq R$; 4. $F : \bar{\Omega}_{\Phi^*} \to \Phi^*$ (acts in Φ^*).

Note that Theorem 1.1. establishes the existence of the invariant et Φ^* for F without the assumption on continuity of F or hat it is single-valued.

This theorem is also applied to the family $F_t : \bar{\Omega} \to X$, $_t(\cdot) = F(t, \cdot)$, $F : \bar{\Omega} \times \Lambda \to X$ of the mappings($t \in \Lambda$, Λ is set of values of the abstract parameter t), the common (for all $_t$) invariant set being of interest. This is achieved by the reducion to Theorem 1.1. by considering the mapping $\Psi(x) = \bigcup_{t\in\Lambda} F_t(x) = F(\Lambda, x)$ and the requirement that Ψ should be coordinated with (in this case we say that the family $\{F_t\}$ is coordinated with).

Proof of Theorem 1.1. Let us define the mapping $Q : C(X) \to \mathfrak{M}$ y the equality $Q(A) = \cap(\Phi_\lambda : \Phi_\lambda \supset A)$. The set $\Phi^1 =$

$Q(R \cup F\overline{\Omega})$ has a property: $R \in \Phi^1, \Omega_{\Phi^1} \neq \emptyset$, $F(\Omega_{\Phi^1}) \subset \Phi^1$ (expanding R if necessary, so that $R \cap \overline{\Omega} \neq \emptyset$). Let us denote the class of all such sets from $\mathfrak{M}$ by $\mathfrak{M}^*$. A set $\Phi^* = \cap \Phi^1$, $\Phi^1 \subset \mathfrak{M}^*$ is minimal in the class $\mathfrak{M}^*$. The mapping $P(\Phi^1) = Q(R \cup F\overline{\Omega}_{\Phi^1})$ acts in $\mathfrak{M}$ and "contracts" Φ^1 : $P(\Phi^1) \subset \Phi^1$. Consequently, a minimal set Φ^* is its fixed point, i.e. $P(\Phi^*) = \Phi^*$. Using axiom 3 and then axiom 2 from the definition of X we obtain

$$\chi(\Phi^*) = \chi(P\Phi^*) = \chi(Q(R \cup F\overline{\Omega}_{\Phi^*})) = \chi(F\overline{\Omega}_{\Phi^*}) \qquad (1)$$

If $\chi(\overline{\Omega}_{\Phi^*}) \neq 0$, then by Definition 1.1, property 2, we have $\chi(F\overline{\Omega}_{\Phi^*}) \not\geqslant \chi(\overline{\Omega}_{\Phi^*})$ and taking into account the beginning of equality (1), $\chi(\Phi^*) \not\geqslant \chi(\overline{\Omega}_{\Phi^*})$, which contradicts axiom 1 for χ. Then $\chi(\overline{\Omega}_{\Phi^*}) = 0$, whence $\chi(F\overline{\Omega}_{\Phi^*}) = 0$ by Definition 1.1., property 1; from equalities (1) follows $\chi(\Phi^*) = 0$ 'i.e. $\Phi^* \in \operatorname{Ker} \chi$ and Φ^* is a required invariant set. Similarly, for the family of the mappings holds

<u>Theorem 1.2</u>. If the family $F_t : \overline{\Omega} \longrightarrow X$, $t \in \Lambda$ is coordinated with X, then there exists $\Phi^* \in \mathfrak{M}^*$ such that properties 1-3 of Theorem 1.1. are fulfilled and $F_t : \overline{\Omega}_{\Phi^*} \longrightarrow \Phi^*$ for all $t \in \Lambda$

3. It is obvious that choosing a compact invariant set of the mapping $F : \overline{\Omega} \longrightarrow X$ it is desirable, however, that the values of F on $\overline{\Omega}_\Phi$, in some specified sense, should "approximate" the values on $\overline{\Omega}$. Let us add the corresponding construction to the result of [17].

<u>Definition 1.2</u>. The set $Q(M \cup N)$ is called the "combination" $[M,N]$ of two sets $M, N \subset X$.

<u>Definition 1.3</u>. The set $\Phi \in \mathfrak{M}$ is called $\mathfrak{M}$-fundamental for the mapping $F : \overline{\Omega} \longrightarrow X$ if $\Omega_\Phi = \Omega \cap \Phi \neq \emptyset$, $F(\Omega_\Phi) \subset \Phi$, and if $x \in [F(x), \Phi]$ when $x \in \overline{\Omega}$ then $x \in \Phi$.

Graphically this means that the image $F(x)$ of the point x beyond Φ is "nearer" to Φ than x.

<u>Theorem 1.3</u>. In the conditions of Theorem 1.2. the set Φ^* may be chosen as $\mathfrak{M}$-fundamental for the family F_t.

<u>Proof</u>. It is sufficient to construct a minimal set $\Phi^* \in \mathfrak{M}^*$ in Theorem 1.3. in a subclass $\mathfrak{N} \in \mathfrak{M}^*$, consisting of $\mathfrak{M}$-fundamental sets from $\mathfrak{M}^*$; the class $\mathfrak{N}$ is not empty (it includes, e.g. $\Phi = X$); $\Phi^* = \cap(\Phi^1 : \Phi^1 \in \mathfrak{N})$ is fundamental and belongs to $\mathfrak{M}^*$. Let us show that the "reducing" operator P acts in $\mathfrak{N}$. In fact, if $\Phi^1 \subset \mathfrak{N}$, then $P(\Phi^1) = [\Psi\overline{\Omega}_\Phi \cup R]$, $R \subset \Phi^1$; if now $x \in [F_t x, P(\Phi^1)]$, then $x \in \Phi^1 \cap \overline{\Omega} = \overline{\Omega}_\Phi$ by the definition of the fundamentality of Φ^1 and

$$F(x) \in F(\overline{\Omega}_{\Phi^1}) \subset \left[\overline{F\Omega}_{\Phi^1} \cup R\right] \subset P(\Phi^1)$$

then it follows that

$$x \in \left[F\, x,\ P(\Phi^1)\right] \subset \left[P(\Phi^1),\ P(\Phi^1)\right] = P(\Phi^1)$$

which means fundamentality $P(\Phi^1)$.

Thus, in the subclass $\mathcal{M}$ the element Φ^*, constructed above, is minimal, i.e. $\Phi^* = P(\Phi^*)$, whence it follows, as in Theorem 1.1., $\Phi^* \in \operatorname{Ker} \chi$.

Remark 1.1. Theorems 1.1. - 1.3. will hold if the closures $\overline{\Omega}$ $\overline{\Omega}_{\Phi^*} = \overline{\Omega} \cap \Phi^*$ (domains of the mappings F, $F|_{\Phi^*}$) are replaced by an arbitrary closed set $Y \subset X$ and $Y_{\Phi^*} = Y \cap \Phi^*$ respectively ($Y_\Phi \neq \emptyset$).

For the first time the notion of a fundamental set was introduced in paper [35] for compact mappings of Banach spaces for another purpose. Yu.I.Sapronov saw the unitily of this notion for the statement of the principle of compact contraction for condensing mappings and in papers [20 , 36] on condensing operators the notion of a fundamental set was applied to the construction of a degree. Later this construction became current (see [37 , 14-16 , 38,39 , 25-27] ; in [39] the reference to it was not made).

4. The principle of compact contraction is extended to the locally convex linear topological spaces (LCS) X with respect to the existence in such spaces of different "measures of non-compactness" which can be used as discriminating mappings and the operation $\overline{co}$ of the convex closure of sets (a systematic theory of measures of non-compactness is developed in [24,25 , 27]). The measure of non-compactness in LCS X is a mapping $\Psi : P(X) \longrightarrow A$ of the collection of all non-empty subsets X onto a partially ordered set A satisfying the condition $\Psi(\overline{co}\, M) = \Psi(M)$ for any $M \subset P(X)$.

Below we consider Hausdorff full LCS X = E and in it the measure of non-compactness Ψ (i.e. "non-negative" function $\Psi(M) \in K'$, $\Psi(\overline{co} M) = \Psi(M)$ given on arbitrary sets $M \subset P(E)$ satisfying additional conditions: c_0) $\Psi(M_1) \leq \Psi(M_2)$, if $M_1 \subset M_2$, c_1) $\Psi(M) = 0$, iff M is pre-compact, c_2). $\Psi(M \cup R) = \Psi(M)$, if R is pre-compact, c_3) $\Psi(M) = +\infty$ only for non-bounded sets in E. Note that the condition $\Psi(\overline{co} M) = \Psi(M)$ together with c_1) implies the compactness of $\overline{co}(M)$ for pre-compact M which is connected with the completeness of LCS E.

Restricting Ψ on closed sets from E, we get a discriminating mapping $\chi = \Psi : C(E) \longrightarrow K' = K \cup \{+\infty\}$. By way of an admissible system $\mathfrak{M}$ we shall choose convex closed sets Φ in E. It is easy to see that χ is a discriminating mapping, $Q(M) = \overline{co}(M)$ is a convex closure, Ker X consists of compact sets in E. The mapping $F : \overline{\Omega} \rightarrow E$ is coordinated with X iff the image of a compact set is precompact and

$$\chi(FM) \not\geq \chi(M) \text{ if } \chi(M) \neq 0,\ M \subset \overline{\Omega} \qquad (2)$$

(the mappings F satisfying (2) are called condensing with respect to the measure of non-compactness χ [27]). If $F_t : \overline{\Omega} \longrightarrow E$ is a family of mappings, $F_t(\cdot) = F(t,\cdot), t \in \Lambda$, then we shall call it condensing, if $\Psi : \overline{\Omega} \longrightarrow E$ ($\Psi(x) = F(\Lambda, x)$) is condensing. From Theorem 1.3. follows

Theorem 1.4. If $F(F_t) : \overline{\Omega} \longrightarrow E$ is a condensing mapping (family) in LCS E and $F(\Psi)$, an image of any compact set in $\overline{\Omega}$, is precompact, then for a given compact $R \subset E$ there exists a fundamental compact set Φ^* for the mapping F (of the family F_t).

For brevity we shall call the set Φ^* a f.c.s. Φ^*. Note that f.c.s. is always convex.

Remark 1.2. The statement of Theorem 1.4. will hold for the mapings $F(F_t) : Y \longrightarrow E$, where $Y \subset E$ is an arbitrary subset with the topology induced from E; a compact contraction takes place onto the set $Y_{\Phi^*} = Y \cap \Phi^* \neq \emptyset$ (instead of $\overline{\Omega}_{\Phi^*}$) closed in a relative topology.

In fact, a discriminating function χ is defined on the relatively closed sets in Y (which are not necessarily closed in E), since it is generated by the measure of non-compactness and this is sufficient for the obtaining of a general scheme where Y is considered in an induced topology.

If the mapping $F(F_t) : Y \longrightarrow E$ allows the restriction onto the compact fundamental set $F(F_t) : Y_{\Phi^*} \longrightarrow \Phi^*$, then they say that $F(F_t)$ is a fundamental compact contractable mapping (f.c.c. mapping). Theorem 1.4. generalizes the result from [36 , 38,14] on the fundamental compact contraction of condensing upper semicontinuous multivalued mappings. Let us state this result but first we shall recall an important

Definition 1.4. Multivalued mapping $F : Y \longrightarrow X$ (the family $F : Y \times \Lambda \longrightarrow X$), where Y, X, Λ are topological spaces, is called upper semicontinuous if for any open $V \subset X$ a small pre-image $F^{-1}(V) = \{y \mid y \in Y, F(y) \subset V$ ($F^{-1}(V) = \{(x,\lambda) \mid (x,\lambda) \in Y \times \Lambda$, $F(x,\lambda) \subset V\}$, repectively) is an open set in Y (in $Y \times \Lambda$).

Denote by K(X) a collection of all compact non-empty subsets in X. f a multivalued mapping $F : Y \longrightarrow X$ with the images in K(X) is up-er semicontinuous and $A \in K(Y)$, then $F(A) \in K(X)$; thus, the mapping : $K(Y) \longrightarrow K(X)$ is induced and similarly the mapping $\hat{F} : K(Y \times \Lambda) \rightarrow$ $\rightarrow K(X)$ is induced for the family $F : Y \times \Lambda \longrightarrow X$.

Theorem 1.5. The condensing mapping $F : Y \longrightarrow E$ (the family $F : Y \times \Lambda \longrightarrow E$) in LCS E, where $Y \subset E$, Λ is a compact topological pace with images from K(E), under the condition of upper semiconti-uity is a f.c.c. mapping.

In fact, considering Y as a topological space with an induced opology from E, we get $F : K(Y) \longrightarrow K(E)$ ($F : K(Y \times \Lambda) \longrightarrow K(E)$). onsequently, all the conditions of Theorem 1.4.(Remark 1.2.) are ulfilled.

5. Applications of the principle of compact contraction. For up-er semicontinuous f.c.c. mappings F with the images from Kv(E)(i.e. ompact and convex), in particular, for single-valued continuous F akes place a "relative" analogue of Leray-Schauder-Krasnosel'skiĭ heory ([29,30 , 14-16 , 26 , 40 , 41,38]). In fact, if s f.c.s. and if $(\mathrm{Fix} F_\Phi) \cap \partial\Omega_\Phi = \emptyset$, then are denoted the rotation f the vector field $I - F_\Phi$ on the boundary $\partial\Omega_\Phi$ with respect o the fundamental compact Φ (denoted by $\gamma(\partial\Omega_\Phi)$), a local degree multiplicity of covering in 0 of the mapping $I - F_\Phi : \bar{\Omega} \longrightarrow E$, lenoted by $d(\bar{\Omega}_\Phi)$), a full Leray index $i(\Omega_\Phi)$ (see a single-alued case in[28], for Kv-mappings define it via ε-approximations).

Theorem 1.6. For the condensing mapping F and f.c.s. Φ we have $\gamma(\partial\Omega_\Phi) = d(\bar{\Omega}_\Phi) = i(\Omega_\Phi)$ To-ological characteristics γ, d, i under the condition $\mathrm{Fix}\, F \cap \partial\Omega =$ $= \emptyset$ do not depend on the choice of f.c.s. Φ .

Proof. The first part of the statement is a direct consequence of he author's works [30 , 29 , 41]. The independence of $\gamma(\partial\Omega_\Phi)$ rom the choice of f.c.s. Φ is proved in [14 , 38], whence ollows the independence of $d(\Omega_\Phi)$, $i(\Omega_\Phi)$ from Φ .

This theorem allows one to define "absolute" rotations, local de-ree, full index of a condensing mapping by the equalities $\gamma(\partial\Omega) = \gamma(\partial\Omega_\Phi)$, $d(\bar{\Omega}) = d(\bar{\Omega}_\Phi)$, $i(\Omega) = i(\Omega_\Phi)$, respecti-ely, if on the boundary $\partial\Omega$ there are no fixed points of the apping F . These characteristics are constant on the condensing fa-ily F_t if the last condition is fulfilled for all t .

Yu.I.Sapronov established an extraordinary fact for the condens-ng mappings: the principle of a bijective correspondence between homotopic classes of quite continuous and condensing vector fields [20].

For the statement of this principle (it will be given in a more general form) we shall introduce the following notation: for the pair of closed subsets $X_1 \subset X \subset E$ denote by $D_\psi(X, X_1)$, $C(X, X_1)$ the sets of all vector fields $I - F$, where $F : X \longrightarrow Kv(E)$, $X_1 \cap \mathrm{Fix} F = \emptyset$, F is upper semi-continuous, condensing or compact, respectively (the latter means that the image $F(X)$ is pre-compact). Let us call the fields from D_ψ and C condensing and compact, respectively. If we consider the families of the fields $I - G_\lambda$, $\lambda \in \Lambda$, generated by the mappings $G : X \times \Lambda \longrightarrow Kv(E)$, condensing or compact, then the collection of such families will be denoted by $D_\psi(X, X_1; \Lambda)$, $C(X, X_1; \Lambda)$, the condition $X_1 \cap \mathrm{Fix}\, G_\lambda = \emptyset$, $\lambda \in \Lambda$ being fulfilled. Let us impose an additional condition (an analogue of semi-additivity [27]) on the measure of non-compactness ψ : c_4) $\psi(M_1 \cup M_2) \leqslant \mathrm{Sup}\{\psi(M_1), \psi(M_2)\}$ $\forall M_1, M_2 \subset E$, assuming that S is a linear structure [42] with $S_+ = K$. This will allow us to denote homotopic classes of a condensing vector field.

The fields $I - F_0$ and $I - F_1$ from $D_\psi(X, X_1)$ (or from $C(X, X_1)$) are called homotopic if there exists a family of fields $I - G_\lambda$ from $D_\psi(X, X_1; \Lambda)$ (from $C(X, X_1; \Lambda)$, respectively) such that $\Lambda = [0,1]$ and $G_{\lambda=0} = F_0$, $G_{\lambda=1} = F_1$. Denote by $[I - F]_\psi$, $[I - F]_C$ the homotopic classes of the field $I - F$ in D_ψ and C, respectively (they are reasonably defined due to the condition of semi-additivity of the measure ψ).

<u>Theorem 1.7</u>. There exists a bijective correspondence $[I - F]_\psi \rightarrow [I - \widetilde{F}]_C$ of the sets of homotopic classes in $D_\psi(X, X_1)$ and in $C(X, X_1)$ where $(I - F) \in D_\psi(X, X_1)$, $(I - \widetilde{F})$ $C(X, X_1)$ and $(I - \widetilde{F}) \in \in [I - F]_\psi$.

We restated Theorem 2.10 from [38] (see also [14]). For the case of Banach space E and single-valued mappings the theorem is proved by Yu.I.Sapronov, when $X = \overline{\Omega}$, $X_1 = \partial\Omega$, and ψ is Hausdorff measure of non-compactness.

This theorem states the equivalence of homotopic structures of the spaces $D_\psi(X, X_1)$ and $C(X, X_1)$, in particular, the spaces $D_\psi(\overline{\Omega}, \partial\Omega)$ and $C(\overline{\Omega}, \partial\Omega)$ of condensing and completely continuous vector fields on $\overline{\Omega}$ in LCS E and, consequently, the equivalence of their topological characteristics. For example, if $I - F \in D_\psi(\overline{\Omega}, \partial\Omega)$ and $I - F$ is a representative of a homotopic class $[I - F]_\psi$ from $C(\overline{\Omega}, \partial\Omega)$, then it is natural to define a local degree of the field $I - F$ by the equality $\deg(I - F, \overline{\Omega}, \mathcal{O}) = \deg(I - \widetilde{F}, \overline{\Omega}, \mathcal{O})$.

where the right-hand side is Leray-Schauder degree; the rotation and full Leray index are defined analogously.

For the Proof of Theorem 1.7. see [38 , 14 , 20]; it follows from Theorem 2.4. of § 2 as a particular case.

We stated the "principle of a bijective correspondence" for condensing vector fields; in [38 , 14] it is proved for a widerclass of f.c.c. mappings.

2. A-condensing mappings and the generalized principle of bijective correspondence.

1. The works by V.G. Zvyagin, V.T. Dmitrienko [1,2] generalize Sapronov's principle of a bijective correspondence (see § 1,5) on the class of A-condensing transformations A - g (by the authors' terminology, g is an "A-condensing perturbation") and its application to the degree theory. Here their result is generalized for the case of linear topological spaces and multivalued mappings g and the construction of the corresponding topological characteristics is outlined. We reduce the problem to the principle of a compact contraction of Yu.G. Borisovich, Yu.I. Sapronov (see § 1, Theorem 1.1.) by the method suggested by the author in [8,5,9] .

Consider LCS E(complete and Hausdorff) and the mappings $A:\overline{\Omega}\longrightarrow E$ (single-valued), $g:\overline{\Omega}\longrightarrow E$ (single- or multivalued) from the closure of an arbitrary set Ω in LCS L. Such pair $(A,g):\overline{\Omega}\longrightarrow E$ gives the mapping (vector field) $f = A - g : \overline{\Omega} \longrightarrow E$.

Definition 2.1. The pair (A,g) is called A-condensing if the mapping $F = g \circ A^{-1}$ (multivalued, generally speaking), acting from the image $A(\overline{\Omega})$ into E, is condensing with respect to the measure of non-compactness. (Here we consider the measure ψ , satisfying the conditions of paragraph 4, Section 1).

Definition 2.2. A-condensing pair is called admissible if the image of F on any compact set in $A(\Omega)$ (having the induced topology from E) is precompact.

Thus, if A-condensing pair (A,g) is admissible, then the mapping $F : Y \longrightarrow E$, where $Y = A(\overline{\Omega})$ is a f.c.c. mapping, since all the conditions of Theorem 1.4.(Remark 1.4.) are fulfilled.

If $g:\Omega\times\Lambda\to E$ gives the family of the mappings $g_t = g(\cdot,t):\overline{\Omega}\longrightarrow E$, then we shall speak about the family of the pairs (A,g_t), $t\in\Lambda$.

Definition 2.3. The family (A, g_t) is called A-condensing, if $F_t = g_t \circ A^{-1} : Y \longrightarrow E$ is a condensing family of mappings and it is called admissible, if an image of Ψ on a compact set, is precompact.

Let us recall that $\Psi(y) = F(y, \Lambda)$, where $F(y,t) = F_t(y)$.

Corollary 2.1. A-condensing admissible family of pairs generates f.c.c. family $F_t = g_t \circ A^{-1}$, $t \in \Lambda$.

Vector field $f = A - g$ (the family $f_t = A - g_t$) will be called admissible A-condensing, if the pair (A,g) (family (A,g_t), respectively) has the analogous property.

If is a f.c.s. for the family $F_t = g_t \circ A^{-1}$, then we shall call it a f.c.s. for the family of the pairs (A,g_t) and the family of the mappings $f_t = A - g_t$.

Let Φ be a f.c.s. for the mapping $F : g \circ A^{-1} : Y \longrightarrow E, Y = A\bar{\Omega}$; let us consider $\tilde{\Phi} = A^{-1}\Phi$, we have $\tilde{\Phi} \subset \bar{\Omega}$.

The set $\mathrm{Fix}(g \circ A^{-1}) = \{y \in Y \mid y \in (g \circ A^{-1})(y)\}$ of the fixed sets of the mapping $g \circ A^{-1}$ lies in $Y_\Phi = Y \cap \Phi$ (from the definition of f.c.s.). The set $S(A,g) = \{x \in \bar{\Omega} \mid A(x) \cap g(x) \neq \emptyset\}$ of "coincidences" of mappings (= the set of solutions to the inclusion $A(x) - g(x) \ni 0$ (zero in E)) lies in $\tilde{\Phi}$. We have $\mathrm{Fix}(g \circ A^{-1}) = AS \subset Y_\Phi$.

Let us give the definition of the approximation of the pair (A,g).

Definition 2.4. The pair $(A,\tilde{g})$ is called "p-approximation on $\tilde{\Phi}$" of the pair (A,g) with f.c.s. Φ, if 1) $\tilde{g} : \bar{\Omega} \longrightarrow \Phi$ ($\tilde{g}$ is compact) 2) $g(x) \subset \overline{co}g(x) + V_p$, $x \in \tilde{\Phi}$, where V_p is an absolutely convex neighbourhood of $0 \subset E$, generated by the continuous semi-norm p.

Definition 2.5. The pair $(A,\tilde{g})$ is called an "approximation on $\tilde{\Phi}$" of the pair (A,g) if the conditions of Definition 2.4. are fulfilled and condition 2) is replaced by condition 2') $\tilde{g}(x) \subset \overline{co}g(x)$, $x \in \tilde{\Phi}$.

Let us introduce the mappings $\overline{co}g$, $\overline{co}g + V$ (V is a set in E) by the equations $(\overline{co}g)(x) = \overline{co}(g(x))$, $(\overline{co}g + V)(x) = \overline{co}(g(x)) + V$.

Consider the linear homotopy

$$G_t(x) = (1 - t)g(x) + t\tilde{g}(x), \qquad t \in [0,1]. \tag{3}$$

Connecting g with $\tilde{g}$, assume that the family $(A,g)_\tau$, $\tau \in [0,1]$ is a linear homotopy connecting the pairs (A,g) and $(A,\tilde{g})$. From Definitions 2.4., 2.5 and equality (3) we get the inclusions

$$\mathrm{co}\;G_{\tau}(x)\subset\overline{\mathrm{co}}(g(x)\cup\phi),\;G_{\tau}(\widetilde{\Phi})\subset\Phi,\;(\overline{\mathrm{co}}G_{\tau})(\widetilde{\Phi})\subset\Phi,$$
$$S(A,\overline{\mathrm{co}}\;G_{\tau})\subset\Phi,\;(x,\tau)\subset\overline{\Omega}\times[0,1] \tag{4}$$

whence follows, in particular, that Φ is a f.c.s. for the family (A,G_{τ})(A,$\overline{co}G_{\tau}$). From conditions 2),2') of the "approximations" (A,g) follow also the corresponding approximational inclusions:

$$G_{\tau}(x)\subset\overline{\mathrm{co}}\;g(x)+Vp \tag{5}$$

$$G_{\tau}(x)\subset\overline{\mathrm{co}}\;g(x),\;(x,\tau)\in\widetilde{\Phi}\times[0,1] \tag{5'}$$

Inclusions (5), (5') characterize "proximity" of the approximation G_{τ} to g on the set $\widetilde{\Phi}$. From (5), (5') we get

$$S(A,\overline{co}G_{\tau})\subset S(A,\overline{co}g+Vp),\mathrm{Fix}((\overline{co}G_{\tau})\circ A^{-1})\subset\mathrm{Fix}\left[(\overline{co}g+Vp)\circ A^{-1}\right] \tag{6}$$

$$S(A,\overline{co}G_{\tau})\subset S(A,\overline{co}g),\;\mathrm{Fix}((\overline{co}G_{\tau})\circ A^{-1})\subset\mathrm{Fix}\left[(\overline{co}g)\circ A^{-1}\right] \tag{6'}$$

If now (A,g_t) is a f.c.c. on Φ family of the pairs, then we shall obtain the definition of the family of "p-approximations"(A,g_t), expanding conditions 1),2),2') to every g_t, $t\in\Lambda$ and by formula (3) we get the family $G_{\tau,t}$ depending on the parameters $(t,\tau)\in\Lambda\times[0,1]$ for every value of which we have analogues of (4),(5),(5'),(6),(6'), where g is replaced by g_t.

There is a theorem on the existence of such approximations.

Theorem 2.1. Let (A,g_t) be an admissible condensing family on $\overline{\Omega}\times\Lambda$ and let Φ be its f.c.s. Then there is a linear homotopy (A,$G_{\tau,t}$) on $\overline{\Omega}\times\Lambda\times[0,1]$, connecting the families (A,g_t) (τ=0), (A,$\widetilde{g}_t$) (τ = 1) and given on $\overline{\Omega}\times\Lambda\times[0,1]$. (A,$g_t$) is a "p-approximation on $\widetilde{\Phi}$" with a given p, when E is separable complete LCS and it is an "approximation on $\widetilde{\Phi}$" when E is a metrizable LCS or a Banach space. The family (A,$G_{\tau,t}$) is an admissible condensing family on $\overline{\Omega}\times\Lambda\times[0,1]$ with a f.c.s. Φ (like the family (A,$\overline{co}G_{\tau,t}$)).

Proof. For any continuous semi-norm p on E it is possible to construct (see [14 , 15]) a quasiretraction $\rho_p : E\longrightarrow\Phi\cap E'$ (E' is a finite-dimensional subspace in E, $E'\cap\Phi\neq\emptyset$) which is continuous and satisfies the condition

$$p(x-\rho_p(x))<1,\quad x\in\Phi \tag{7}$$

Then assume

$$\widetilde{g}_t(x) = \rho_p \circ g_t(x),\ (x,t) \in \Omega \times \Lambda \times [0,1], \qquad (8)$$

It is easy to check that $\widetilde{g}_t$ satisfies the conditions of Definition 2.4. at every $t \in \Lambda$. In the case of a metrizable LCS E assume

$$\widetilde{g}_t(x) = \rho \circ g_t(x),\ (x,t) \in \overline{\Omega} \times \Lambda \qquad (9)$$

where ρ is Dugundji retraction on Φ (i.e. $\rho : E \longrightarrow \Phi$ is continuous, $\rho|_\Phi = 1_\Phi$); then $\widetilde{g}_t(x) = g_t(x)$, $(x,t) \in \widetilde{\Phi} \times \Lambda$ by virtue of a f.c.c. F_t on $\widetilde{\Phi}$, whence follows 2') of Definition 2.5.

Denote $G_\tau(x,t)$ by formula (3), replacing $g, \widetilde{g}$ for $g_t, \widetilde{g}_t$. It remains to check the admissibility and condensability of the families $(A, G_{\tau,t})$, $(A, coG_{\tau,t})$. Let $F_{\tau,t} = G_{\tau,t} \circ A^{-1} : Y \longrightarrow E$, $Y = A\overline{\Omega}$ and let

$$\Psi(y) = \bigcup_{(t,\tau)\in\Lambda\times[0,1]} F_{\tau,t}(y) = \bigcup_{\tau\in[0,1]} G_\tau(A^{-1}y, \Lambda) \qquad (10)$$

Then we get

$$G_\tau(A^{-1}y,\Lambda) \subset \bigcup_{(x,t)} co\,\{g_t(x) \cup \widetilde{g}_t(x)\} \subset co\,\{\Psi_g(y) \cup \Phi\} \qquad (11)$$

where $\Psi_g(y)$ is a corresponding mapping Ψ for a condensing family (A, g_t) and $(x,t) \in (A^{-1}y) \times \Lambda$. For $M \subset Y$ from (11) we get

$$\Psi(M) \subset \bigcup_{y\in M} co\,\{\Psi_g(y) \cup \Phi\} \subset co\,\{\Psi_g(M) \cup \Phi\},$$

whence with the help of axioms $c_0) - c_2)$ of the measure of non-compactness ψ we deduce

$$0 \leq \psi(\Psi(M)) \leq \psi[co\,\{\Psi_g(M) \cup \Phi\}] = \psi[\Psi_g(M) \cup \Phi] = \psi[\Psi_g(M)] \qquad (12)$$

From (12) it follows that the condensability of Ψ_g involves the condensability of Ψ and the admissibility of the family (A, g_t) involves pre-compactness of $\Psi(K)$ for every compact $K \subset Y$. Since the last statement does not depend on the choice of g_t (8) or (9), then Theorem 2.1. is proved (for the family $(A, co\,G_{t,\tau})$ the reasonings are analogous and (11) is filfilled).

2. Let us investigate a homotopic structure of condensing pairs.

We shall assume that ψ is semi-additive (i.e. it satisfies axiom Φ_4)). Usually the homotopies of the pairs (A,g) are considered under the additional condition $S(A,g) \cap X_1 = \emptyset$, where $X_1 \subset \overline{\Omega}$ is a fixed closed subset (e.g. $X_1 = \partial\Omega$). When analysing multivalued mappings one has to consider **four** types of such conditions (a bar over g means the closure of images, $\overline{co}$ is a convex closure):

$$(A - \overline{g})\, X_1 \cap \{0\} = \emptyset \qquad (13),$$

$$(A - \overline{co}g)\, X_1 \cap \{0\} = \emptyset \qquad (14)$$

called "0-separability on X_1", "convex 0-separability on X_1" respectively and for the neighbourhood of zero in L generated by a continuous semi-norm q on E ;

$$(A - \overline{g})\, X_1 \cap V_q(0) = \emptyset \qquad (15),$$

$$(A - co g)\, X_1 \cap V_q(0) = \emptyset \qquad (16)$$

called "V-separability on X_1", "convex V-separability on X_1". We are also interested in the fulfilment of conditions (13)-(16) for the homotopy $G_{\tau,t}$ constructed in Theorem 2.1. (which is a f.c.c. on Φ).

Lemma 2.1. Let (A,g_t) be an admissible condensing family on $\overline{\Omega}\times\Lambda$, at every t satisfying one of the conditions (13)-(16) and let Φ be its f.c.s. and let $(A,G_{\tau,t})$ be a linear homotopy from Theorem 2.1. Then

1. (the case of metrizable E) a) if the mapping g_t has convex values, then conditions (13), (14) are identical and hold for the family $(A,G_{\tau,t})$ on X_1; conditions (15)≃(16) hold for $(A, G_{\tau,t})$ on $X_1 \cap \Phi$ for $(t,\tau) \in \Lambda \times [0,1]$); b) otherwise condition (14) holds for $(A, G_{\tau,t})$ on X_1.

2. (the case of non-metrizable E). a) g_t has convex values: condition (15) holds for $(A,G_{\tau,t})$ on $\Phi \cap X_1$, but with another neighbourhood Vp, such that $\overline{V}p + \overline{V}p \subset Vq$ and it guarantees condition (14) for $(A,G_{\tau,t})$ on the whole X_1; b) images g_t are not necessarily convex: condition (16) holds for $(A,G_{\tau,t})$ on $\Phi \cap X_1$ with the neighbourhood Vp and guarantees condition (14) for $(A,G_{\tau,t})$ on the whole X_1.

The proof of the lemma's statements is based on the relations (4), (5), (5'), (6),(6') and on Theorem 2.1. and is presented to the reader as a technical exercise.

Consider admissible condensing mappings **with continuous** g

$$f = A - g : (\bar{\Omega}, X_1) \longrightarrow (E, E \setminus 0) \qquad (17)$$

satisfying condition (14) of "convex $()$ -separability" on X_1 in the case of metrizable LCS E and the condition of "convex V-separability" **for** (A, G_τ) **on** $X_1' = X_1 \cap (A^{-1}\Phi_1)$ **for compact** $\Phi_1 \subset E$ **in non-metrizable case.** In the case of the family $f_t = A - g_t$ of such mappings conditions (14), (16) are supposed to be fulfilled for all (t, τ) (with Vq independent from (t, τ)).

Denote the class of such mappings f (of the families f_t **with continuous** g_t) by $D_\psi(\bar{\Omega}, X_1)$ (**by** $D_\psi(\bar{\Omega}, X_1, \Lambda)$); here ψ is a symbol of the measure of non-compactness. If $X_1 = \emptyset$, then denote $D_\psi(\bar{\Omega}, X_1; \Lambda)$, $D_\psi(\bar{\Omega}, X_1)$ by $D_\psi(\bar{\Omega}, \Lambda)$, $D_\psi(\bar{\Omega})$, respectively. If $\Lambda = 0,1$, then $D_\psi(\bar{\Omega}, X_1, [0,1])$ is a class of homotopy families f_t, $t \in [0,1]$. The product of two homotopies f_t', f_t'' from $D_\psi(\bar{\Omega}, X_1; [0,1])$ is again a homotopy by virtue of semi-additivity of the measure of non-compactness ψ, that is why the class $D_\psi(\bar{\Omega}, X_1)$ is divided into homotopic classes $[A - g; \bar{\Omega}, X_1]_\psi$ (with representatives $f = A - g$). In $D_\psi(\bar{\Omega}, X_1)$ we shall point out a subclass $D_c(\bar{\Omega}, X_1)$ of the mappings $f = A - k$ with compact $k : \bar{\Omega} \longrightarrow E$; similarly in $D_\psi(\bar{\Omega}, X_1; \Lambda)$ we shall distinguish a subclass $D_c(\bar{\Omega}, X_1; \Lambda)$ of the families $f_t = A - k_t$ with compact $k : \bar{\Omega} \times \Lambda \rightarrow E$; $D_c(\bar{\Omega}, X_1)$ is also divided into homotopic classes $[A - k; \bar{\Omega}, X_1]_c$. There is a natural inclusion $i : D_c(\bar{\Omega}, X_1) \rightarrow D_\psi(\bar{\Omega}, X_1)$; the mapping $j : D_\psi(\bar{\Omega}, X_1) \rightarrow D_c(\bar{\Omega}, X_1)$ is defined by the equality $j(A - g) = A - G_{\tau=1}$, where G_τ is a linear homotopy (3) between g and $k = \tilde{g}$ constructed with respect to the chosen f.c.s. Φ for $f = A - g$.

<u>Lemma 2.2</u>. The mappings i, j induce the mappings of homotopic classes

$$\left\{[A - g; \bar{\Omega}, X_1]_\psi\right\} \underset{i_*}{\overset{j_*}{\rightleftarrows}} \left\{[A - k; \bar{\Omega}, X_1]_c\right\} \qquad (18)$$

<u>Proof</u>. It is obvious that if $f_t = A - k_t$ is a homotopy in D_c, $t \in [0,1]$, then it is a homotopy in D_ψ, which means the correctness of the **map** i_*. If $f_t = A - g_t$ is a homotopy in D_ψ, then by Theorem 2.1. we shall construct the homotopy $A - G_{\tau,t} = f_{\tau,t}$ from f_t to $\tilde{f}_t = A - \tilde{g}_t$, $(t, \tau) \in [0,1]^2$. In the metrizable case condition (14) holds for the homotopy $f_{\tau,t}$ (case 1 of Lemma 2.1) and, consequently, for the family $f_{\tau=1,t} = \tilde{f}_t$. In a non-metrizable case condition (16) of a "convex V-separability" on $\Phi \cap X_1$ holds for

the homotopy $(A, G_{\tau,t})$ by Lemma 2.1.(Φ is f.c.s. of f_t), in particular, for $\tilde{f} = (A, G_{\tau=1,t}), t \in [0,1]$. One can show for each compact $\Phi_1 \subset E$ that $\tilde{f}_t$ satisfies condition (16) on $X_1' = X_1 \cap (A^{-1}\Phi_1)$. Then it follows that the homotopy f_t belongs to a single homotopic class in D_c.

It remains to establish that the class from the image j_* does not depend on the choice of f.c.s. Φ. This is stated by

Lemma 2.3. If Φ_0 and Φ_1 are two f.c.s. for $f_t \in D_\psi(\bar{\Omega}, X_1; \Lambda)$ and ${}^0G_{\tau,t}$, ${}^1G_{\tau,t}$ are the corresponding linear homotopies, then ${}^0\tilde{f}_t = A - {}^0G_{\tau=1,t}$, ${}^1\tilde{f}_t = {}^1G_{\tau=1,t} - A$ lie in a single homotopic class in $D_c(\bar{\Omega}, X_1, \Lambda)$.

For the proof it is sufficient to construct a f.c.s. Φ for the family f_t, containing $\Phi_0 \cup \Phi_1$, the corresponding approximation $G_{\tau,t}$ with a sufficiently small semi-norm p, its restriction $G_{\tau=1,t}$ and linear homotopies ${}^iG_{s,t} = (1-s)G_{\tau=1,t} + s\,{}^iG_{\tau=1,t}$, $0 \leq s \leq 1$, $i = 0,1$, for which on $X_1' = X_1 \cap (A^{-1}\Phi_1)$ a "convex V-separability" of the mapping ${}^if = A - {}^iG_{s,t}$, $i = 0,1$ with a sufficiently small neighbourhood $V \subset V_p$ at any s,t is checked.

Remark 2.1. The condition of "convex V-separability" on $\tilde{\Phi} \cap X_1$, where Φ is a f.c.s. for a f.c.c. of the mapping $f = A - g$, involves the condition of "convex $\mathcal{O}$-separability" on X_1 (condition (14)).

In fact, by Lemma 2.1.(2.) condition (16) on $\tilde{\Phi} \cap X_1$ for f is preserved for $f_{t,t}$ on $\tilde{\Phi} \cap X_1$, and condition (14) on X_1 follows from the fundamentality of Φ for the pair $(A, \overline{co}G_{\tau,t})$ (see (4)).

Lemma 2.4. The mappings i_*, j_* are mutually inverse.

Proof. If $j(A - g) = A - g = A - G_{\tau=1}$, then $A - G_\tau$, $\tau \in [0,1]$ is a homotopy in D_ψ, whence $[A - \tilde{g}]_\psi = [A - g]_\psi$ and $i_*[A - g]_c = [A - g]_\psi$, consequently $i_* \circ j_* = 1$. Let then $i_*[A - k]_c = [A - k]$ and $A - g_0 \in [A - k]_\psi$. For the homotopy $A - g_t$, connecting $A - g_0$ and $A - k$ in D_ψ, we shall construct, by Theorem 2.1., the homotopy $f_{\tau,\tau} = A - G_{\tau,t}$. Let $G_{\tau=1,t=0} = k_0$, $G_{\tau=1,t=1} = k_1$ ($G_{\tau=0,t=1} = k$ by the definition of $G_{\tau,t}$).

Then $j_*[A - k]_\psi = [A - k_0]_c$ by the definition of j_*, but $A - k_0$, $A - k_1$ are homotopic in D_c (homotopy $G_{\tau=1,t}$), $A - k_1$, $A - k$ are also homotopic in D_c, whence $j_*[A - k]_\psi = [A - k]_c$, i.e. $j_* \circ i_* = 1_{D_c}$. The lemma is proved.

Lemma 2.4 is a "weakened variant" of the principle of a bijective correspondence of the homotopic classes in D_ψ and D_c. Without additional topological conditions on the mappings $f = A - g$ the separations of D_ψ, D_c into homotopic classes may prove to be trivial. It depends on the homotopic properties of the space $P(E) \setminus 0$, in which acts the mapping $f : X_1 \longrightarrow E \setminus 0$. Even in the finite-dimensional situation $E = R^n$, $\partial\Omega = S^{n-1}$, in the class of the mappings $f : S^{n-1} \longrightarrow R^n \setminus 0$ under the condition $f : S^{n-1} \longrightarrow K(R^n \setminus 0)$ there is only one homotopic class because $\pi_m[K(R^n \setminus 0)]$ is trivial for all $m \geqslant 0$; here $K(R^n \setminus 0)$ is considered in an exponential topology. If $f : S^{n-1} \longrightarrow Kv(R^n \setminus 0)$, then homotopic classes of such mappings bijectively correspond to $\pi_{n-1}(S^{n-1}) \cong Z$. This follows from the equality $\pi_m(S^{n-1}) = \pi_m(Kv(R^n \setminus 0))$, $m \geqslant 0$. $Kv(R^n \setminus 0)$ may be considered both in the exponential topology $\mathrm{Exp}\, Kv$ and in the upper semi-finite topology $\varkappa Kv$ (naturally, f is considered continuous in the chosen topology). Since the problem under consideration is important for the theory of mappings, we suggested that it should be included in the thesis by N.Benkafadar who, together with B.D.Gel'man, obtained the following results (see [16 , 50]):

Theorem 2.2. Let X be a subset of a Banach space E. Then: 1) $\mathrm{Exp}\, Kv(X)$ and X are homotopically equivalent; 2) $\pi_m[\varkappa Kv(X)] = \pi_m[X]$, $m \geqslant 0$, if $X \subset T$ (T is a closed convex set in E) is both open and pathwise connected in T.

If X is a topological Hausdorff space, pathwise connected and locally contractible, then $\mathrm{Exp}\, K(X)$ (a space of compact not necessarily convex set in X) is pathwise connected and $\pi_m(\mathrm{Exp}\, K(X))$ are trivial for all m 0.

There exist several classes of mappings $f = A - g$ for which the set D_c has a non-trivial homotopic structure. We shall consider them in the next item. Here we shall give an abstract statement of the principle of bijection.

We say that the mapping $f : \overline{\Omega} \times J^k \longrightarrow E$, where $J = [0,1]$ is α - mapping if it satisfies some condition (α) having the following properties:

α_0) restriction $f_{|Q}$ on any subspace $Q \subset X_1 \times J^k$ satisfies condition (α);

α_1) the property of α-homotopy of α-mappings is equivalence;

α_2) linear homotopy between two α-homotopies is α - homotopy;

α_3) "approximation" or "V-approximation" of α - mapping f (α - family f_t) is an α - mapping (α - family).

Denote by $D^{\alpha}_{\psi}(\overline{\Omega}, X_1)$, $D^{\alpha}_{c}(\overline{\Omega}, X_1)$ the collections of α - mappings from the classes D_ψ , D_c, respectively.

Theorem 2.3. (principle of bijection of A - condensing vector fields). The mappings i_* , j_* between homotopic classes in $D_\psi(\overline{\Omega}, X_1)$, $D_c(\overline{\Omega}, X_1)$ induce the mappings $\hat{i}_*$, $\hat{j}_*$ of α - homotopic lasses

$$\left\{\left[A - g, \overline{\Omega}, X_1\right]^{\alpha}_{\psi}\right\} \underset{\hat{i}_*}{\overset{\hat{j}_*}{\rightleftarrows}} \left\{\left[A - K, \overline{\Omega}, X_1\right]^{\alpha}_{c}\right\} \quad ; \quad (19)$$

he mappings $\hat{i}_*$, $\hat{j}_*$ are mutually inverse, i.e. they effect a bi-ective correspondence of α- homotopic classes in $D^{\alpha}_{\psi}(\overline{\Omega}, X_1)$ and $c^{\alpha}(\overline{\Omega}, X_1)$.

The statement follows immediately from the weakened principle of ijection (Lemma 2.4) and the properties of condition α .

One of the main problems in the theory of nonlinear mappings is he problem of the construction of topological characteristics of the ector fields $f = A - g$ preserved under homotopies and connected ith the existence of the solutions to inclusion (20) with a non-com-act operator g

$$A(x) - g(x) \ni 0 \qquad (20)$$

n our case can be reduced to the analogous problem with a compact perator g .

Corollary 2.2. (principle of compact contraction). If the condi-ions of the principle of bijection are fulfilled for the class of $^{\alpha}_{\psi}(\overline{\Omega}, X_1)$ of A - condensing vector fields and if for the class $c^{\alpha}(\overline{\Omega}, X_1)$ some topological characteristic was constructed, then it s extended to the class D^{α}_{ψ} by the equality

$$\gamma(A - g, \overline{\Omega}, X_1) = \gamma(j(A - g), \overline{\Omega}, X_1) \qquad (21)$$

n fact, we get an invariant of a homotopic class

$$\left[A - g, \overline{\Omega}, X_1\right]^{\alpha}_{\psi}$$

3. Let us realize various variants of the principle of bijection. onsider at first Hausdorff LCS L and E, where E is complete. Let $\tilde{\Omega}$ e a closure of an open set in L, $X_1 = \partial\Omega$. Let the measure of on-compactness ψ on E satisfy axioms $c_0) - c_4)$.

Consider, as condition (α), upper semicontinuity of the multi-alued mapping $f : X \longrightarrow K(Y)$ of the topological spaces X, Y .

The validity of axioms $\alpha_0)$, $\alpha_1)$ is obvious; $\alpha_2)$ follows rom the upper semicontinuity of a linear combination of vectors as a unction of these vectors and coefficients (see, e.g. [15] , Section .3).The property $\alpha_3)$ means upper semicontinuity of $\tilde{g}_t(x) = \rho \circ g_t(x)$ on

$\bar{\Omega} \times \Lambda$, which follows from the continuity of a quasiretraction (retraction) ρ(see[14, item 1.2.37])and semicontinuity of a superposition([15 , item 1.3.13]).

Consider the mappings (vector fields) $f = A - g$ from $D_\psi(\bar{\Omega})$, $D_c(\bar{\Omega})$, satisfying condition (α); denote by $D^*_\psi(\bar{\Omega}, \partial\Omega)$, $D^*_c(\bar{\Omega}, \partial\Omega)$ the classes of **upper** semicontinuous A-condensing (A-compact, respectively) vector fields, satisfying the condition of "convex O -separability" on the boundary $\partial\Omega$:

$$(A - \overline{co}\, g)(\partial\Omega) \cap \{0\} = \emptyset \tag{14'}$$

and then fix a number of properties of the mapping A.

Lemma 2.5. $A : \bar{\Omega} \longrightarrow E$ be a proper and continuous mapping, $f = A - g \in D^*_\psi(\bar{\Omega}, \partial\Omega)$. Then on every intersection $(\partial\Omega) \cap (A^{-1}\Phi_1)$ where $\Phi_1 \in E$ is compact, condition (16) of "convex V-separability" for (A, G_t) is fulfilled.

Proof. By virtue of the completeness of E, the mapping $\overline{co} g : \Omega \rightarrow E$ is also upper semi-continuous ([15 , 1.3.21]) ; the set $\tilde{\Phi}_1 = A^{-1}\Phi_1$ is compact by virtue of the properness of A, the mappings A, $(\overline{co}\, g) : (\partial\bar{\Omega} \cap \tilde{\Phi}) \rightarrow \Phi_1$, respectively, are continuous and upper semi-continuous; consequently, $A - \overline{co}\, g : (\partial\bar{\Omega}) \cap \tilde{\Phi}_1 \rightarrow E$ is upper semi-continuous and its image lies in K(E). Then there exists Vp-neighbourhood of O , separating O from the compact image $(A - \overline{co}\, g)[(\partial\Omega) \cap \tilde{\Phi}_1]$, i.e. condition (16) on $X_1 = \tilde{\Phi}_1 \cap (\partial\Omega)$ is fulfilled. The case of (A, G_t) is analogous. Lemma is proved.

Corollary 2.3. Under the condition of properness and continuity of the mapping A the classes $D^*_\psi(\bar{\Omega}, \partial\Omega)$, $D^*_c(\bar{\Omega}, \partial\Omega)$ are α - classes, figuring in Theorem 2.3.

Let us strengthen property (α) and consider upper semi-continuous mappings g with the images from Kv(x) (convex compact). If "V - approximation" $\tilde{g}$ has the form $\tilde{g}(x) = \overline{co}(\rho \circ g(x))$, then axioms $\alpha_0) - \alpha_3)$ hold also in the class of the mappings with convex values; it is easy to see that the classes $D^{*V}_\psi(\bar{\Omega}, \partial\Omega)$, $D^{*V}_c(\bar{\Omega}, \partial\Omega)$ of the mappings with convex values from D^*_ψ, D^*_c are also α - classes (under the condition of properness and continuity of A) and condition (14') transforms into the condition

$$(A - g)(\partial\Omega) \cap \{0\} = \emptyset \tag{13'}$$

Consider in $D^{*V}_\psi(\bar{\Omega}, \partial\Omega)$, $D^{*V}_c(\bar{\Omega}, \partial\Omega)$ the homotopic classes $[A - g, \bar{\Omega}, \partial\Omega]^{*V}_\psi$, $[A - k, \bar{\Omega}, \partial\Omega]^{*V}_c$, respectively.

From Theorem 2.3. immediately follows

Theorem 2.4. (principle of bijection for the mappings with convex values). Let A be proper and continuous. Then the homotopic classes of upper semicontinuous A-condensing and A - compact vector fields $\tilde{f} = A - k$ (where k is upper semi-continuous with compact convex values) are in a natural bijective correspondence

$$\left\{\left[A - g, \overline{\Omega}, \partial\Omega\right]^{*V}_{\Psi}\right\} \underset{\hat{i}_*}{\overset{\hat{j}_*}{\rightleftarrows}} \left\{\left[A - k, \overline{\Omega}, \partial\Omega\right]^{*V}_{c}\right\} \quad (19')$$

where $\hat{i}_*$, $\hat{j}_*$ are mutually inverse mappings induced by the mappings i, j.

Theorem 2.5. For the mappings $f = A - g : \overline{\Omega} \to E$ into a complete metrizable LCS or Banach space E ($\Omega \subset L$ - LCS) the statement of Theorem 2.4. holds without the condition of properness and continuity of the mapping A **for admissible f.**

In fact, when constructing "approximation" $\tilde{g}(x) = \overline{co}\,(\rho \circ g(x))$ with Dugundji retraction one can assume that $Vp = 0$ in all approximational valuations which makes us free from checking V-separability of the image (A - g) on $X_1 \cap \tilde{\Phi}_1$ (Lemma 2.5.). The rest constructions do not change.

Remark 2. Theorem 2.5. generalizes Theorem 3 of V.G.Zvyagin and V.T.Dmitrienko [1] on A-condensing mappings f = A - g from a Banach space L into a Banach space E, where A,g are single-valued mappings.

On the class $D^{*V}_{\Psi}(\Omega, \partial\Omega)$ of vector fields $f = A - g$ with convex values one can also define, by equality (21), topological characteristics in the conditions of Theorem 2.4. and the additional condition on A: A_1) $A : \overline{\Omega} \longrightarrow E$ is Fredholm of the class $\Phi_n C^1$, $n \geqslant 0$, the mapping of Banach spaces L,E; or A_2) $A : \overline{\Omega} \longrightarrow E$ is a demicontinuous bounded operator, satisfying condition α) of I.V.Skrypnik [49].

In fact, in case A_1) for the operators A - k the theory of a topological characteristic is developed [22,23,6,7,47] which generalizes earlier theories with single-valued k and linear operator A ([45, 46]). The theory constructed here on the class $D^*(\overline{\Omega}, \partial\Omega)$ generalizes the results of [1-4, 5, 9,10, 12, 13].

In case A_2) the rotation is constructed [49] for the vector field A - k and can be extended naturally onto the class D^{*V}_{c}.

A detailed analysis of degree theories on the class D_{ψ}^{*v} will be presented in Part II.

Remark. All results of Section 2 (except the case A_2) are valid for the mappings $f = A - g$ acting from an infinite-dimensional manifold modelled by a Banach space or LCS.

REFERENCES

1. Dmitrienko V.T., Zvyagin V.G. Homotopic classification of a single class of continuous mappings. - Mat. zametki, 1982, 31, issue 5, p. 801-812.(in Russian)

2. Zvyagin V.G. To the theory of generalized condensing perturbations of continuous mappings. - In: Topological and geometrical methods in mathematical physics. Voronezh: VGU, 1983, p.42-62.(in Russian)

3. Dmitrienko V.T., Zvyagin V.G. Homotopic classification of generalized condensing perturbations of mappings. - IV Tiraspol' Symposium on General Topology and its Applications. Kishinev: Shtiintsa, 1979, p.32-33.(in Russian)

4. Dmitrienko V.T. Homotopic classification of a single class of multi-valued mappings. - Voronezh University, Voronezh, 1980, Deposited in VINITI, No.2091-80 Dep.(in Russian)

5. Borisovich Yu.G., Zvyagin V.G. Perturbations of nonlinear Fredholm operators and their application to boundary value problems. - In:General theory of boundary problems, Kiev: Nauk.dumka, 1983,. p.35-43.

6. Borisovich Yu.G. To the theory of topological degree of nonlinear Fredholm mappings perturbed by a multi-valued operator. Voronezh University, Voronezh, 1980. Deposited in VINITI No.5026-80 Dep.,7p.

7. Ratiner N.M. To the degree theory of Fredholm mappings of non-negative index. - Voronezh University, Voronezh, 1981, Deposited in VINITI, No.1493-81(in Russian)

8. Borisovich Yu.G. Topology and nonlinear functional analysis. - Uspekhi mat. nauk, 1979, 34, issue 6, p.14-22.(in Russian)

9. Borisovich Yu.G. On topological methods in the problem of solvability of non-linear equations. Trudy.Leningr. mezhdunar. topol. konf., 23-27 August, 1982, L.: Nauka, 1983, p.39-49.(in Russian)

10. Borisovich Yu.G. On topological characteristics of nonlinear Fredholm operators and their perturbations. - In: Optimal management. Geometry and analysis. Tezisy dokl. vsesoyuznoi shkoly, Kemerovo, izd-vo Kem.GU, 1968, p.68.(in Russian)

11. Borisovich Yu.G. Global analysis and solvability of nonlinear boundary value problems. - In: Application of functional methods of the function theory to the problems of mathematical physics. Tezisy dokl. IX Sovetsko-Chekhoslovatskogo soveshchaniya. - Donetsk, 1986, p.17. (in Russian)

12. Borisovich Yu.G.,Ratiner N.M. On new constructions of topological characteristics of nonlinear Fredholm operators. In: XI All-Union Seminar on the Operator Theory in Functional Spaces. Tezisy dokl.p.I, Chelyabinsk, 1986, p.17. (in Russian)

.3. Lecture Notes in Math. 1108. Global Analysis - Studies and Appli-
ations I. Ed. by Yu.G.Borisovich, Yu.E.Gliklikh, Springer-Verlag,
984.

.4. Borisovich Yu.G., Gel'man B.D., Myshkis A.D., Obukhovskiĭ V.V.
opological methods in the theory of fixed points of multivalued mapings. - Uspekhi mat. nauk, 1980, v.35, issue I, p.59-126.(In Russian)

.5. Borisovich Yu.G., Gel'man B.D., Myshkis A.D., Obukhovskiĭ V.V.
ultivalued mappings. - In: Results of science and technology. Math.
nalysis, Moscow, 1982, v.19, p.127-230.(In Russian).

.6. Borisovich Yu.G., Gel'man B.D., Myshkis A.D., Obukhovskiĭ V.V.
ultivalued analysis and operator inclusions. - In: Results of science
nd Technology. Modern problems of mathematics. Latest achievements.,
oscow, 1986, v.29, p.127-230.(IN Russian).

7. Borisovich Yu.G., Sapronov Yu.I.To the topological theory of com-
act contracted-mappings. - In: Trudy seminara po funkts.analizu, Vo-
onezh, VGU, 1969, issue 12, p.43-68. (In Russian).

8. Borisovich Yu.G., Sapronov Yu.I. To the topological theory of con-
ensing operators. - DAN SSSR, 1968, 183, No.1, p.18-20. (In Russian).

.9. Sapronov Yu.I. To the algebraic index theory. - Trudy seminara po
unkts.analizu, Voronezh, VGU, 1969, issue 12, p.143-154. (In Russian)

0. Sapronov Yu.I. To the homotopic classification of condensing mapings. - Trudy matem. fak. VGU, Voronezh, VGU, 1972, issue 6, p.78-
0. (In Russian).

1. Sapronov Yu.I. To the degree theory of nonlinear Fredholm mappings.
In: Trudy NIIM, Voronezh, VGU, 1973, issue XI, p.92-101. (In Rus-
ian).

2. Borisovich Yu.G., Zvyagin V.G., Sapronov Yu.I. Nonlinear Fredholm
appings and Leray-Schauder theory. - Uspekhi mat.nauk, 1977, 32, No.4,
.3-54.(In Russian).

3. Borisovich Yu.G., Zvyagin V.G. On a single topological principle
f solvability of equations with Fredholm operators. - Dokl.AN USSR,
976, ser.A, No.3 (In Russian).

4. Sadovskiĭ B.N. Limiting compact and condensing operators. - Us-
ekhi matem- nauk, 27, issue I(1972), p.81-146.(In Russian).

5. Akhmerov R.R., Kamenskiĭ M.I., Potapov A.S., Sadovskiĭ B.N. Con-
ensing operators. - Itogi nauki i tekhniki, v.18, Moscow, VINITI,
980, p.185-250. (In Russian).

6. Krasnosel'skiĭ M.A., Zabreĭko P.P. Geometrical methods of nonli-
ear analysis, Moscow, Nauka, 1975.(In Russian).

7. Akhmerov R.R., Kamenskiĭ M.I., Potapov A.S.,Rodkina A.E., Sadov-
kiĭ B.N. Measures of non-compactness and condensing operators. - No-
osibirks, 1986. - 265 p.(In Russian).

8. Leray J. Sur les equations et les transformations. - Jour. de Math.
ures et Appl. 9e Serie, 24(1945); pp.201-248.

9. Borisovich Yu.G. On one application of the notion of vector field
otation. - Dokl.AN SSSR, 1963, 153, No.I, p.12-15. (In Russian).

0. Borisovich Yu.G. On relative rotation of compact vector fields in
inear spaces. - In: Trudy seminara po funkts. analizu, Voronezh,
GU, 1969, issue 12, p.3-27.(In Russian).

1. Sadovskiĭ B.N. On measures of non-compactness and condensing ope-
ators. - IN: Problemy matem. analiza slozhnykh sistem, Voronezh,
GU, 1968, issue 2. (In Russian).

32. Nussbaum R.D. The fixed point index and asymptotic fixed-point theorem for k-set-contractions. - Bull. Amer. Math. Soc., 1969, v.75, No.3, p.490-495.

33. Nussbaum R.D. The fixed point index for local condensing maps. - Ann. mat. pura ed appl. , 1971, t.89, p.217-258.

34. Nussbaum R.D. Degree theory for local condensing maps.- J. Math. Analysis and Applic., 1972, v.37, No.3, p.741-766.

35. Zabreĭko P.P., Krasnosel'skiĭ M.A., Strygin V.V. On the principles of rotation invariance. - Izv. vuzov, Matematika, 1972, v.5(120), p.51-57. (In Russian).

36. Obukhovskiĭ V.V. On some principles of fixed point for multivalued condensing operators. - In: Trudy mat. fak., Voronezh, VGU, 1971, Issue 4, p.70-79. (In Russian).

37. Obukhovskiĭ V.V., Gorokhov E.V. To the definition of rotation of a single class of compact contracted multivalued vector fields. - Trudy mat. fak., Voronezh, VGU, 1974, issue 12, p.45-52.(In Russian).

38. Borisovich Yu.G., Obukhovskiĭ V.V. Homotopic properties, theory of rotation and theorems on fixed point for a single class of noncompact multivalued mappings. Voronezh, 1980. Deposited in VINITI, No.5033-80 Dep., 34 p. (In Russian).

39. Evert J. Homotopical properties and the Topological Degree for γ - contraction Vector Fields. - Bull. Acad. Polon. Sci., 1980, 28, No.5-6, p.273-274.

40. Izrailevich Ya.A. On the notion of relative rotation of multivalued vector field. - Trudy seminara po funkts. analizu, Voronezh, VGU, 1969, issue 12, p.111-115. (In Russian).

41. Borisovich Yu.G. Relative rotation of compact vector fields and Lefshetz number. - Trudy seminara po funkts. analizu, Voronezh, VGU, 1969, issue 12, p.28-42 (In Russian).

42. Vulikh B.Z. Introduction into the theory of semi-ordered spaces.- Moscow, 1961, 407 p. (In Russian).

43. Smale S. An infinite dimensional version of Sard's theorems. - Amer. J. Math. , 1965, 87, p.861-866.

44. Elworthy K.D., Tromba A.J. Differential structures and Fredholm maps on Banach manifolds. - Proc. Sympos. Pure Math., AMS, 1970, 15, p.45-94.

45. Shvarts A.S. Homotopic topology of Banach spaces. - Dokl. AN SSSR, 154 (1964), p.61-63. (In Russian).

46. Nirenberg L. Generalized degree and nonlinear problems. - Contribution to Nonlinear Functional analysis. - Acad. Press, New York/London, 1971, p.1-9.

47. Zvyagin V.G. On the existence of continuous branch of eigenfunctions of nonlinear elliptic boundary value problem. - Differentsial'nye uravneniya, 1977, issue 8, p.1524-1527. (In Russian).

48. Zvyagin V.G. On one topological method of investigation of boundary value problems, nonlinear with respect to highest derivative. - In: Granichnye zadachi matematicheskoi fiziki. Kiev: Nauk.dumka, 1981, p.35-37.(In Russian).

49. Skrypnik I.V. Nonlinear elliptic equations of highest order. - Kiev:Nauk.dumka, 1973 (in Russian).

QUALITATIVE GEOMETRICAL THEORY OF INTEGRABLE SYSTEMS. CLASSIFICATION OF ISOENERGETIC SURFACES AND BIFURCATION OF LIOUVILLE TORI AT THE CRITICAL ENERGY VALUES

A.T.Fomenko
Department of Mechanics and Mathematics
Moscow State University
119899, Moscow, USSR

This paper briefly describes the main results of a new theory developed by the author recently. This theory permitted a classification of isoenergetic surfaces of intergrable systems, as well as a rearrangement of Liouville tori at critical Hamiltonian levels.

1. How is the complete integrability (in Liouville's sense) of a Hamiltonian system related to the topology of a phase- or configurational space? A more-or-less complete qualitative description of the behaviour of integral trajectories of a system in the neighbourhood of a single isolated Liouville torus can be obtained from a complete set of integrals in the involution. In appropriate coordinates (action-angle variables), the system trajectories describe a quasiperiodic motion on the torus. However the following question defied explanation until recently: what is the relative arrangement of Liouville tori in a phase space? In which way do they abut one another and how are they rearranged in the neighbourhood of the critical energy values? In other words, how to construct a qualitative theory of topological arrangement, interaction and rearrangement of Liouville tori in integrable Hamiltonian systems?

In this paper, we shall briefly describe a new concept developed by the author to answer some of these questions. We shall give a complete topological classification (and indicate the canonical representation) of all three-dimentional constant-energy (isoenergetic) surfaces of Hamiltonian systems which can be integrated with the help of the Bott integral on four-dimentional symplectic manifolds. The concept of the Bott integral, i.e. the "general position" integral, is introduced in a natural way. This is an integral whose critical points on an isoenergetic surface are arranged in nondegenerate critical submanifolds. Further, we shall prove that in some cases the existence of at least two stable periodic solutions can be guaranteed on three-dimensional isoenergetic surfaces simply on the basis of the information on one-dimensional homologies of these surfaces or the data about the fundamental group. We shall completely describe the structure of singular constant-energy surfaces (i.e. the surfaces corresponding to the critical energy values). As an application of this theory, we shall completely classify all the rearrangements of the general position of Liouville tori in the vicinity of bifurcation diagrams mapping the moment of integrable systems. It turns out that all such rearrangements of tori can be described in an explicit and effective manner. The analogs of these results will be also proved for an arbitrary multidimensional case, [1-3].

To solve these problems, we had to develop a new specific theory of rearrangements of the level surfaces for Hamiltonian system integrals. This theory differs from Morse's normal theory and from Bott's theory for functions with degenerate singularities. We shall then prove that the isoenergetic surfaces of integrable systems have well-defined specific properties which effectively distinguish them from the class of all three-dimensional closed orientable manifolds. In particular, the class of isoenergetic surfaces of integrable systems forms in a certain sense a "meager" subset in the class of all three-dimensional manifolds.

Together with Zieschang and Matveev, we have discovered four ways of representing a class of isoenergetic surfaces, i.e. four different methods of their description (among all three-dimensional manifolds). Deep-rooted links with the problems in the theory of three-dimensional manifolds were revealed in this work.

While constructing this theory, we have used some basic ideas put forth, among others, by S.Smale, R.Bott, S.Novikov, D.Anosov, V.Kozlov, V.Arnold, J.Moser, J.Marsden and A.Weinstein (see [6-20]).

2. Let (M^{2n}, w) be a smooth symplectic manifold (compact or noncompact) and $v = \operatorname{sgrad} H$ be a Hamiltonian system (vector field) on M^{2n} with a smooth single-valued Hamiltonian H. Let the system v be integrable in Liouville's sense. This means that v has a complete set of integrals in involution, i.e. there exist functions $f_1 = H$, $f_2, \ldots, f_n$ such that $\{f_i, f_j\} = 0$ and all functions are functionally independent (almost everywhere) on M. Here $\{f, g\}$ denotes

Poisson's bracket, i.e.

$$\{f, g\} = w\,(\mathrm{sgrad}\ f,\ \mathrm{sgrad}\ g)$$

By $F : M^{2n} \to R^n$ we denote the map of the moment of an integrable system, defined by

$$F(x) = (f_1(x), \ldots, f_n(x)).$$

The point $x \in M^{2n}$ is called the critical point for the map F if the rank $dF(x) < n$. The set of all critical points is denoted by K. Then its image, i.e. the set

$$\Sigma = F(K) \subset R^n$$

is called the bifurcation diagram of the moment map. The points $c \in \Sigma$ are called the critical values of the moment map, while the points $a \in R^n \setminus \Sigma$ are called the noncritical values. Let a be a noncritical value. Then, in view of Liouville's theorem, each connected component of the complete inverse image $F^{-1}(a)$ is homeomorphic to the n-dimensional torus T^n (and is called Liouville's torus). If points $a \in R^n$ slide along a smooth curve, the Liouville tori $F^{-1}(a)$ are deformed inside the manifold M^{2n} and are somenow rearranged when the point a "pierces" the bifurcation set Σ. We shall describe all such rearrangements of general position and classify them.

We shall first construct the theory for the case $n = 2$ and then generalize the results for an arbitrary n.

3. Let M^4 be a four-dimensional (compact or noncompact) symplectic manifold, and v be a Hamiltonian system with a Hamiltonian H, having a second supplementary independent smooth integral f. We shall study the integrability of the system on a single isolated constant-energy surface $Q^3 = \{H = \mathrm{const}\}$. This is associated with the fact that in mechanics, physics and geometry, one frequently encounters situations in which a system is integrable only on one isoenergetic surface and is not integrable on the remaining (even close) surfaces. This effect is encountered in smooth systems but is not observed in analytic systems. Hence the smooth integral f may satisfy a less stringent condition which we shall formulate now. Suppose that a fixed isoenergetic surface Q is given by the equation $H=0$.

We shall actually study below the properties of the function f satisfying the equation $\{H, f\} = \lambda(H)$, where $\lambda(0) = 0$ and the function $\lambda(H)$ is quadratic in H in the vicinity of the value H = 0. For example, we can put $\lambda(H) = \varepsilon H^2$, $\varepsilon \neq 0$. Thus, the integral f must commute with H only on the given surface Q, and must not commute with H outside Q. Henceforth, we shall assume that $Q = Q^3$ is a compact nonsigular isoenergetic surface, i.e. grad $H \neq 0$ everywhere on Q. Since M^4 is always orientable (as a symplectic manifold), the surface Q^3 is automatically orientable in all cases. Let us confine the integral f to Q. The smooth function f obtained in this way will also be denoted by the same letter f.

Definition 1. We shall call the integral f a Bott integral on the isoenergetic surface Q, if its critical points form critical nondegenerate smooth submanifolds in Q.

This means that the Hessian d^2f of the function f is nondegenerate on the planes normal to the critical submanifolds of the function f.

Definition 2. Let γ be a closed integral trajectory of the system v on the surface Q^3 (i.e., γ is a periodic solution). We shall say that γ is stable if some tubular neighbourhood of γ is completely (without gaps) foliated on concentric two-dimensional tori that are invariant relative to the system v.

In other words, all integral trajectories of the system that are close to γ can be "fit into" invariant two-dimensional tori whose common axis is the closed trajectory γ (circle). In particular, a normal two-dimensional disc of a small radius is completely foliated without gaps into concentric circles that are invariant relative to the system v. A study of the number of periodic stable solutions is an important problem of Hamiltonian mechanics.

4. We denote by (M) the class of all closed compact orientable three-dimensional manifolds. We shall denote by (H) the class of all closed compact nonsingular isoenergetic surfaces of Hamiltonian systems that can be integrated with the help of Bott integrals. As was mentioned above, $(M) \supset (H)$. It should be emphasized that it is absolutely not clear a priori whether or not the class (H) coincides with the class (M). In other words, it can be asked whether there are any topological restrictions which forbid any three-dimensional manifolds to act as isoenergetic surfaces of integrable systems. If such restrictions do exist, then the class (H) is smaller than the

class (M) and hence we encounter obstructions to the integrability of Hamiltonians systems in the class of smooth Bott integrals. Such obstructions are indeed known to exist. At first, it should be remarked that the class (H) admits another important representation which we shall describe below.

Let us consider five simple three-dimensional manifolds with a boundary, i.e. five "building blocks" which will be used to construct a new class of closed three-dimensional manifolds by gluing these "building blocks". Let D^n denote an n-dimensional disc.

Type 1. We shall call the direct product $S^1 \times D^2$ a complete torus. Its boundary shall be the two-dimensional torus T^2.

Type 2. We shall call the direct product $T^2 \times D^1$ a cylinder. Its boundary will be two tori T^2.

Type 3. We shall call the direct product $N^2 \times S^1$ an oriented saddle (or "trousers" in a more figurative manner of speech), where N^2 is a two-dimensional disc with two holes (in other words, N^2 is obtained from the sphere S^2 by discarding three non-intersecting discs). The manifold N^2 is homotopically equivalent to a figure of 8 (a combination of two circles), and hence $N^2 \times S^1$ is homotopically equivalent to the direct product of a figure of 8 and a circle. The boundary of $N^2 \times S^1$ is formed by three tori T^2. The manifold $N^2 \times S^1$ can be realized in the three-dimensional space R^3 as follows. We consider a standard complete torus and discard from it two complete tori which circumvent the axis of the basic "large" torus exactly once.

Type 4. Let us realize the manifold N^2 as a disc with two holes which are fixed and labelled as 1 and 2. We consider the nontrivial fibration $A^3 \xrightarrow{N} S^1$ with the circle S^1 as its base and N^2 as a fibre. Obviously, only two nonequivalent fibrations exist over the circle with a fibre, such that the boundary of these fibrations is a union of two-dimensional tori. The first fibration is the direct product $N^2 \times S^1$ (i.e. a manifold of type 3, see above). The second fibration A^3 is characterized by the fact that when the fibre N^2 is translated along the base S^1, it returns to its original position with holes 1 and 2 transposed. Since N^2 is homotopically equivalent to a figure of 8 formed by the two circles 1 and 2, we can also represent types 3 and 4 as follows. From the homotopic point of view, type 3 is a direct product of a figure of 8 and a circle, while in type 4 the figure of 8 moves along the circle (remaining orthogonal to it) in such a way that the two circles interchange places after a complete circumvention. The small neighbourhood of the base circle S^1 is homeomorphic in this case to two Mobius strips which intersect transversally along a common axis. Two tori T^2 form the boundary of A^3. The manifold A^3 can be realized in R^3 as follows. Let us consider in R^3 a complete torus imbedded in the standard way. We drill a thin complete torus inside this large torus in such a way that it coils twice round the generator of the large torus. We shall call the manifold A^3 an unoriented saddle. It is a space of the oriented skew product $N^2 \times S^1$.

Type 5. Let K^2 be a two-dimensional Klein's bottle and K^3 be the space of the oriented skew product of K^2 and a segment, i.e.

$$K^3 = K^2 \tilde{\times} D^1 .$$

The boundary of the manifold K^3 is a torus. Graphically, the manifold K^3 can be represented as follows. We consider the embedding of Klein's bottle K^2 into R^3 and take a tubular normal neighbourhood of this embedded manifold. This gives a three-dimensional manifold with a boundary T^2, embedded in R^3. This is the required manifold K^3. The boundary of the normal tubular neighbourhood of Klein's bottle in R^3 is a two-dimensional torus which is the boundary of K^3.

For brevity, we shall denote the manifolds of types 1-5 by the numbers I, II, III, IV and V.

Statement 1. Manifolds of type 4 and 5, i.e.

$$A^3 = N^2 \tilde{\times} S^1 \quad \text{and} \quad K^3 = K^2 \tilde{\times} D^1$$

are not new from a topological point of view. This means that IV = I + III, i.e. A^3 is obtained by gluing the complete torus $S^1 \times D^2$ with the oriented saddle $N^2 \times S^1$ along some diffeomorphism of the boundary tori. Further, V = 2 I + III, i.e. K^3 is obtained by gluing two complete tori $S^1 \times D^2$ and the oriented saddle $N^2 \times S^1$ along some diffeomorphisms of the boundary tori. Besides, V = I + IV.

Let us consider a new class (Q) of closed compact orientable three-dimensional manifolds obtained by piecing together an arbitrary number of elementary manifolds of types I, II, III, IV and V along arbitrary diffeomorphisms of their two-dimensional boundary tori. In other words, each manifold in class (Q) can be represented (generally, not unambiguously) as a gluing of the type mI + pII + qIII + sIV + rV, where m, p, q, s, r are nonnegative integers, and the sign

"+" indicates gluing along diffeomorphisms of torus boundaries. It is remarkable that the classes (H) and (Q) coincide. This is not a trivial fact, as we shall see later.

Statement 2. Let f be a Bott integral on some three-dimensional compact nonsingular isoenergetic surface Q. Then it can have only three following types of critical submanifolds: (a) circles, (b) two-dimesional tori, and (c) two-dimensional Klein bottles.

Let us recall the concept of the separatrix diagram of the critical submanifold T for a Bott function. Let $x \in T$ be an arbitrary point and $N_x(T)$ be a disc of small radius normal to the submanifold T at the point x. In view of the Bott integral f, its restriction to the normal disc $N_x(T)$ is a normal Morse function with the critical point x having a certain index λ. The separatrix of the critical point x is the integral trajectory of the field grad f, which is entering or leaving x. Let us consider a set of separatices entering the critical point x and leaving it (in the normal disc). The union of all the separatices entering the point x gives a disc of dimensionality λ and is called the incoming separatrix diagram (incoming separatrix disc). The union of outgoing separatrices gives a disc of additional dimensionality and is called the outgoing separatrix diagram (outgoing separatrix disc). Varying the point $x \in T$ and contstructing the incoming and outgoing separatrix discs for each point x, we obtain the incoming and outgoing separatrix diagrams of the entire critical submanifold T.

Let f be a Bott integral on a compact nonsingular isoenergetic surface Q. We denote by m the number of critical circles of the integral f, on which it attains its local minimum or maximum. Clearly, each such circle is a stable periodic solution of the Hamiltonian system. Let p be the number of two-dimensional critical tori of the integral f. Each of them is necessarily a local minimum or maximum of the integral f. Let q be the number of the critical saddle circles of the integral f (the number of unstable periodic trajectories of the system v) with the orientable separatrix diagram. Let s be the number of the critical saddle circles of the integral f (the number of unstable periodic solutions of the system) with the nonorientable separatrix diagram. Let r be the number of the critical Klein bottles for the integral f. All Klein's bottles are necessarily minima or maxima of the integral f. Taking into account Statement 2, we see that we have listed all possible connected critical manifolds of the Bott integral f on an isoenergetic surface.

Theorem 1. (The theorem on the topological classification of three-dimensional isoenergetic surfaces of integrable systems). Let M^4 be a smooth symplectic manifold (compact or noncompact), and let $v = \text{sgrad } H$ be a Hamiltonian system integrable in Liouville's sense on some nonsingular compact three-dimensional constant-energy surface Q with the help of the Bott integral f. Then the manifold Q is represented by gluing (along certain diffeomorphisms of boundary tori) the following elementary manifolds:

$$Q = mI + pII + qIII + sIV + rV = m(S^1 \times D^2) + p(T^2 \times D^1) + q(N^2 \times S^1) + s(N^2 \tilde{\times} S^1) + r(K^2 \tilde{\times} D^1).$$

Here, the nonnegative numbers m, p, q, s, r have the sense indicated above, and consequently reveal the number of critical manifolds of each type in the given Bott integral f.

This decomposition of the manifold Q will be called the Hamiltonian decomposition since each "building block" corresponds to some critical manifold of the integral f. If we neglect this exact interpretation of the numbers m, p, q, s, r and seek the simplest topological representation of the surface Q, we need the following theorem.

Theorem 2.(Theorem on the topological decomposition of isoenergetic surfaces.) Let Q be a compact nonsingular constant-energy surface of the Hamiltonian system $v = \text{sgrad } H$ on Q, which is integrable with the help of the smooth Bott integral f. Then Q admits the following topological decomposition:

$$Q = m'I + p'II + q'III = m'(S^1 \times D^2) + p'(T^2 \times D^1) + q'(N^2 \times S^1),$$

where m', p', q' are nonnegative integers, connected to the numbers m, p, q, s, r in Theorem 1 as follows:

$$m' = m + s + 2r, \quad p' = p, \quad q' = q + s + r.$$

If at least one of the numbers m' or q' is nonzero, the topological decomposition can be further simplified as

$$Q = m'I + q'III$$

In other words, it can be assumed that $p' = 0$.

Thus, any isoenergetic surface can be represented (generally, not unambiguously) in either of the following two forms:

(a) $Q = m'I + q'III$,

(b) $Q = p'II$.

Here, m', p' and q' are nonnegative integers.

Thus, we have obtained two decompositions, viz. Hamiltonian and topological, for each isoenergetic surface of an integrable system. The Hamiltonian decomposition is more "comprehensive", and "remembers" the structure and the number of critical manifolds of the integral. The topological decomposition is simpler but less exact.

5. Thus, Theorem 1 (and Theorem 2) states that the following inclusion is valid: $(H) \subset (Q)$.

It was shown by the author and A.V.Brailov [1] that the inverse inclusion is also true:

$$(H) \supset (Q),$$

i.e., we finally get $(H) = (Q)$.

Statement 3. The classes (H) and (Q) coincide. In other words, any closed compact orientable three-dimensional manifold obtained by piecing together complete tori, cylinders and trousers can be realized as an isoenergetic surface (with the help of Bott integral) of a Hamiltonian system on an appropriate symplectic four-dimensional manifold M^4 (which may be non-compact).

Thus, we have described a remarkable class of three-dimensional manifolds, viz. the class of isoenergetic surfaces of integrable systems.

6. The integral f on Q can be considered as a smooth mapping

$$f : Q \longrightarrow R$$

Let $a \in R$ be a noncritical value of the mapping f, i.e. the complete inverse image $f^{-1}(a)$ does not have any critical point of the function f. Then, in accordance with Liouville's classical theorem, each compact connected nonsingular component of the inverse image $f^{-1}(a)$ is homeomorphic to a two-dimensional torus. It is well known that we can choose in the neighbourhood of each such torus (we shall call it the Liouville torus) regular coordinates (action-angle variables) for which the Hamiltonian system v (with respect to which the Liouville tori are invariant) will describe a quasiperiodic motion over the torus. If. however, $c \in R$ is a critical value, the inverse image $f^{-1}(c)$ need not be composed of Liouville tori and, generally speaking, may not be a manifold. Two important questions arise in this case: (a) how are the "singular foils" of $f^{-1}(c)$ arranged, and (b) how are the integral trajectories of the system v arranged on these singular fibres ? We shall obtain an answer to these questions for the case of Bott integrals (i.e. general position integrals).

Theorem 3. Let v = sgrad H be a Hamiltonian system on M^4, which is integrable in Liouville's sense on some constant energy surface Q^3 with the help of the Bott integral f. Then each critical level surface of the integral f , being a smooth manifold, must be diffeomorphic to the torus T^2, the circle S^1, or the Klein bottle K^2. Further, each singular saddle level surface $f^{-1}(c)$ of the integral f, which is not a manifold (this will be the case if the critical points fill a certain critical saddle circle), is obtained by piecing together two two-dimensional tori T_1^2 and T_2^2 along nontrivial cycles (circles) γ_1 and γ_2 lying respectively on tori T_1^2 and T_2^2. Moreover, the integral trajectory of the system v on such a singular fibre (singular level surface of the integral) will asymptotically coil (in the case of general position) around this nontrivial cycle or coincide with it.

It was assumed in Theorem 3 that a given singular level surface contains exactly one critical manifold of the integral. This is not a restriction since it can be proved that the general case can be reduced to a small movement of the original integral.

Thus, each of the tori T_1^2 and T_2^2 constituting the singular level surface $f^{-1}(c)$ has a "locking cycle" $\gamma = \gamma_1 = \gamma_2$ which "screens" the torus. Each integral trajectory lies on one of the tori T_1^2 and T_2^2 and moves along the torus in such a way that as it approaches the locking cycle, it coils round the torus. When moving in the opposite direction, the same trajectoty approaches the locking cycle from the other side and again coils round it.

7. We can now give a complete classification of all the possible rearrangements of Liouville tori resulting from a change in the value of the integral f. Transposing H and f, we could speak of the bifurcation of Liouville tori when they cross the critical energy le-

vel of H for a fixed value of the integral f.

We shall consider the following five rearrangements of the tori T^2 corresponding to the basic manifolds I, II, III, IV and V mentioned earlier. The torus is realized as one of the boundary components of the appropriate basic manifold. Then the torus obtained as a result of a change in the integral f is transformed into a union of Liouville tori which are the remaining boundary components of the same basic manifold. It can be verified that these rearrangements have the following form.

(1) The torus T^2 contracts to the axial circle of the complete torus and then vanishes from the level surface of the integral f. This rearrangement can be represented as follows:

$$T^2 \longrightarrow S^1 \longrightarrow 0.$$

It occurs in the neighbourhood of the maximal or minimal critical circle of the integral, i.e. in the manifold $S^1 \times D^2$.

(2) Two tori T^2 move towards each other along a cylinder, merge together, and then "vanish". This takes place in the neighbourhood of the maximal or minimal critical torus of the integral, i.e. in the manifold $T^2 \times D^1$. This rearrangement can be represented as follows:

$$2T^2 \longrightarrow T^2 \longrightarrow 0.$$

(3) The torus T^2 splits into two tori when passing through the centre of an oriented saddle. These tori then "remain" on the level surfaces of the integral f. The rearrangement can be denoted by T^2 $2T^2$. This event occurs in the neighbourhood of the critical saddle circle of an integral with an oriented separatrix diagram, i.e. inside the manifold $N^2 \times S^1$.

(4) A torus T^2 coils round the torus T^2 twice (following the topology of the unoriented saddle $A^3 = N^2 \tilde{\times} S^1$), and remains on the surface of the level integral f. This rearrangement is denoted by

$$T^2 \longrightarrow T^2$$

and takes place inside the manifold $N^2 \tilde{\times} S^1$.

(5) The torus T^2 is transformed into a Klein's bottle, covering it twice. After this, the torus "vanishes" from the level surface of the integral f. This event is denoted by

$$T^2 \to K^2 \to 0$$

and takes place in the manifold $K^2 \tilde{\times} D^1$.

The five rearrangements obtained from the above-mentioned events by reversing the direction of the arrows will not be considered new rearrangements.

Theorem 4. (Theorem on the classification of rearrangements and bifurcation of two-dimensional Liouville tori.) Let f be a Bott integral on a nonsingular constant-energy surface Q^3 . Then any rearrangement of a Liouville torus due to its transition through the critical level surface of the integral f is a composition of a certain number of the above-mentioned elementary (canonical) rearrangements 1, 2, 3, 4, 5. Moreover, only the first three of these five rearrangements are actually independent (from a topological point of view). The rearrangements 4 and 5 decompose into compositions of rearrangements of the type 1 and 3.

A similar theorem also applies to multidimensional Liouville tori (see [1-3] for details). Thus, during the motion of a Liouville torus (induced by a change in the value of the integral f), only the following canonical events can take place: decomposition into two tori, merger with another torus, transformation into a circle, or transformation into a Klein's bottle. All the remaining rearrangements are just compositions of these rearrangements. We shall now describe a simple method which graphically illustrates the isoenergetic surface with a given integral on it.

8. The integral f on Q can be considered as a smooth mapping

$$f : Q \to R .$$

If $a \in R$ is a noncritical value of the mapping f, the complete inverse image $f^{-1}(a)$ of the value a consists of Liouville tori. Consequently, mapping each such torus by an individual point, we find that the inverse image of the value a will be represented by a set of points. By varying a, we make these points move. These points sweep the edges of a one-dimensional graph. When the value of a becomes critical, some of these edges meet (i.e. the corresponding Liouville tori begin to interact and get rearranged). As a result, we obtain a certain one-dimensional graph which we denote by $\Gamma(Q, f)$. Obviously, to construct this graph, we must know not only the surface Q, but also the integral f. For different values of f, the corresponding graphs $\Gamma(Q, f)$ on the same surface Q may be quite different. Theorem 1, which was proved above, not only leads to a construction of the graph $\Gamma(Q)=\Gamma(Q, f)$, but also gives a complete description of all its vertices. This would be impossible if we had not classified all the elementary manifolds in whose union we can decompose any isoenergetic surface with integral f. In other words, only after proving Theorem 1 can we classify the graphs $\Gamma(Q, f)$ according to their vertices and the type of interaction of their edges. In order to depict the graphs $\Gamma(Q, f)$ conveniently, we use the notation shown in Fig.1. (1) The large dark spot with one outgoing (incoming) graph edge represents the complete torus generated by the minimal (resp. maximal) the critical circle of the integral f. The dark spot represents the critical circle, while the edge represents a one-dimensional family of tori foliating a complete torus, i.e. a tubular neighbourhood of the critical circle (minimal or maximal).

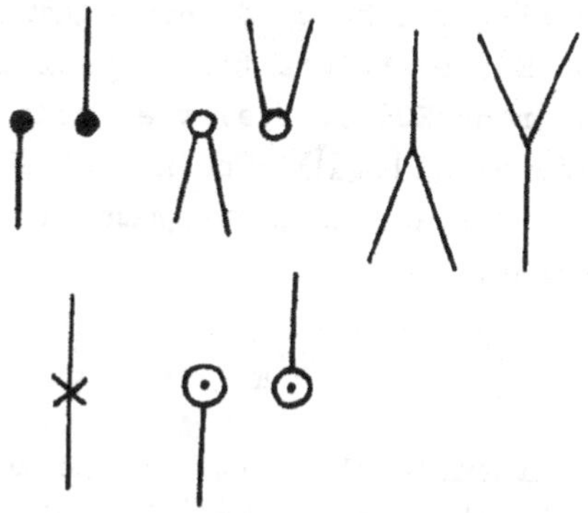

Fig.1

(2) The light spots with two outgoing (incoming) graph edges represent the minimal (resp.maximal) critical torus and two families of nonsingular Liouville tori emerging from it. (3) The trifolium, i.e. the point with three graph edges converging at it, represents a connected neighbourhood of the critical saddle circle with oriented separatrix diagram (in this case, one Liouville torus is decomposed into two tori). (4) The asterisk (with an incoming and an outgoing edge) represents the circumference of the critical saddle circle with an unoriented separatrix diagram (in this case, the Liouville torus is "reconstructed" during its motion into a single torus "coiled" twice round it. (5) The spot with a dot and one outgoing (incoming) graph edge represents the minimal (resp. maximal) Klein's bottle.

Each trifolium (see type 3) describes either the decomposition of one torus into two tori, or a merger of two tori into one torus depending on whether the trifolium is oriented with two edges pointing upwards or downwards.

The vertices of the graphs are labelled 1-5 not accidentally. As a matter of fact, there is a one-to-one correspondence between these five types of vertices of the graph $\Gamma(Q)$ and the five types of elementary manifolds described in Theorem 1. It is found that a quite small "neighbourhood" of an i-type vertex on the graph (i = 1, 2, 3, 4, 5) is homeomorphic (from the point of view of surface Q) to an i-type manifold described in Theorem 1.

9. By way of an important illustration of the general theory, we shall now give a complete description of the isoenergetic surfaces for an integrable system representing the motion of a four-dimensional rigid body fixed at its centre of mass (Euler's case). It is well known that this problem can be solved with the help of four quadratic integrals. The corresponding system of Euler differential equations is realized on the Lie algebra of the group SO(4), i.e. the group of rotations of the space R^4 (see the review in [4,5]). The Lie algebra so(4) of the group SO(4) is realized as a space of skew-symmetric matrices $X = (x_{ij})$. Two independent integrals

$$h_1 = \sum_{i<j} x_{ij}^2 \quad \text{and} \quad h_2 = x_{12}x_{34} - x_{13}x_{24} + x_{14}x_{23}$$

isolate on the Lie algebra so(4) four-dimensional orbits of the adjoint representation of the group SO(4). This means that the joint level surfaces of these two integrals are invariant to the transformations $x \to g x g^{-1}$, where $X \in$ so(4) and $g \in SO(4)$. The general position orbits are homeomorphic: $S^2 \times S^2 = \{ h_1 = p_1^2 , h_2 = p_2 \}$, $p_i = \text{const}$, $|2p_2| < p_1^2$. The equations of motion of a rigid body for the so-called normal series (see [4]) correspond to the Hamiltonian

$$H = \sum_{i<j} \lambda_{ij} x_{ij}^2 , \quad \lambda_{ij} = \frac{b_i - b_j}{a_i - a_j} , \quad \sum a_i = \sum b_i = 0.$$

The remaining two integrals of this system have the form

$$h_3 = \sum_{i<j} (a_i + a_j)x_{ij}^2 \quad \text{and} \quad h_4 = \sum_{i<j} (a_i^2 + a_i a_j + a_j^2)x_{ij}^2 .$$

These integrals are in involution on the $S^2 \times S^2$ orbits and are functionally independent. Thus, for the four-dimensional symplectic manifold M^4 we can take $S^2 \times S^2$, while for H we use the general Hamiltonian mentioned above. For the second additional integral f (on Q^3), we can use either of the integrals h_3 and h_4. In accordance with our general theory, the vertices of the graph Γ (Q, f) are split into five types of vertices. It was found that for the Hamiltonian equations of motion of a rigid body (case so(4)), only two of these rearrangements, viz. 1 and 3 (Fig.1) are encountered on isoenergetic surfaces, i.e. all critical submanifolds of the second integral f are circles with oriented separatrix diagrams (maxima, minima or saddles). Sometimes, a rearrangement of two Liouville tori into two Liouville tori on the same critical level takes place. The critical level surface of the integral f is homeomorphic to $P \times S^1$, where P is the cross section of the sphere S^2 (standardly embedded in R^3) by a pair of planes passing through its centre. This rearrangement is a composition of two rearrangements of type 3. The following two theorems were proved by A.Oshemkov who applied our general theory to this particular case.

Statement 4. Let a normal series Hamiltonian system of a rigid body be given on the Lie algebra so(4) with a Hamiltonian h_3. Then the non-singular orbits $S^2 \times S^2$ in the Lie algebra so(4) are foliated into isoenergetic surfaces $Q^3 = \{ h_3 = \text{const} \}$, such that with the exception of perhaps a finite number of these surfaces Q^3, the function

$$f = h_4 \mid S^2 \times S^2$$

is a Bott integral for all of them. Depending on the numerical values of the parameters P_1 and P_2 defining the general position orbit $S^2 \times S^2 = \{ h_1 = p_1^{\,2}, h_2 = p_2 \}$ the bifurcation parameters for the moment map

$$F = h_3 \times h_4 : S^2 \times S^2 \longrightarrow R^2$$

can be only of three types. The explicit form of these parameters is computed and shown in Fig.2.

The numbers in Fig.2 indicate the number of Liouville tori constituting the complete inverse image $F^{-1}(y)$ for the points y of a given region on R^2. The reconstructions along the arrows shown in Fig.2 have the form A, B, C shown in Fig.3.

Statement 5. Let a normal series Hamiltonian system of a rigid body be given on a Lie algebra so(4) with a general type Hamiltonian H indicated above. For the second integral f, which is functionally independent of H almost everywhere, we can take h_3 or h_4. Then the non-singular orbits (i.e. general position orbits) $S^2 \times S^2$ in the Lie al-

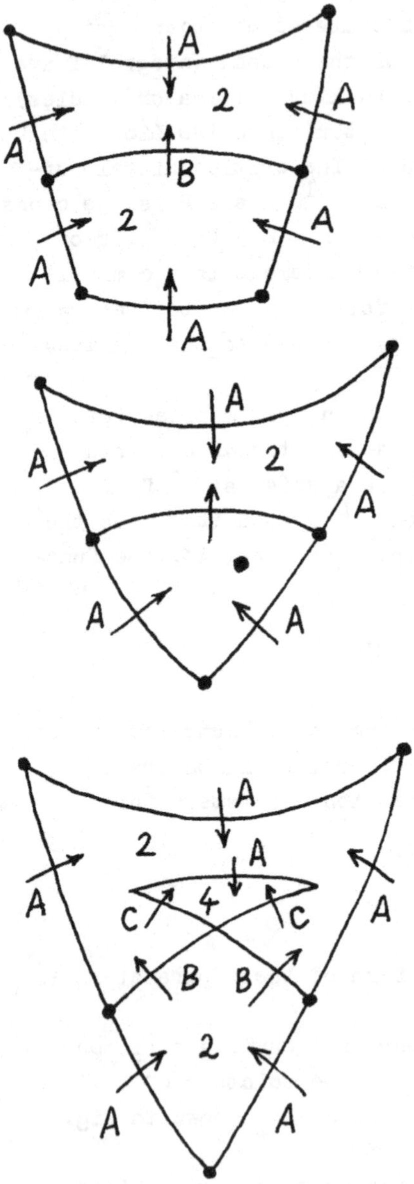

Fig.2

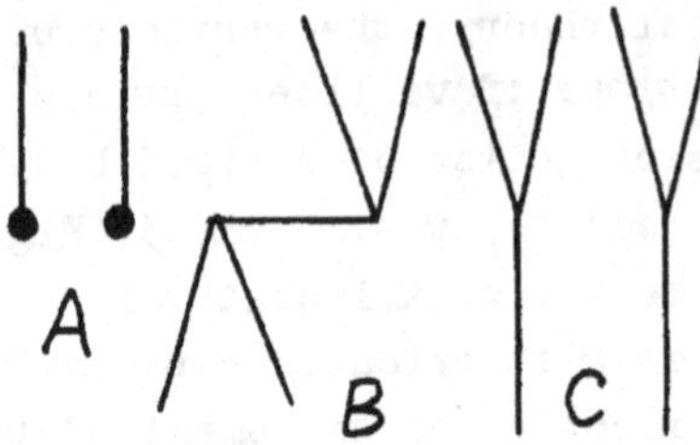

Fig.3

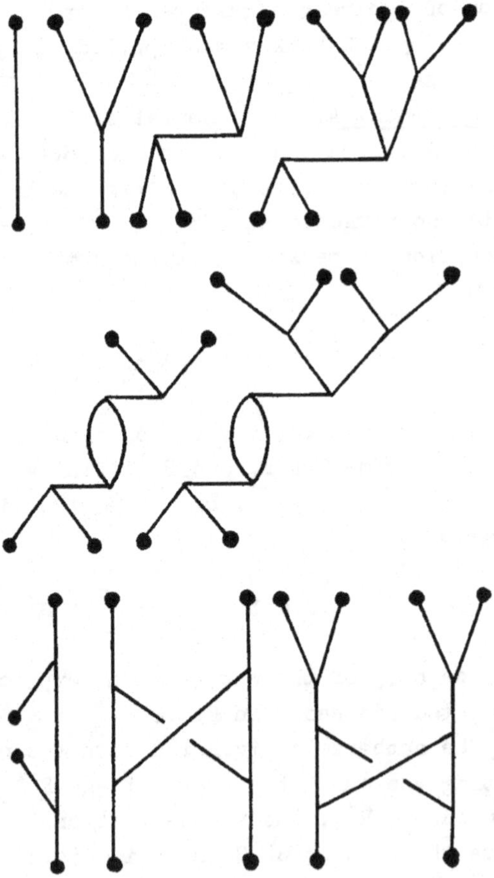

Fig.4

gebra so(4) are foliated into isoenergetic surfaces $Q^3 = \{H = \text{const}\}$, such that the integral f is a Bott integral for all Q except, perhaps, a finite number of these surfaces. The complete list of all the nine connected Q graphs $\Gamma(Q, f)$ describing all nonsingular surfaces encountered in this problem is given in Fig.4. All these nine graphs are indeed realized for some common Hamiltonian (with an appropriate choice of the parameters defining it and the surface of the second integral).

It should be noted that an arbitrary normal-series Hamiltonian H can be represented in the form

$$H = k_1h_1 + k_2h_2 + k_3h_3 ,$$

where the constants k_1, k_2 and k_3 are independent of x_{ij}. Hence the bifurcation sets for a system with a normal series Hamiltonian H are obtained from the bifurcation sets for a system with Hamiltonian h_3 with the help of some nondegenerate linear transformation. Some of the above-mentioned canonical rearrangements of the Liouville tori were earlier discovered by many authors in specific examples of important mechanical systems (see, for example, the works of Kharlamov and Pogosyan [17,18]). In particular, some of the reconstructions of tori in the Kovalevskaya and Goryachev-Chaplygin cases are of this type. It is found that some of the rearrangements obtained in [17,18] can be split into compositions of our elementary rearrangements (we can calculate and find out the arrangements for which this statement is valid). An interesting circumstance is worth noting here. While in the integrable case of the equations of motion of a four-dimensional rigid body fixed at the centre of mass the Bott integrals have arrangements of types 1 and 3 only (see above), the integrable cases considered in [17,18] also include, say, a rearrangement of type 4, generated by the elementary manifold $N^2 \tilde{\times} S^1 = A^3$. This points towards a curious topological distinction between these integrable cases.

Let us chalk out the programme of further investigations in this direction. Since it has been established that we can graphically represent the topology of the isoenergetic surface Q^3 and the qualitative behaviour of the integral f on it with the help of a one-dimensional graph $\Gamma(Q, f)$ (with five types of vertices), we must compile the most exhaustive list possible for all known integrable cases of systems on four-dimensional manifolds M^4. After this, we can use the scheme described above and construct the graphs $\Gamma(Q, f)$. This gives a visual and qualitatively important information on the topology of the integrable cases.

10. It turns out that the class (H) = (Q) of manifolds introduced by us can also be represented in two other important ways. Proceeding from the three-dimensional topology problems, Waldhausen [11] introduced a class (W) of three-dimensional manifolds, such that the manifold W contains a set of non-intersecting two-dimensional tori whose omission leads to a manifold in which each connected component is foliated with a circular foil over a certain two-dimensional manifold (possibly with an edge).

Recently, on the advice of the author, Matveev and Burmistrova [6] investigated a class (S) of three-dimensional manifolds on which there exists a smooth function g, such that all its critical points are arranged in nondegenerate critical circles and all nonsingular level surfaces of the function g are unions of two-dimensional tori. In other words, the manifold M belongs to the class (S) if and only if there exists a Bott function on it, such that all its critical manifolds are circles and all nonsingular level surfaces are composed of tori. It is important to note that the function g need not be an integral of any Hamiltonian system. It was proved by Fomenko and Zieschang that the classes (W) and (Q) coincide. Later, it was proved by Matveev and Burmistrova that the classes (S) and (Q) also coicide. We can combine all these results into the following important (and nontrivial) theorem.

Theorem 5. All four classes of three-dimensional manifolds described above coincide, i.e. (H) = (Q) = (W) = (S).

This leads to the following important corollary.

Statement 6. Not every compact closed oriented manifold can serve as a constant energy surface of some Hamiltonian system which can be integrated with the help of Bott's smooth integral. The class (H) = = (W) = (Q) = (S) is a "meager" subset (in a certain sense, a "zero-measure" subset) in the class M of all three-dimensional manifolds.

Thus, the topology of the isoenergetic surface under consideration may serve as an obstruction to the integrability of a Hamiltonian system (in the class of smooth Bott integrals). The new topological obstructions to the integrability which were discovered by us are useful in the sense that the computation of the required topological surface characteristics is often simpler than the investigation of the analytical properties of the Hamiltonian system under investigation. However, it turns out that from the point of view of homology groups, the class (H) cannot be "recognized" in the class (M). It was shown by Mamedov that we can glue trousers, complete tori and cylinders to form a manifold with any groups of integral homologies (for three-dimensional closed oriented manifolds). In other words, we can effectively compute the obstructions to integrability on the basis of the homology groups alone, although we cannot effectively isolate the class of isoenergetic surfaces among all three-dimensional manifolds. We shall explicitly present the topological obstructions which prevent an "overwhelming majority" of three-dimensional manifolds to be realized as isoenergetic surfaces of integrable systems.

Statement 6 follows from Theorem 5 as well as Waldhausen's theorem [11], according to which the class (W) does not coincide with the class of all three-dimensional manifolds and is a "meager subset" in it. We shall also indicate another class (R) of three-dimensional manifolds which is being actively studied at present. These are manifolds that can be decomposed into a sum of round handles. The Morse circular function is a Bott function whose all critical manifolds are circles. (Unlike the manifolds of the class (S), however, it is not necessary in this case that the nonsingular level surfaces be tori.) Morse circular functions are closely related to the decomposition of a manifold into a sum of round handles. It is proved in [12] that the manifold M^n (of arbitrary dimension n) can be decomposed as a sum of round handles if and only if M^n admits a circular Morse function. It is found in [14] that the manifold M^n of dimension $n \geqslant 4$ can be decomposed into a sum of round handles if and only if the Euler characteristic $\chi(M)$ is equal to zero. The situation is more complicated for three-dimensional manifolds [12-15]. Obviously, (R) $\supset$ (H), i.e. any isoenergetic surface of an integrable system admits a circular function. This follows from Statement 6, since the class (S) actually consists of manifolds with circular functions (moreover, all their nonsingular level surfaces are tori). It can be proved that the class (R) is strictly larger than the class (H), i.e. there exist manifolds which admit circular functions but are not realizable as isoenergetic surfaces. The distinction between the classes (R) and (H) is due to the difference in their behaviour to the operation of taking the connected sum $\#$ of two manifolds. According to [13], if M is any three-dimensional manifold (we shall not stipulate each time that we are dealing with a class of closed compact oriented manifolds), let us consider the connected sum with a sufficiently large number of manifolds $S^1 \times S^2$. This will eventually lead to the class (R). In other words, we shall have for a sufficiently large m

$$M \# (\#_{i=1}^{m} S^1 \times S^2) .$$

We shall show in the next article that the class (H) has the following remarkable property: if

$$M = M_1 \# M_2 \in (H)$$

the manifolds M_1 and M_2 also must belong to the class (H) (Matveev) . This means that the class (R) is strictly larger than the class (H). Indeed, it is sufficient to take a manifold M_0 not belonging to the class (H) (we know that there are no manifolds of this kind). Taking its connected sum with a very large number of manifolds $S^1 x S^2$, we arrive at the class (R). The manifold thus obtained, however, does not belong to the class (H), since otherwise the original manifold M_0 must belong to (H), which is contrary to the choice of M_0. If we consider the class of irreducible manifolds (which cannot be represented as the connected sum of manifolds $M_1 \# M_2$, where M_1 and M_2 differ from the sphere S^3), the following remarkable identity is found to be valid:

$$(H) = (Q) = (W) = (S) = (R) .$$

11. Let us now estimate from below the number of stable periodic solutions of an integrable Hamiltonian system. It should be noted that a system may be integrable, and yet not possess any closed stable trajectory (although there may be many unstable closed trajectories). Let us consider a simple example, i.e. the geodesic flow of a two-dimensional plane torus T^2 with a local Euclidean metric $g_{ij} = \delta_{ij}$. It can easily be seen that this flow has an additional integral, but all the closed trajectories of the system are unstable (!). It turns out that there exists a simple qualitative connection between the number of stable periodic solutions of a system on a given isoenergetic surface and the topology of this surface, viz. the group of one-dimensional integral homologies $H_1(Q, Z)$ or the fundamental group $\pi_1(Q)$. Before precisely formulating the results obtained by us, let us consider a natural definition which we shall be requiring later.

12. Definition 3. We shall call a Bott integral on the isoenergetic surface Q orientable if all its critical submanifolds are orien-

table. If even one of the critical submanifolds is not orientable, the integral f will be called unorientable.

It turns out that we can investigate only orientable integrals without any significant loss of generality.

Statement 7. Let Q^3 be a nonsigular compact constant energy surface in M^4 anf f be an unorientable Bott integral on Q. Then all unorientable critical manifolds are homeomorphic to Klein's bottle and the integral f attains a local minimum or maximum on them. Let U(Q) be a quite small tubular neighbourhood of the surface Q in M. Then there exists a two-sheeted covering

$$\pi : (\tilde{U}(\tilde{Q}), \tilde{H}, \tilde{f}) \to (U(Q), H, f)$$

(with a fibre Z_2), where $\tilde{U}(\tilde{Q})$ is a symplectic manifold with a Hamiltonian system $\tilde{v}$ = sgrad $\tilde{H}$ (the Hamiltonian $\tilde{H}$ has the form $\tilde{H} = \pi^* H$), which is integrable on $\tilde{Q} = \pi^{-1}(Q)$ with the help of the orientable (!) Bott integral $\tilde{f} = \pi^* f$. All the critical Klein bottles are "unfolded" into critical tori T^2 in Q (being the minima or maxima of the integral f). The manifold $\tilde{U}(\tilde{Q})$ is a tubular neighbourhood of the manifold $\tilde{Q}$.

This means that if f is an unorientable integral on Q, then $\pi_1(Q) \neq 0$ automatically, and the group $\pi_1(Q)$ contains a subgroup of index 2. If, for example, Q is homeomorphic to the sphere S^3 (a frequent case in mechanics), then any Bott integral f is always orientable on Q. We denote by m = m(Q) the number of stable periodic solutions of the system v on Q. Let r = r(Q) be the number of critical Klein bottles for the integral f on Q. If the integral is orientable, then r = 0.

Theorem 6. Let M^4 be a smooth symplectic four-dimensional manifold (compact or noncompact) and v = sgrad H be a Hamiltonian system on M^4, where H is a smooth Hamiltonian. We assume that the system is integrable in Liouville's sense on some nonsingular three-dimensional compact level surface Q^3 of the Hamiltonian H, the second smooth integral f commuting with H on Q and being a Bott integral on Q. Then the number m = m(Q) of stable periodic solutions of the system v on Q is estimated from below as follows in terms of the topological invariants of the isoenergetic surface:

(1) If the integral f is orientable on Q, we have

(a) $m \geq 2$ if the homology group $H_1(Q, Z)$ is finite;

(b) $m \geq 2$ if the fundamental group $\pi_1(Q) = Z$.

(2) If the integral f is unorientable on Q, we have

(a) $m + r \geqslant 2$ if the homology group $H_1(Q, Z)$ is finite;

(b) $m \geqslant 2$ if $H_1(Q, Z) = 0$ (in this case, the group $\pi_1(Q)$ may be infinite;

(c) $m \geqslant 1$ if the group $H_1(Q, Z)$ is cyclic and finite;

(d) $m \geqslant 1$ if $\pi_1(Q) = Z$ or if $\pi_1(Q)$ is a finite group;

(e) $m \geqslant 2$ if the group $H_1(Q, Z)$ is finite and cyclic, and the surface Q does not belong to a small series of manifolds of the type

$$Q_0 = I + sIV + rV$$

where s and r are nonnegative integers.

In both the cases (1) and (2), the integral f attains a local minimum or maximum on each of these stable periodic solutions of the system v (or on the Klein bottles). If the homology group $H_1(Q, Z)$ is infinite, i.e. the rank of H_1 is equal to or greater than unity, the system v may not have any stable periodic solutions on the surface Q (such examples can be easily constructed).

The criterion obtained above is quite effective, since the rank of a homology group can be usually calculated without any difficulty. For many integrable mechanical systems, the isoenergetic surfaces are diffeomorphic to the sphere S^3 or to the projective space RP^3, or to $S^1 \times S^2$. For example, while composing the equations of motion for a heavy rigid body in the region of high velocities, we can assume after an appropriate factorization that $Q \approx RP^3$. In the Kovalevskaya integrable case, the isoenergetic surfaces (after factorization) are homeomorphic to $S^1 \times S^2$. If the Hamiltonian H has an isolated minimum or maximum (i.e. an isolated equilibrium position) on M^4, all the neighbouring level surfaces $Q^3 = \{ H = \text{const} \}$ are spheres S^3. By $L_{p,q}$ we denote the lens space (cyclic group factors of the sphere S^3). Let us consider some cases of interest from the point of view of mechanics in the form of statements.

Statement 8. Let the system v be integrable with the help of the Bott integral f on the surface Q which is homeomorphic to one of the following manifolds: S^3, RP^3, $S^1 \times S^2$, $L_{p,q}$.

(1) If the integral f is orientable, we always have $m \geqslant 2$, i.e. the system must have at least two stable periodic solutions.

(2) If the integral f is unorientable, we have $m \geqslant 2$ for S^3, while for the manifolds RP^3, $S^1 \times S^2$, $L_{p,q}$, we have $m \geqslant 1$. In particular, the integrable system always has on the sphere S^3 (irrespective of the orientability of f) at least two stable periodic solutions.

The criterion of Theorem 6 is exact in the following sense. There exist cases in which the integrable system has exactly one (exactly two) stable periodic solutions on the surface $Q = RP^3$ (resp S^3).

It follows from the results ontained by Anosov, Klingenberg and Takens (see [7,16]) that in the set of all geodesic flows on smooth Riemannian manifolds there exists a dense and everywhere open subset of flows which do not have closed stable integral trajectories. This means that the property of geodesic flow not to have stable periodic solutions is a property of the general position. Together with our results, this leads to the following theorem.

Theorem 7. Suppose that a two-dimensional smooth surface is homeomorphic to a sphere and is supplied with a smooth general position Riemannian metric, i.e. there is no closed stable geodesic on the surface. Then the smooth geodesic flow corresponding to this metric is nonintegrable (on each nonsingular isoenergetic surface) in the class of smooth Bott integrals.

This theorem states the nonintegrability of the given type of systems just on the basis of topological restrictions which, as was shown by us, are imposed on the isoenergetic surfaces of integrable systems. In this respect, we shall indicate new topological obstructions to integrability.

It is found that the number of critical manifolds of the integral f on Q (and sometimes the number of stable periodic solutions) can be estimated from below by a certain universal constant which depends only on the first homology group $H_1(Q, Z)$. Let $\beta = \operatorname{rank} H_1(Q, Z)$, (i.e. a one-dimensional Betti number) and ε be the number of elementary multipliers in a finite part Tor H_1 of the group $H_1(Q, Z)$. If Tor H_1 is decomposed into an ordered sum of subgroups, where the order of each subgroup is a divisor of the order of the preceding subgroup, then ε is the number of terms. The following theorem was proved by Fomenko and Zieschang

Statement 9. Let Q^3 be a compact nonsingular isoenergetic surface of an integrable (with the help of the Bott integral) system. Let m be the number of stable periodic solutions of the system, s be the number of unstable periodic solutions of the system with an unorientable separatrix diagram, and r be the number of critical Klein bottles. Then the following inequality is satisfied:

if $q + s > 0, \quad p + m > 0$

then $m' = m + s + 2r \geqslant \varepsilon - 2\beta + 1, \quad q' \geqslant m' - 2,$

and $q \geqslant m + r - 2$

where the numbers q' and q have been defined above. For $p = 1$ and $m = r = 0$, we have

$$0 \geqslant \varepsilon - 2\beta .$$

If the integral f is orientable and all separatrix diagrams of its critical submanifolds are also orientable, we get $s = r = 0$, i.e. in this case we obtain an estimate from below for the number m of stable periodic solutions of the system:
$m \geqslant \varepsilon - 2\beta + 1$, and $q \geqslant m - 2$.

This theorem supplements Theorem 6. Besides, if the integral f is orientable, all separatrix diagrams of the critical manifolds are orientable and the homology group $H_1(Q, Z)$ is finite, then we have $\beta = 0$ and $m \geqslant \varepsilon + 1$, i.e. $m \geqslant 1$. In this particular case, we "come across" one of the statements of Theorem 6. However, the statement of Theorem 6 is stronger, since it contains no assumption concerning the orientability of the separatrix diagrams. In the general case, Theorem 6 and Statement 9 are independent.

REFERENCES

1. A.T.Fomenko. Topology of Constant-energy Surfaces of Integrable Hamiltonian Systems and Obstructions to Integrability. - Izv. Akad. Nauk SSSR, 1986, v.50, No.6.

2. A.T.Fomenko. Topology of Three-dimensional Manifolds and Integrable Mechanical Systems. - in: Proceedings of V Tiraspol Symposium on General Topology and Its Applications. Kishinev, 1985.

3. A.T.Fomenko. The Morse Theory of Integrable Hamiltonian Systems. - Dokl. Akad. Nauk SSSR, 1986, v.287, No.5.

4. V.V.Trofimov and A.T.Fomenko. Liouville Integrability of Hamiltonian Systems on Lie Algebras. - Uspekhi Mat. Nauk , 1984, v.39, No.2.

5. A.V.Brailov. Complete Integrability of Some Geodesic Flows and Integrable Systems with Noncommuting Integrals. - Dokl. Akad. Nauk SSSR, 1983, v.271, No.2.

6. A.B.Burmistrova and S.V.Matveev. The Structure of S-functions on Orientable 3-manifolds. - in: Abstracts of XX All-Union School on the theory of Operators in Functional Spaces. Chelyabinsk, 1986.

7. D.V.Anosov. On Typical Properties of Closed Geodesics. - Izv. Akad. Nauk SSSR, Ser. Matem., 1982, v.46, No.4.

8. V.V.Kozlov. Integrability and Nonintegrability in Hamiltonian Mechanics. - Uspekhi Mat. Nauk, 1983, v.38, No.1.

9. V.I.Arnold. Mathematical Methods in Classical Mechanics. Nauka,

Moscow 1974 (in Russian).

10. S.P.Novikov. Variational Methods and Periodic Solutions of Kirchhoff-type Equations, II. - Funkts. Analiz i Ego Prilozh., 1982, v.15, No.4.

11. F. Waldhausen Eine-Klasse von 3-dimensional Mannigfaltigkeiten, I. - Invent. Math., 1967, v.3, No.4.

12. S.Miyosh. Foliated Round Surgery of Codimension-one Foliated Manifolds. - Topology, 1982, v.21, No.3.

13. J.Morgan. Non-singular Morse-Smale Flows on 3-dimensional Manifolds. - Topology, 1979, v-18, No.1.

14. D.Asimov. Round Handles and Non-singular Morse-Smale Flows. - Ann. of Math., 1975, v.102, No.1.

15. W.Thurston. Existence of Codimension-one Foliation. - Ann. of Math., 1976, v.104, No.2.

16. W.Klingenberg. Lectures on Closed Geodesics. Springer-Verlag, 1978.

17. M.P.Kharlamov. Topological Analysis of Classical Integrable Systems in Dynamics of a Rigid Body. - Dokl. Akad. Nauk SSSR, 1983, v.273, No.6.

18. T.I.Pogosyan and M.P.Kharlamov. Bifurcation Set and Integral Manifolds in the Problem on the Motion of a Rigid Body in a Linear Force Field. - Prikl. Matem. i Mekh., 1979, v.43, No.3.

19. S.Smale. Topology and Mechanics. - Uspekhi Mat. Nauk, 1972, v.27, No.2.

20. P.Orlik, E.Vogt, H.Zieschang. Zur Topologie gefarserter dreidimensionaler Mannigfaltigkeiten. - Topology, 1967, v.6, No.1.

GEOMETRICAL ASPECTS OF NELSON's STOCHASTIC QUANTIZATION

Yu.E. Gliklikh

Department of Mathematics,
Voronezh State University,
394693, Voronezh, USSR

The present paper is devoted to stochastic mechanics - the theory which is, in principle, a classical one, but gives the same results as quantum mechanics. In this theory the motion of a particle is described by a diffusion process which obeys a certain analogue of Newton's second law. Using the probability density of the diffusion process, we can construct a particle quantum-mechanical wave function satisfying the Schrödinger equation. Imre Fényes [14] was, seemingly, the first to describe diffusion processes satisfying the Schrödinger equation; for historical details the reader is referred to [6,24,26] . In Nelson's works [23,24] this approach has become quite natural, and later it gained wide application.

At present, stochastic mechanics has been extended to describe particles with spin [6,9,13,26] and relativistic particles in the theory of special relativity (for example, it has been established in [9,20] that the Klein-Gordon equation is related to stochastic mechanics, and a similar relationship for the Dirac equation has been found in [1]); promising results have been obtained for quantum fields, etc. The theory has been extended to Riemannian manifolds, using interesting geometrical constructions. Nelson's book [26] (published in 1985) describes in detail the modern state of (non-relativistic) stochastic mechanics.

In those cases where both quantum and stochastic mechanics are applicable they give similar predictions, though they frequently use essentially different explanations (for example, the description of electron interference in [26]). In Nelson's oppinion [26] it is not yet clear whether stochastic mechanics reflects real physics or it is only a suitable mathematical apparatus. There exists another point of view (see [30]), according to which stochastic mechanics employs a third quantization method differing from the Hamiltonian and Lagrangian approaches.

The task of the paper is to outline a clear introduction to the mathematical apparatus of stochastic mechanics, which is intended for specialists in global analysis. In this connection, special attention is paid to geometrical constructions in stochastic mechanics on manifolds. Less attention is paid to the probabilistic concepts, except for the cases where these concepts acquire a geometrical meaning (as for details, the reader is referred to the references).

The exposition is based on the Belopolskaya-Daletsky approach to stochastic differential equations on manifolds [2,3,5] . A new topic in the paper is the description of the current velocity hodograph in stochastic mechanics (using a modification of the stochastic integral with Riemannian parallel displacement [18]), as well as an attempt at describing stochastic mechanics with non-holonomic constraints, which leads to equations with degenerate diffusion.

1. Introduction to stochastic mechanics

In this section we outline all the necessary preliminaries and survey the basic mathematical constructions of stochastic mechanics in a linear space.

In what follows we shall consider stochastic processes with continuous time $t \in [0,T]$ defined on a probabilistic space (Ω, F, P) on which one can specify a Wiener process $w(t)$ assuming values in R^n. By $E(\ \ |B)$ we denote conditional expectation with respect to the σ-subalgebra B of the σ-algebra F. The subalgebra B may be generated either by a **random variable** η (via inverse images of Borel sets) or by a certain condition u; the corresponding notation is as follows: $E(\ \ |\eta)$ and $E(\ \ |u)$.

Any stochastic process $\xi(t)$ defines three families of σ-subalgebras of the σ-algebra F: "the past" P_t^ξ generated by $\xi(s)$ for $s \leq t$, "the future" F_t^ξ generated by $\xi(s)$ for $s \geq t$, and "the present" N_t^ξ generated by $\xi(t)$. These families are assumed to be completed with all sets of zero probability. Hereinafter, instead of $E(\quad | N_t^{\xi})$ we shall simply write E_t^{ξ}.

Let $c(t,x)$ be a continuous (in both variables) vector field on R^n and $C(t,x): R^n \longrightarrow R^n$, a linear operator continuously dependent on the parameters $t \in [0,T]$ and $x \in R^n$. We shall consider stochastic differential equations written in the Ito form:

$$d\xi(t)=c(t,\xi(t))dt+C(t,\xi(t))dw(t). \tag{1.1}$$

Equation (1.1) implies that the following relations are valid for $t \in [0,T)$ and $\Delta t > 0$: $E_t^{\xi}(\xi(t+\Delta t)) = c(t, \xi(t))\Delta t + O(\Delta t)$ and $E_t^{\xi}(\xi(t+\Delta t)^2) = (\mathrm{tr}\, CC^*)\Delta t + O(\Delta t)$, where $O(\Delta t)$ is an infinitesimal of higher order than Δt in the norm of the space $L_{\infty}(\Omega, R^n)$. The use of symbol d (forward differential) means that all the increments Δt are calculated in the direction where t increases. Equation (1.1) has an equivalent integral form in terms of the Ito integral with respect to dw (see, for example, [15,22]).

For a solution $\xi(t)$ of Eq. (1.1), the field $c(t,x)$ is called the drift coefficient and the field of linear operators $\frac{1}{2}C(t,x)C^*(t,x)$, the diffusion coefficient; the process ξ is called a diffusion process. If not stated otherwise, $c(t,x)$ and $C(t,x)$ are assumed to be smooth with respect to both variables.

We recall that under smooth changes of coordinates in R^n equation (1.1) is not transformed according to a tensor law: let $\varphi: R^n \longrightarrow R^n$ be a diffeomorphism (change of coordinates), then $\varphi(\xi)$ satisfies the following equation (the Ito formula; see, for example, [15]):

$$d\varphi(\xi(t))= \varphi'(c(t,\xi(t))dt+\frac{1}{2}\mathrm{tr}\, \varphi''(C,C)dt+ \varphi'(C(t,\xi(t))dw(t)$$

Stochastic mechanics in R^n deals with diffusion processes characterized by the diffusion coefficient $\frac{1}{2}\sigma^2 I$ (σ is a constant and I is the identity operator). In other words, these processes are defined by equations of the form

$$d\xi(t)=b(t,\xi(t))dt+\sigma dw \tag{1.2}$$

Hereinafter, the constant $\frac{1}{2}\sigma^2$ is set equal to $\hbar/2m$, where m is the particle mass and $\hbar = h/2\pi$, Planck's constant. For simplicity, we use the system of units in which $m = 1$. Thus, the less is σ , the greater is the mass.

Remark 1. The case of equations with non-degenerate positive definite diffusion coefficient of general form corresponds to stochastic mechanics on a Riemannian manifold where the Riemannian metric is defined by the diffusion coefficient. This case is considered in the next section. In Section 4 we shall consider the case of degenerate diffusion coefficient, which corresponds to stochastic mechanics with constraints.

A sample trajectory of the process $\xi(t)$ is almost surely (a.s.) non-differentiable, i.e. the derivative $d\xi/dt$ does not exist. Following Nelson, we define the "mean forward derivative" by

$$D\xi(t)=\lim_{\Delta t\to +0} E_t^{\xi}\left(\frac{\xi(t+\Delta t)-\xi(t)}{\Delta t}\right) \tag{1.3}$$

where $\Delta t \to +0$ means that $\Delta t \to 0$ and $\Delta t > 0$. Since $w(t)$ is a martingale with respect to P_t^{ξ}, and $b(t, \xi(t))$ is measurable with respect to N_t^{ξ} , we obtain $D\xi(t) = b(t, \xi(t))$. It can easily be seen that the relation

$$\xi(b)-\xi(a)=\int_a^b D\xi(t)dt+\sigma(w(a)-w(b))$$

holds for any $a \leqslant b \in [0,T]$.

The "mean backward derivative", $D_*\xi(t)$, is defined by

$$D_*\xi(t)=\lim_{\Delta t\to +0} E_t^{\xi}\left(\frac{\xi(t)-\xi(t-\Delta t)}{\Delta t}\right) \tag{1.4}$$

It is known that for the solution $\xi(t)$ of Eq. (1.2) there exists a vector field $b_*(t,x)$ such that $D_*\xi(t) = b_*(t, \xi(t))$, and, generally, $b(t,m) \neq b_*(t,m)$. If the trajectories of ξ were smooth, these field would, evidently, be identical.

Let the process $w_*(t)$ be defined by

$$\xi(b)-\xi(a)=\int_a^b D_*\xi(t)dt+\sigma(w_*(b)-w_*(a))$$

for any $a \leq b \in [0,T]$. If time is reversed, this process becomes a Wiener process relative to F_t. We shall call w_* a reverse Wiener process. It can easily be seen that the process $\xi(t)$ also satisfies

$$d_*\xi(t)=b_*(t,\xi(t))d_*t+\sigma d_*w(t) \tag{1.5}$$

Equation (1.5) means that the following relations are valid for $t \in (0,T]$ and $\Delta t > 0$: $E_t^\xi(\xi(t-\Delta t)) = -b_*(t,\xi(t))\Delta t + O(\Delta t)$ and $E_t^\xi(\xi(t-\Delta t)^2) = \sigma^2 n\Delta t + O(\Delta t)$. Note that (1.5) has an equivalent integral form in terms of the anticipating stochastic integral with respect to d_*w_* (see, for example, [22]). The use of the symbol d_* (backward differential) means that all the increments Δt are calculated in the direction where t decreases. Note that under coordinate transformation φ equation (1.5) is transformed according to the law

$$d_*\varphi(\xi(t))=\varphi'(b_*(t))d_*t-\frac{\sigma^2}{2}\operatorname{tr}\varphi''_\xi d_*t+\sigma\varphi' d_*w_*(t)$$

Remark 2. It can easily be seen that solutions $\xi(t)$ of equations (1.2) and (1.5) satisfy the equalities $b(t,x) = \lim_{\Delta t\to+0} E(\frac{\xi(t+\Delta t)-\xi(t)}{\Delta t}|\xi(t)=x)$ and $b_*(t,x) = \lim_{\Delta t\to+0} E(\frac{\xi(t)-\xi(t-\Delta t)}{\Delta t}|\xi(t)=x)$.

Following Nelson, we shall call the quantities $v = \frac{1}{2}(D+D_*)\xi = \frac{1}{2}(b+b_*)$ and $u = \frac{1}{2}(D-D_*)\xi = \frac{1}{2}(b-b_*)$ current and osmotic velocity of a process ξ , respectively. Introducing the current $D_c = \frac{1}{2}(D+D_*)$ and osmotic $D_o = \frac{1}{2}(D-D_*)$ derivatives, we obtain $v = D_c\xi$ and $u = D_o\xi$. The osmotic velocity can be described as follows. It is known that for $\xi(t)$ there exists on $[0,T]\times R^n$ a probability density $\rho(t,x)$ such that for any continuous function $f(x,t)$ on $[0,T]\times R^n$ we have

$$\int_{[0,T]\times R^n} f\rho\, d\mu = \int_{[0,T]\times\Omega} f(\xi(t))dPdt \tag{1.6}$$

where μ is the Lebesgue measure on $[0,T]\times R^n$. Finally, one obtains $u=\frac{1}{2}\sigma^2 \text{grad} \ln\sqrt{\rho}$. The proof of this formula can be found in [23,24,26] .

For a vector field $Z(t,x)$ the forward DZ and backward D_*Z derivatives along ξ are defined by the relations

$$DZ=\lim_{\Delta t\to+0} E_t\left(\frac{Z(t+\Delta t,\xi(t+\Delta t))-Z(t,\xi(t))}{\Delta t}\right),$$
$$D_*Z=\lim_{\Delta t\to+0} E_t\left(\frac{Z(t,\xi(t))-Z(t-\Delta t,\xi(t-\Delta t))}{\Delta t}\right) \tag{1.7}$$

Using the methods, which are customary in the theory of diffusion processes, one finds

$$DZ=\left(\frac{\sigma^2}{2}\Delta+b\cdot\nabla+\frac{\partial}{\partial t}\right)Z,$$
$$D_*Z=\left(-\frac{\sigma^2}{2}\Delta+b_*\cdot\nabla+\frac{\partial}{\partial t}\right)Z \tag{1.8}$$

where $\Delta=\left(\frac{\partial^2}{(x^1)^2}+\ldots+\frac{\partial^2}{(x^n)^2}\right)$ is Laplacian, $\nabla=\left(\frac{\partial}{\partial x^n},\ldots,\frac{\partial}{\partial x^1}\right)$, and point denotes scalar product in R^n.

Acceleration of a process $\xi(t)$ is said to be the vector

$$a=\frac{1}{2}(DD_*+D_*D)\xi=\frac{1}{2}(Db_*+D_*b)=(D_cD_c-D_oD_o)\xi=D_cv-D_ou \tag{1.9}$$

Apparently, $D_c v = \frac{\partial}{\partial t} v + (v \cdot \nabla)v$, $D_o u = \frac{1}{2}\sigma^2 \Delta u + (u \cdot \nabla)u$, i.e.

$$a=(\frac{\partial}{\partial t} +v\cdot \nabla)v-(\frac{1}{2}\sigma^2\Delta +u\cdot \nabla)u. \tag{1.10}$$

Let on R^n there be given a mechanical system with a force vector field $F(t,x,\dot{x})$. The trajectory of this system is described by Newton's second law

$$\ddot{x}=F(t,x,\dot{x}) \tag{1.11}$$

The basic postulate of Nelson's stochastic mechanics reads: the trajectory of a quantum-mechanical system is a diffusion process with the diffusion coefficient $\sigma^2 I$, which is described by the equation

$$a=F(t,\xi,v) \tag{1.12}$$

This equation is an obvious analogue of Newton's law (1.11), and turns to this law as $\sigma \longrightarrow 0$. The transition from deterministic processes and equation (1.11) to diffusion processes and equation (1.12) is called stochastic quantization.

Let us consider the simplest case where the classical system is natural, i.e. $F = -\text{grad } V$ is a conservative force. Suppose also that $v = \frac{1}{2}\sigma^2 \text{grad } S$ (note that S has a meaning of action). Recall that we always have $u = \sigma \text{ grad } R$, where $R = \ln\sqrt{\rho}$. Let us consider a complex-valued function $\psi = e^{R+iS}$. It is a simple matter to prove (see [23,24,26]) that for an appropriate gauge of S (which does not alter grad S) the function ψ satisfies the Schrödinger equation

$$\frac{\partial\psi}{\partial t} = i\frac{\hbar}{2m}\Delta\psi - i\frac{I}{\hbar} V\psi \tag{1.13}$$

where $\hbar/2m = \frac{1}{2}\sigma^2$ (see above).

The additional condition $v = \frac{1}{2}\sigma^2 \text{grad } S$ is not very severe, since

The solution of (1.12) for which this condition is fulfilled can be reconstructed from any solution of (1.13). Indeed, let ψ be a solution of (1.13) for which $\int |\psi|^2 d\mu = 1$. Let us represent this solution in the form $\psi = e^{R+iS}$, and specify $u = \sigma \operatorname{grad} R$, $v = 2\sigma \operatorname{grad} S$, and $b = v + u$. Let ξ be a solution of (1.2) with $b = v + u$. Then, it satisfies (1.12) with $F = -\operatorname{grad} V$, and the wave function constructed from ξ coincides with ψ.

Note that for more complicated forces a relationship between v and S is more intricate. For example, let 1-form of the force be $\tilde{F} = -dV + d\omega(\ \ ,\dot{x})$, where ω is a 1-form. Then, $\tilde{v} + \omega = \frac{1}{2}\sigma^2 dS$, where $\tilde{v}$ is the 1-form which is physically equivalent to the vector v.

Remark 3. Here we "quantize" directly Newton's second law, while in ordinary quantization methods we proceed from the classical Lagrange and Hamilton equations, so that the scopes of various quantization methods differ. However, in all the cases where both quantization methods are applicable their scopes are equivalent. Among the results known to the author, the most general cases (with velocity-dependent forces) are considered in [26,30]. Note that variational stochastic principles have been developed (see [19,31]) which lead to (1.12) under the quantization of systems obeying the Lagrangian formalism. In the description of purely quantum effects, which do not have a classical analogue, additional terms (forces) can (naturally) appear on the right-hand side of (1.12), "spin" forces arising in the stochastic mechanics of systems with spin being an example [26]. In the limiting process to classical equation (1.11) these forces vanish.

Remark 4. Let us explain why definition (1.9) of the acceleration is natural. E. Nelson [24] demonstrated that among physically correct definitions of acceleration, which are symmetric with respect to time inversion and usually lead to a conventional definition for smooth trajectories, only relation (1.9) gives a correct result in certain particular cases. Later it was shown that acceleration a , as defined by (1.10), naturally arises from the variational principles mentioned in Remark 3 (see [26]).

2. Stochastic mechanics on Riemannian manifolds

The systems describing stochastic mechanics are closely related to the configuration space geometry. For example, the theory of stochastic differential equations on manifolds deals (because of the non-tensor character of these equations) with interesting geometrical constructions. The Riemannian metric on the configuration space defines kinetic energy (just as in ordinary geometrical mechanics), and also a field of Wiener processes which describes the motion of the system under study. The curvature of the configuration space is taken into account in Newton's law of stochastic mechanics, etc. These topics are briefly considered in this Section.

As was already mentioned in Remark 1, a diffusion process with positive definite diffusion coefficient defines on a manifold the Riemannian metric (see, for example, [26]). In the present paper we shall use an essentially different approach: proceeding from the Riemannian metric, we shall construct a standard field of Wiener processes in spaces tangent to the manifold, and then consider the diffusion processes as solutions of the corresponding stochastic differential equations.

Let M be an n-dimensional Riemannian manifold. Consider the bundle $\pi : O(M) \longrightarrow M$ of orthonormal bases on M and the Levi-Civita connection (i.e. connection without torsion) H on O(M) (see [4]). It is known that H, as a subbundle of TO(M), is a trivial bundle over O(M), trivialization being given by the smooth (and linear on fibers) mapping $E: O(M) \times R^n \longrightarrow H$ which is described as follows. Let $x = (x^1, \ldots, x^n) \in R^n$ and $\bar{b} \in O_{\pi\bar{b}}(M)$ be an orthonormal basis in $T_{\pi\bar{b}}M$. The basis $\bar{b}$ may be considered as a linear operator $\bar{b}: R^n \longrightarrow T_{\pi\bar{b}}M$; in this case $\bar{b}x$ is a vector in $T_{\pi\bar{b}}M$ with the coordinates $(x^1, \ldots, x^n)$ in the basis $\bar{b}$. By definition (see [4]), we have $E(\bar{b},x) = T\pi^{-1}(\bar{b}x)|_{H_{\bar{b}}}$. Fixing $x \in R^n$, we obtain on O(M) a vector field E(x), which is called a basis field; a vector of this field in $\bar{b} \in O(M)$ is denoted by $E_{\bar{b}}(x)$.

Let us consider in R^n a Wiener process w(t). At a certain point $m \in M$ we consider two bases $\bar{b}_1, \bar{b}_2 \in O_m(M)$. Apparently, the processes $T\pi E_{\bar{b}_1}(w(t))$ and $T\pi E_{b_2}(w(t))$ in T_mM are obtained from each other through the action of an orthogonal operator. Since a Wiener process is invariant relative to the action of an orthogonal group, we obtain in each tangent space T_mM the Wiener process constructed from the Wiener process w in R^n. These processes in tangent spaces will also be denoted by w.

Choosing a fixed chart on M, we can describe the field of Wiener processes in tangent spaces as images of w from R^n under the action of a certain section of $O(M)$ over the chart (recall that sections of $O(M)$ do exist over a chart). Thus, a Wiener process in T_mM is defined as $A_m w$, where $A_m: R^n \longrightarrow T_mM$ is a linear (smooth in m) operator such that for $X,Y \in T_mM$ we have $\langle X,Y \rangle = (A_m^{-1}X)\cdot(A_m^{-1}Y)$.

Let $b(t,m)$ be a vector field on M. By $(b(t,m)dt \,\widetilde{+}\, \sigma\, dw(t))$ we denote the class of stochastic processes $\eta(t)$ in T_mM such that $\eta(t) = 0$ and the equalities $E_t^{\eta}(\eta(t+\Delta t)) = b(t,\eta)\Delta t + o(\Delta t)$, $E_t^{\eta}(\eta(t+\Delta t)^2) = \sigma^2 n\Delta t + o(\Delta t)$ hold true for $\Delta t > 0$; $n = \dim M$. The expression

$$d\xi(t)=\exp_{\xi(t)}(\widetilde{b(t,\xi(t))dt+\sigma dw(t)}) \tag{2.1}$$

where $\exp_m: T_mM \longrightarrow M$ is exponential mapping of the Levi-Civita connection, means that for each t the process $\xi(t+\Delta t)$, for sufficiently small $\Delta t > 0$, a.s. belongs to the class $\exp_{\xi(t)}(b(t,\xi(t))dt \,\widetilde{+}\, \sigma dw(t))$. In local coordinates expression (2.1) is equivalent to the stochastic differential equation

$$d\xi(t)=b(t,\xi(t))dt-\frac{\sigma^2}{2}\operatorname{tr}\Gamma_{\xi}(A_{\xi},A_{\xi})dt+\sigma A_{\xi}dw(t) \tag{2.2}$$

where $\Gamma_m(X,Y)$ is a local connector of the Levi-Civita connection in the given chart (see [10,11]). Under change of coordinates, Eq. (2.2) is transformed covariantly. A detailed description of this approach to stochastic differential equations on manifolds is presented in [2,3,5] (see also [18]). Apparently, expression (2.1) is an analogue of (1.2).

We now describe an analogue of equation (1.5). Let $b_*(t,m)$ be a vector field on M. By $(b_*(t,m)d_*t + \widetilde{\sigma d_*w_*(t)})$ we denote the class of stochastic processes $\eta(t)$ in T_mM for which $\eta(0) = 0$ and the equalities $E_t^{\eta}(\eta(t-\Delta t)) = -b_*(t,\eta)\Delta t + O(\Delta t)$ and $E_t^{\eta}(\eta(t-\Delta t)^2) = \sigma^2 n\Delta t + O(\Delta t)$ hold true for $\Delta t > 0$. The expression

$$d_*\xi(t)=\exp_{\xi(t)}(b_*(t,\xi)\widetilde{d_*t+\sigma d_*w_*(t)}) \tag{2.3}$$

implies that for $t\in[0,T]$ the process $\xi(t-\Delta t)$ on M, for sufficiently small $\Delta t>0$, a.s. belongs to the class $\exp_{\xi(t)}(b_*(t,\xi(t))d_*t \widetilde{+ \sigma d_*w_*(t)})$. It can easily be seen that in local coordinates expression (2.3) is equivalent to the following stochastic differential equation:

$$d_*\xi(t)=b_*(t,\xi(t))d_*t+\frac{\sigma^2}{2}\operatorname{tr}\Gamma_{\xi}(A_{\xi},A_{\xi})d_*t+\sigma A_{\xi}d_*w_*(t) \tag{2.4}$$

where A_m is the operator appearing in (2.2).

As previously, for the diffusion process $\xi(t)$ (i.e. for the solution of (2.1) with a certain $b(t,m)$) there exists a field $b_*(t,m)$ such that $\xi(t)$ satisfies (2.3). Defining D and D_* by relations similar to (1.3) and (1.4), we obtain: $D\xi(t) = b(t,\xi(t))$ and $D_*\xi(t) = b_*(t,\xi(t))$. Introduce the objects : $D_c=\frac{1}{2}(D + D_*)$, $D_o = \frac{1}{2}(D - D_*)$, $v = D_c\xi = \frac{1}{2}(b + b_*)$, and $u = D_o\xi = \frac{1}{2}(b - b_*)$. Then, the following relation is valid: $u = \sigma\operatorname{grad}\ln\sqrt{\rho}$, where is defined by formula (1.6). The proof of these assertions (though in another notation) is presented in [26].

Let Z be a vector field on M. Riemannian parallel displacement of vectors is defined along the process $\xi(t)$ (see [21]). The result of parallel displacement of a (random) vector Y from $\xi(s)$ to $\xi(r)$ will be denoted by $\Gamma_{s,r}Y$. Then, DZ and D_*Z are defined, similarly to ordinary covariant derivative, by the relations

$$DZ=\lim_{t\to +0} E_t^{\xi}\left(\frac{\Gamma_{t+\Delta t,t} Z(t+\Delta t, \xi(t+\Delta t))-Z(t,\xi(t))}{\Delta t}\right), \tag{2.5}$$

$$D_* Z=\lim_{t\to +0} E_t^{\xi}\left(\frac{Z(t,\xi(t))-\Gamma_{t-\Delta t,t} Z(t-\Delta t, \xi(t-\Delta t))}{\Delta t}\right).$$

The following analogue of relation (1.8) is valid:

$$\begin{aligned} DZ &= \frac{\sigma^2}{2}\nabla^2 Z+ \nabla_b Z+ \frac{\partial}{\partial t} Z \\ D_* Z &= -\frac{\sigma^2}{2}\nabla^2 Z+ \nabla_{b_*} Z+ \frac{\partial}{\partial t} Z \end{aligned} \tag{2.6}$$

where $\nabla_X Y$ is covariant derivative of the Levi-Civita connection of the vector field Y with respect to the vector field X, ∇^2 is the Laplace-Beltrami operator (in local coordinates $\nabla^2 = \sum_{i,j} g^{ij}\nabla_i\nabla_j$, where $(g^{ij})^{-1} = (g_{ij})$ is the Riemannian metric matrix and ∇_k is covariant derivative with respect to the k-th coordinate). Similarly to Section 1, we introduce acceleration: $a = \frac{1}{2}(DD_* + D_*D)\xi = \frac{1}{2}(Db_* + D_*b)$. Apparently, $D_c v = \frac{\partial}{\partial t} v + \nabla_v v$, $D_o u = \frac{1}{2}\sigma^2\nabla^2 u + \nabla_u u$, and

$$a=(D_cD_c-D_oD_o)\xi = D_c v-D_o u=\left(\frac{\partial}{\partial t}v+ \nabla_v v\right)-\left(\frac{\sigma^2}{2}\nabla^2 u+\nabla_u u\right) \tag{2.7}$$

It should be emphasized that the expression for $D_c v$ resembles an ordinary formula for covariant derivative, with respect to time, for a non-atonomous vector field along its flow.

Let M be the configuration space of a mechanical system with kinetic energy $K(\dot m) = \frac{1}{2}\langle \dot m, \dot m\rangle$, $\dot m \in TM$, and with a force vector field $F(t,m,\dot m)$. Then (see, for example, [16]), the trajectory of this mechanical system $m(t)$ is described by the equation (an analogue of Newton's second law)

$$\nabla_{\dot m(t)}\dot m(t)=F(t,m,\dot m) \tag{2.8}$$

The equation

$$a = F(t, \xi, v) - \frac{1}{2}\sigma^2 \hat{R}(\xi) \cdot u \tag{2.9}$$

is an analogue of Eq. (1.12), i.e. of Newton's law in stochastic mechanics. Here the linear operator $\hat{R}(m): T_m M \longrightarrow T_m M$ is the Ricci (1,1) tensor of the Riemannian metric $\langle \, , \, \rangle$ at point m. It is the presence of $\hat{R} \cdot u$ on the right-hand side of Eq. (2.9) that makes its solution naturally related to the solution of the corresponding Schrödinger equation (similarly to Section 1; see [6,26]).

Remark 5. As is known, for any vector field X the following relation (Weitzenböck's formula) is valid: $\nabla^2 X - \hat{R} \cdot X = \Delta X$, where $\Delta = (d\delta + \delta d)$ is the Laplace-de Rham operator (here d is exterior differential and δ , codifferential). Since the Riemannian metric there exists on M, we need not distinguish between vector fields and 1-forms. If we defined acceleration $\bar{a}$ by the formula $\bar{a} = (\frac{\partial}{\partial t} v + \nabla_v v) - (\frac{\sigma^2}{2}\Delta u + \nabla_u u)$, we would obtain Newton's law of stochastic mechanics in the conventional form: $\bar{a} = F$. Precisely in this form this law is described in [6,26] , and precisely this form of Eq. (2.9) is used to prove the aforementioned relationship with the Schrödinger equation. To make the definition of acceleration $\bar{a}$ natural, the concept of parallel displacement is modified in [7,8,26] in such a manner that in relations (2.6) Δ is substituted for ∇^2 (modified parallel displacement allows for the deviation of geodesics). In this case, $\bar{a} = \frac{1}{2}(\bar{D}\bar{D}_* + \bar{D}_*\bar{D})\xi$, where $\bar{D}$ and $\bar{D}_*$ are defined by relations (2.5) using modified parallel displacement.

3. The hodograph of current velocity and the stochastic integral with Riemannian parallel displacement

In this section we modify the construction of the curvilinear stochastic integral with Riemannian parallel displacement and demonstrate, using this modified integral, that equation (2.9) is reduced to a stochastic equation in the tangent space in the initial configuration of the system, i.e. to the so-called equation of the current velocity hodograph. For the sake of simplicity, we shall consider a compact manifold M and deterministic initial conditions for the solution of (2.9). We shall also choose the system of units in which $\sigma = 1$.

Let, on M, there be given a vector field $p(t,m)$. Consider the equation of the form

$$D_c \xi(t)=p(t,\xi(t)) \tag{3.1}$$

Definition 1. A solution of equation (3.1) is said to be a stochastic process $\xi(t)$ on M which is a solution of equation (2.1) with a certain $b(t,m)$ for which (3.1) is a.s. valid for any t.

To analyse equation (3.1), we construct the stochastic integral. Let $y \in H_b$, $b \in O(M)$, and $E(x)$ be a basis vector field on $O(M)$ such that $E_b(x) = y$. Consider the integral curve $\gamma_y(t)$ of the field $E(x)$ with the initial condition $\gamma_y(0) = b$.

Definition . Let $e:H \longrightarrow O(M)$ denote the mapping which sends any vector $y \in H$ into the point $\gamma_y(1)$.

Since $O(M)$ is a compact manifold (we have assumed above that M is compact), the point $\gamma_y(1)$ exists for any $y \in H$. It is known that $\pi e(y) = \exp T\pi\, y$ (see, for example, [4]). The mapping e may be considered as restriction to H of the exponential mapping of a certain connection on $TO(M)$.

Let us fix a point $m_o \in M$ and a basis $\hat{b} \in O_{m_0}(M)$. Let in $T_{m_0}M$ there be given a Markovian process $v(t)$ with a.s. continuous trajectories. Consider the families P_t^v, F_t^v, and N_t^v corresponding to this process. Let in $T_{m_0}M$ there exist a Wiener process $w(t)$ adapted to P_t^v. We define the process $D_*^v w(t)$ by the relation $D_*^v w(t) = \lim_{t \to +0} E(\frac{w(t)-w(t-\Delta t)}{\Delta t} | P_t^v)$.

Let us consider on $O(M)$ the stochastic differential equation

$$d\xi(t)=e\cdot E_{\xi(t)}(\hat{b}^{-I}\widetilde{(v(t)-\tfrac{I}{2}D_*^v w(t))dt+dw(t))} \tag{3.2}$$

Theorem 1. For any initial condition $\xi(0) = \bar{b} \in O(M)$ there exists a unique strong solution of Eq. (3.2), which is defined on $[0,T]$ and is Markovian with respect to P_t.

Theorem 1 is proved similarly to Theorem 2 of Ref. [18] (see also the proof of the existence of stochastic development [12]). It is important in this case that the process $\int_0^t (v(\ )-\frac{1}{2}D_*^v w(\ ))d\tau \ +w(t)$ satisfies the conditions under which its development can be constructed as in [12] . Local existence and uniqueness follow from the smoothness of the mapping E, and global existence follows from the compactness of O(M).

Remark 6. Compactness of M may be rejected if we require that the Riemannian metric on M should possess a Riemannian uniform atlas [17,18] (in particular, in this case M is a complete manifold) in the charts of which there are satisfied some conditions of uniform boundedness for **local connector** of the Levi-Civita connection. It is shown in [17] that for any Riemannian metric there exists the corresponding conformal metric which possesses a Riemannian uniform atlas.

Let us denote the solution of Eq. (3.1) with the initial condition $\xi(0) = \hat{b}$ by $\hat{\xi}(t)$, and introduce the notation $Sv(t) = \pi\hat{\xi}(t)$.

Lemma 1. The process $Sv(t)$ on M does not depend on the choice of the initial basis $\hat{b}$ in $O_{m_0}(M)$.

The proof of Lemma 1 is similar to the proof of independence of Cartan development of the choice of the initial basis [4] . The stochastic case is considered in [12,18] .

Theorem 2. For any $t \in [0,T]$ the stochastic vector $D_c Sv(t)$ is parallel along Sv to the stochastic vector $v(t) \in T_{m_0}M$, and the vector $D_0 Sv(t)$ is parallel to $-D_* w(t)$.

To prove the theorem, suffice it to notice that $\hat{\xi}(t)$ is parallel displacement of the basis $\hat{b}$ along Sv. Thus, $T\pi E_{\hat{\xi}(t)}(b^{-1}v(t))$ is parallel along Sv to the vector $v(t)$. Since the process Sv is Markovian, it follows from (3.2), with an allowance for the definition of D_c and D_0, that the vectors $D_0 Sv$ and $-D_* w$ are parallel and that the relation $D_c Sv(t) = T\pi E_{\xi(t)}(b^{-1}v(t))$ is valid.

The operator S thus constructed is similar to the stochastic integral with variable upper limit in R^n. A specific construction of this operator on a Riemannian manifold leads to the fact that the integrand belongs to the tangent space, while the integral belongs to the manifold.

Let, on M, there be given a vector field $p(t,m)$ and a stochastic process $\xi(t)$, $\xi(0) = m_0$, along which parallel displacement is defined. By $\Gamma p(t,\xi(t))$ we denote a stochastic process in $T_{m_0}M$ obtained by parallel displacement of vectors $p(t, \xi(t))$ along ξ

Theorem 3. A process $v(t)$ in $T_{m_0}M$, $t \in [0,T]$, a.s. satisfies the relation

$$v(t)=\Gamma p(t,Sv(t)) \tag{3.3}$$

if and only if the process $Sv(t)$ is a solution of (3.1).

Theorem 3 is a consequence of Theorem 2 and of parallel displacement properties.

The solution of Eq. (3.3) has a mechanical meaning of the current velocity hodograph for the solution of (3.1).

We now describe the equation of the current velocity hodograph for a trajectory of a stochastic mechanical system. For any process of the type Sv, on M, we can find D_cSv, D_oSv, and D_oD_oSv. The stochastic vector

$$B(t,Sv)=F(t,Sv(t),D_cSv(t))-\sigma\hat{R}(Sv(t))\cdot D_oSv(t)-D_oD_oSv(t) \tag{3.4}$$

is called current force acting on Sv at a moment t.

Theorem 4. The equation of the current velocity hodograph for the solution of Eq. (2.9) is of the form

$$v(t)=v(0)+\int_0^t(\Gamma B(\tau,Sv)-\tfrac{1}{2}D_*w(\tau))d\tau+w(t). \tag{3.5}$$

If v is a weak solution of (3,5), then Sv is a weak solution of (2.9).

Indeed, let on a certain probabilistic space there exist a process $v(t)$ in $T_m M$ and a Wiener process $w(t)$ in $T_m M$; these processes are assumed to be adapted to P_t^v and be such that for them relation (3.5) holds almost surely for any t. Then, $v(t)$ has a.s. continuous trajectories and is a Markovian process. According to Theorem 2, in this case the vector $D_c Sv$ is parallel along Sv to the vector $v(0) + \int_0^t (\Gamma B(\tau,Sv)-\frac{1}{2}D_* w(\tau))d\tau + w(t)$. Then, by definition of the vector $D_c D_c Sv$ we obtain $D_c D_c Sv(t) = B(t,Sv)$, i.e. Sv satisfies (2.9).

4. Stochastic mechanics with constraints

In this section we describes one of the possible approaches to stochastic mechanics with linear (generally non-holonomic) constraints, proceeding from the geometrical definition of a mechanical system with constraints [27-29].

Let us consider a mechanical system with configuration space M, and let the kinetic energy of the system be given by the Riemannian metric $\langle\ ,\ \rangle$ (see Section 2). A smooth distribution β on M (a subbundle of the tangent bundle) is called a linear constraint imposed on the system. The condition imposed by the constraint β is that the velocity vectors along a trajectory of the mechanical system must belong to the distribution β.

Let $P:TM \longrightarrow \beta$ denote the operator of orthogonal (relative to the Riemannian metric) projection of fibers of TM on their subspaces ($P_m : T_m M \longrightarrow \beta_m$ for each $m \in M$). Let also $F(t,m,\dot{m})$ be a force vector field. The equation of motion of a system with constraint is represented by the following analogue of Newton's law:

$$\overline{\nabla}_{\dot{m}}\dot{m} = PF(t,m,\dot{m}) \tag{4.1}$$

where $\overline{\nabla}_X Y = P\nabla_X Y$ (∇ is covariant derivative of the Levi-Civita connection on M). If the distribution β is involutary (i.e. integrable, according to the Frobenius theorem), the constraint β is called holonomic. In this case, making restriction to an integral constraint submanifold, we obtain a mechanical system without constraint, but on a manifold of lower dimension. The case of a non-involutary (i.e. non-integrable) distribution, where the constraint is called non-holonomic, requires the introduction of auxiliary constructions.

Let us consider the principal bundle $\pi : O^{\beta}(M) \longrightarrow M$ of "partial" orthonormal frames (i.e. $\tilde{b} \in O_m(M)$ is an orthonormal frame in β_m) with the stuctural group $O(k)$, $k = \dim \beta$. It is shown in [29] that $\overline{\nabla}$ is covariant derivative on $O^{\beta}(M)$, i.e. it is generated by a certain connection $\bar{H}$ (and by the corresponding parallel displacement) on $O^{\beta}(M)$. This connection has been called truncated. Let us define on $O^{\beta}(M)$ basis vector fields just as it was done in the classical case: for $x \in R^k$ and $\tilde{b} \in O^{\beta}(M)$ we define the basis vector $\bar{E}_{\tilde{b}}(x)$ by the relation $\bar{E}_{\tilde{b}}(x) = T\bar{\pi}^{1}\tilde{b}x_{|\bar{H}_{\tilde{b}}}$. It can easily be seen that $\bar{E} : O^{\beta}(M) : R^k \longrightarrow TO^{\beta}(M)$ is a smooth mapping which is linear on the fibers.

We now consider a Wiener process $w(t)$ in R^k. Let $b_1, b_2 \in O_m^{\beta}(M)$. Apparently, the processes $T\pi\bar{E}_{\tilde{b}_1}(w)$ and $T\pi\,\bar{E}_{\tilde{b}_2}(w)$ differ only by the action of an operator from $O(k)$. Since w is invariant relative to the action of $O(k)$, we have obtained in each space β_m a Wiener process, which is again denoted by w. In local coordinates, the field of Wiener processes in spaces of the constraint β can be represented as $A_m w$, where $A_m : R^k \longrightarrow \beta_m$ is a linear operator such that $(A_m^* A_m)^{-1}$ defines the Riemannian metric in β_m, i.e.
$\langle X,Y \rangle = (A_m^{-1}X)\cdot(A_m^{-1}Y)$; $X,Y \in T_m M$.

<u>Remark 7</u>. One may assume that A_m acts from R^n ($n = \dim M$) into $T_m M$ and that β_m is the image of A_m; in other words, A_m is a degenerate operator. Thus, the introduction of constraint β leads to processes with degenerate diffusion coefficient (cf. Remark 1).

It is more convenient to describe the equations of stochastic mechanics with constraint β in terms of integral-type operators which are analogues of the operators S and Γ appearing in Section 3. For simplicity, we assume, as before, that M is compact, $\xi(0) = m_0$, and $\sigma = 1$. The case of non-compact manifolds is similar to the case with no constraints involved.

Let us fix a point $m_0 \in M$ and a truncated basis $\hat{b} \in O_{m_0}(M)$. Suppose a Markovian process $v(t)$ is given in β_{m_0}, which has a.s. continuous trajectories. Let also a Wiener process $w(t)$ be defined in β_{m_0}, which is adapted to P_t^v. Similarly to Definition 2, we define the mapping $\bar{e}:\bar{H} \longrightarrow O^{\beta}(M)$ and consider on $O^{\beta}(M)$ the following stochastic differential equation:

$$d\xi(t)=\bar{e}\cdot\bar{E}_{\xi(t)}(\hat{b}^{-1}\widetilde{(v(t)-\tfrac{1}{2}D_*^v w(t))dt+dw(t))} \tag{4.2}$$

Equation (4.2) differs from (3.2) only by the manifold on which the former is defined. Exactly as in Section 3, we can prove the existence of solution $\xi(t)$, which is Markovian with respect to P_t^v, on the entire interval $[0,T]$ and for any initial condition $\xi(0) = \bar{b}$. Let $\hat{\xi}(t)$ be a solution with the initial condition $\hat{\xi}(0) = \hat{b}$. The process $\bar{\pi}\hat{\xi}(t)$ will be denoted by $\bar{S}v(t)$; it does not depend on the choice of the initial basis $\hat{b}$ (cf. Section 3). Note that $\hat{\xi}$ is parallel displacement of the truncated basis $\hat{b}$ along Sv, i.e. parallel displacement of vectors from the distribution β is defined along Sv.

Let a vector field $p(t,m)$, lying in the distribution β, be given on M. We shall consider the process $\bar{\Gamma}p(t,\bar{S}v)$ obtained via parallel displacement of vectors $p(t,\bar{S}v(t))$ along Sv to the point m_0. If $v(t) = \bar{\Gamma}p(t,\bar{S}v)$, the process $\bar{S}v$ is a solution of an analogue of (3.1) for the case with a constraint involved.

Let, on M, there be given a force field $F(t,m,\dot{m})$. For a process of type $\bar{S}v$ we can find $D_c\bar{S}v$, $D_0\bar{S}v$, and $D_0D_0\bar{S}v$ following an ordinary definition of Section 2, and define the current force $B(t,\bar{S}v)$ by relation (3.4). The hodograph of the velocity of a trajectory of a stochastic mechanical system with constraint is said to be the solution of the equation

$$v_0(t)=v_0(0)+\int_0^t (\bar{\Gamma}PB(\tau,\bar{S}v_0(\tau))-\tfrac{1}{2}D_*w(\tau))d\tau+w(t) \tag{4.3}$$

in β_{m_0}. Then, the process $\bar{S}v_0$ on M describes the trajectory of a stochastic mechanical system with constraint β.

5. On relativistic stochastic mechanics

An attempt at constructing relativistic stochastic mechanics encounters difficulties of essentially new character (see, for example, [20,25]). In particular, the conventional apparatus of stochastic differential equations is inapplicable in this case. Papers [9,20] describe, without using stochastic differential equations, a system of stochastic mechanics on the Minkowski space, which corresponds to the motion of a scalar charged particle in an external electromagnetic field, and establish natural relations with the Klein-Gordon equation. In this section we briefly outline the main idea of the approach just mentioned and try to modify the stochastic equations in such a manner that they could be used in the relativistic case.

Let us denote the Minkowski space with the signature (+ - - -) by M^4 and decompose M^4 into an orthogonal direct sum $M^4 = R^1 + R^3$, where R^1 is timelike and R^3, spacelike. Changing the sign of the scalar product in R^3, we may consider R^1 and R^3 as Euclidean spaces. In this case we can define independent Wiener process: one-dimensional process in R^1, $w^0(\tau)$, and three-dimensional in R^3, $\overline{w}(\tau)$, where τ is an invariant parameter acting as proper time. Finding $D_* w^0(\tau)$ and $D_*\overline{w}(\tau)$, we can express $w^0_*(\tau)$ and $\overline{w}_*(\tau)$, for any $a \leq b$, as follows:

$$w^0(b)-w^0(a)= \int_a^b D_* w^0(\tau)d\tau \quad +(w^0_*(b)-w^0_*(a)),$$

$$\overline{w}(b)-\overline{w}(a)= \int_a^b D_*\overline{w}(\tau)d\tau \quad +(\overline{w}_*(b)-\overline{w}_*(a)) .$$

For simplicity, we shall use the system of units in which $\sigma = 1$.

Let $\xi(\ )$ be a stochastic process in M^4. Following [9,20] , we shall modify the definitions of the forward and backward derivatives in such a fashion that they become covariant relative to the Lorentz group on M^4. Namely, we define $D\xi(\tau)$ at a point $m \in M^4$ by the relation

$$D\xi(\tau)\underset{m}{=} \lim_{\Delta\tau\to+0} \Big(E\Big(\frac{\xi(\tau+\Delta\tau)-\xi(\tau)}{\Delta\tau}\Big|\, \xi(\tau)=m,\ (\xi(\tau+\Delta\tau)-\xi(\tau))^2 \geq 0\Big) +$$

$$+ E\Big(\frac{\xi(\tau)-\xi(\tau-\Delta\tau)}{\Delta\tau}\Big|\, \xi(\tau)=m,\ (\xi(\tau)-\xi(\tau-\Delta\tau))^2 \leq 0\Big)\Big). \tag{5.I}$$

The derivative $D_* \xi(\tau)$ at point $m \in M^4$ is defined analogously to (5.1) with the corresponding change of signs.

Let $b(m)$ be a timelike vector field on M^4. Consider, on M^4, the process $\xi(\tau)$ described by the covariant equation (an analogue of Eq. (1.2))

$$\xi(\tau)=\xi(0)+\int_0^\tau b(\xi(s))ds+w^o(\tau)+\overline{w}_*(\tau). \qquad (5.2)$$

If $\xi(\tau)$ is a solution of (5.2), the independence of w^o and $\overline{w}$ leads to the equality $D\xi(\tau)_m=b(m)$.

An analogue of (1.5) is

$$\xi(\tau)=\xi(0)+\int_0^\tau b_*(\xi(s))ds+w^o_*(\tau)+\overline{w}(\tau).$$

As before, $D_* \xi(\tau)_m = b_*(m)$. The quantities v, u, D_c, and D_o are defined exactly as in Section 2 with a natural modification. The field u is related to the density ρ, with respect to the invariant Lebesgue measure on M^4, by the ordinary formula (see [9,20]).

For a vector field Z on M^4 we define DZ and D_*Z along $\xi(\tau)$ by transforming relations (1.7) just in the same way as (1.3) has been transformed into (5.1). We obtain (cf. relations (1.8))

$$DZ=(\tfrac{1}{2}\Box + b\cdot\nabla)Z, \qquad D_*Z=(-\tfrac{1}{2}\Box + b_*\cdot\nabla)Z$$

where $\Box$ is d'Alembertian; point denotes scalar product in M^4.

Further constructions are similar to those of Section 1. The equation of motion is analogous to (2.12). The case where the force is given by a 2-form of electromagnetic field is exactly the same as that considered in [9,20] .

Transition to a non-linear Lorentz manifold can be effected via simple modification of the methods described in Section 2.

Assistance of A.A. Talashev, translation editor, is highly appreciated.

REFERENCES

1. de Angelist G.F., Jona-Lasino G., Servo M., and Zanghi N., Stochastic mechanics of a Dirac particle in two spacetime dimensions.- Journ. Phys., Ser. A, 1986, v. 19.

2. Belopol'skaya Ya.I. and Daletskiĭ Yu.L., Ito equations and differential geometry.- Usp. Matem. Nauk, 1982, v. 37, No. 3 (in Russian).

3. Belopol'skaya Ya.I. and Daletskiĭ Yu.L., Stochastic equations and differential geometry.- In: An analysis on manifolds and differential equations. Voronezh, 1986 (in Russian; English translation: Lect. Notes Math., 1986, v. 1214).

4. Bishop R.L. and Crittenden R.J., Geometry of manifolds.- Academic Press, 1964.

5. Daletskiĭ Yu.L., Stochastic differential geometry.- Usp. Matem. Nauk, 1983, v. 38, No. 3 (in Russian).

6. Dankel T.G. (Jr.), Mechanics on manifolds and the incorporation of spin into Nelson's stochastic mechanics.- Arch. Rat. Mech. Anal., 1970, v. 37, No. 3.

7. Dohrn D. and Guerra F., Nelson's stochastic mechanics on Riemannian manifolds.- Lettere al Nuovo Cimento, 1978, v.22, No.4.

8. Dohrn D. and Guerra F., Geodesic correction to stochastic parallel displacement of tensors.- Lect. Notes Phys., 1979, v. 93.

9. Dohrn D., Guerra F., and Ruggiero P., Spinning particles and relativistic particles in the framework of Nelson's stochastic mechanics.- Lect. Notes Phys., 1979, v. 106.

10. Dombrowski P., On the geometry of tangent bundle.- Journ. reine und angew. Math., 1962, Bd. 210, Heft 1/2.

11. Eliasson H.I., Geometry of manifolds of maps.- Journ. Diff. Geometry, 1967, v. 1, No. 2.

12. Elworthy K.D., Stochastic differential equations on manifolds.- Cambridge, 1982.

13. Farris W.G., Spin correlation in stochastic mechanics.- Foundations of Physics, 1982, v. 12, No. 1.

14. Fényes I., Eine wahrscheinlichkeitstheoretische Begründung und Interpretation der Quantenmechanik.- Zeitschrift für Physik, 1952, v. 132.

15. Gikhman I.I. and Skorokhod A.V., The theory of stochastic processes.- Moscow, 1975, v. 3 (in Russian).

16. Gliklikh Yu.E., Riemannian parallel translation in non-linear mechanics.- In: Topological and geometrical methods in mathematical physics. Voronezh, 1983 (in Russian; English translation: Lect. Notes Math., 1984, v. 1108).

17. Gliklikh Yu.E., On Riemannian metrics possessing a Riemannian uniform atlas.- In: Differential geometry of manifolds of figures. Kaliningrad, 1985, No. 16 (in Russian).

18. Gliklikh Yu.E., Riemannian parallel translation, Ito integral, and stochastic equations on manifolds.- In: An analysis on manifolds and differential equations. Voronezh, 1986 (in Russian; English translation: Lect. Notes Math., 1986, v. 1214).

19. Guerra F. and Morato L.M., Quantization of dynamical systems and stochastic control theory.- Phys. Reviews, Ser. D, 1983, p. 1774-1786.

20. Guerra F. and Ruggiero P., A note on relativistic Markov processes. Lettere al Nuovo Cimento, 1978, v. 23, No. 16.

21. Ito K., Stochastic parallel displacement.- Lect. Notes Math., 1975, v. 451.

22. Ito K. Extension of stochastic integrals.- In: Proc. of Intern. Symp. SDE (Kyoto, 1976), New York, 1978.

23. Nelson E., Derivation of Schrödinger equation from Newtonian mechanics.- Phys. Rev., 1966, v. 150, No. 4.

24. Nelson E. Dynamical theories of Brownian motion. Princeton, 1967.

25. Nelson E. "Le mouvement brownien relativiste" by J.-P. Caubet (book review).- Bull. Amer. Math. Soc., 1979, v. 84, No. 1.

26. Nelson E., Quantum fluctuations.- Princeton, 1985.

27. Vershik A.M. and Faddeev L.D., Differential geometry and Lagrangian mechanics with constraints.- Dokl. Akad. Nauk SSSR, 1972, v. 202, No. 3 (in Russian).

28. Vershik A.M. and Faddeev L.D., Lagrangian mechanics in an invariant representation.- In: Problems in theoretical physics. Leningrad, 1975, No. 2 (in Russian).

29. Vershik A.M. Classical and non-classical dynamics with constraints. In: Geometry and topology in global non-linear problems. Voronezh, 1984 (in Russian; English translation: Lect. Notes Math., 1984, v. 1108).

30. Yasue K., Quantum mechanics and stochastic control theory.- J. Math. Phys., 1981, v. 22, No. 5.

31. Yasue K., Stochastic calculus of variations.- J. Funct. Anal., 1981, v. 41.

SINGULARITIES OF SOLUTIONS OF DIFFERENTIAL EQUATIONS ON COMPLEX MANIFOLDS (CHARACTERISTICAL CASE)

B.Yu. Sternin
Moscow Institute of
Electronic Building
Moscow, 109208, URSS

V.E. Shatalov
Moscow Institute of Civil
Aviation Engineers
Moscow, 135838, USSR

The paper continues the study of equations on complex-analytical manifolds with the help of maslov's canonical operator method [1]. In our previous papers [2] - [4] an apparatus was developed allowing to construct explicit solutios of equations with constant coefficients as well as to investigate solutions of equations in non-characteristical case (constructing, in particular, their asymptotics by smoothness). We suggest in this paper a method of constructing asymptotic solutions of Cauchy problem in characteristical case, i.e. in case the initial manifold contains a characteristical submanifold of (complex) codimension 1.

The results of authors' investigations in the field of equations on complex-analytical manifolds and, in particular, the results of present paper have been the topic of a series of lectures delivered at Voronezh Winter Mathematical School (1986). The authors acknowledge the opportunity provided by the organizers of the School to presenting their lectures. Fruitful discussions which accompanied the lectures have been of great use for the authors.

We are most grateful to Academician V.P. Maslov for the statement of the problem of asymptotic investigation of the equations on complex-analytical manifolds. We are also thankful to Prof. Yu.G. Borisovich for his support and to Prof. A.S.Mishchenko for useful discussions and great help.

I. <u>Statement of the problem.</u> Let us consider the equation

$$\hat{H}u = f(x), \tag{1}$$

where

$$\hat{H} = H(x, -\frac{\partial}{\partial x}) = \sum_{|\alpha| \le m} a_{\alpha}(x) (-\frac{\partial}{\partial x})^{\alpha}$$

is a differential operator of order m. The homogeneous component of
order j of the symbol we denote by

$$H_j(x,p) = \sum_{|a|=j} a_\alpha(x)p^\alpha,$$

the function $H_m(x,p)$ being denoted simply by $H(x,p)$.

Let X be an analytical manifold of codimension 1 in "n-dimentio-
nal space $\mathbb{C}^n_x$. Throughout what follows we denote arithmetical spaces
listing their coordinates in square brackets; for example $[x] = [x^1,...,x^n] = \mathbb{C}^n_x$.

We suppose the right-hand side f(x) of the equation (1) be a
multiple-valued analytical function defined in a neighbourhood $U(x_0)$
of a point $x_0 \in X$, singular no more than on the set X and satisfying
the inequality

$$|f(x)| \le c|s(x)|^q.$$

Here $X = \{x \mid s(x) = 0\}$, $ds|_x \ne 0$, q is a real number, $q > -1$. The
set of described functions is denoted by $A_q(X)$.

Let $[x,p]$ be a phase (symplectic) space endowed with the structure form

$$dx \wedge dp = dp_1 \wedge dx^1 + ... + dp_n \wedge dx^n,$$

which is homogeneous with respect to the action of the group   :

$$\lambda(x,p) = (x, \lambda p), \quad \lambda \ne 0.$$

Suppose the manifold X contains a subset of points characteristic
with respect to $H(x,p) = H_m(x,p)$, i.e. points where

$$H\left(x, \frac{\partial s(x)}{\partial x}\right) = 0.$$

The set of such points we denote by char X; we suppose $x_0 \in$ char X.

Any analytical manifold X determines a Lagrangian manifold

$$L_X = \left\{(x,p) \mid p = \lambda \frac{\partial s(x)}{\partial x}, \lambda \in \mathbb{C}_* \right\} \subset [x,p]$$

in the space $[x,p]$. The quotient space $L_x / \mathbb{C}_* = l_x$ is a Legendre
manifold in $[x,p]/\mathbb{C}_*$ biholomorphic to X. The map $\alpha: x \longmapsto$
$\longmapsto (x, cls(\lambda \frac{\partial s}{\partial x}))$ is a biholomorphism X on l_x.

Let

$$V(H) = H_p(x,p) \frac{\partial}{\partial x} - H_x(x,p) \frac{\partial}{\partial p}$$

be a Hamulton vector field defined by $H(x,p)$. The corresponding contact vector field we denote by X_H. The contact vector field trajectory emerging from the point $\alpha(x)$, corresponding to the point $x \in X$ we shall call for shortness <u>the trajectory with origin in $x \in X$</u> and its projection we shall call <u>the ray with origin in $x \in X$</u>. The ray with origin in x is evidently a projection of every trajectory of Hamylton vector field with origin in a point in L_x projecting on x (i.e. in a point of a form $(x, \lambda\, \partial s(x)/\partial x), \lambda \in \mathbb{C}_*$).

<u>Condition 1</u>. Char X is a submanifold in X of codimension 1 transversal to the ray with origin in $x_0 \in X$.

Analytically this condition means that

$$H_p\left(x_0, \frac{\partial s}{\partial x}(x_0)\right) \frac{\partial}{\partial x}\left[H\left(x, \frac{\partial s(x)}{\partial x}\right)\right]\Big|_{x=x_0} \neq 0.$$

The union of rays with origins in points $x \in$ char X we denote by Y.

<u>Lemma 1</u>. The set Y is an analytical submanifold in $U(x_0)$ of codimention 1. This submanifold is characteristical; it intersects with X only in points of char X and it is tangent to X in such points.

The assertion of the Lemma is valid, of course, in a sufficiently small neighbourhood of the point x_0. Such refinements will be omitted below.

<u>Proof</u>. Due to Condition 1 the submanifold α(char X) has the dimension n-2 and is transversal to the vector field X_H. Therefore its inverse image $\tilde{\alpha}$ (char X) in the space $[x,p]$ has the dimension n-1 and is transversal to V(H). The manifold $\tilde{\alpha}$(char X) lies on the zero level surface of the Hamiltonian H due to the definition of the set char X. Consequently (see [5]) the phase flow L' of the manifold $\tilde{\alpha}$ (char X) along V(H) is a Lagrangian manifold lying on the zero level surface of the Hamiltonian H. From the other hand, the projection of L' on $[x]$ evidently coincides with Y. Thus, $L' = L_Y$ and consequently Y is characteristic:

$$Y = \left\{ x \,\middle|\, s_1(x) = 0 \quad, ds_1(x)\big|_Y \neq 0;\ H\left(x, \frac{\partial s_1(x)}{\partial x}\right) \equiv 0\right\}.$$

Since L_x coincides with L_Y in points lying over char X, X and Y are tangent in such points, i.e.

$$\frac{\partial s(x)}{\partial x^i} = \lambda \frac{\partial s_1(x)}{\partial x^i} \quad, i = 1, \ldots, n,\ \lambda \in \mathbb{C}_* \text{ on char X.}$$

For $\lambda = 1$ we have

$$\dot{s}(x) = H_p\left(x, \frac{\partial s_1(x)}{\partial x}\right)\frac{\partial s(x)}{\partial x} = H_p\left(x, \frac{\partial s(x)}{\partial x}\right)\frac{\partial s(x)}{\partial x} \neq 0$$

on char X (the natural derivative of s with respect to natural parameter of the ray being denoted by $\dot{s}$). We have also

$$\ddot{s}(x_0) = H_p\left(x_0, \frac{\partial s}{\partial x}(x_0)\right)\frac{\partial}{\partial x}\left\{H_p\left(x, \frac{\partial s_1(x)}{\partial x}\right)\frac{\partial s(x)}{\partial x}\right\}\Big|_{x=x_0} =$$

$$= H_{p_i}\left(x_0, \frac{\partial s(x_0)}{\partial x}\right)\left\{H_{p_j x^i}\left(x_0, \frac{\partial s}{\partial x}(x_0)\right)\frac{\partial s}{\partial x^j}(x_0) + H_{p_j p_k}\left(x_0, \frac{\partial s}{\partial x}(x_0)\right)\times\right.$$

$$\left.\times\frac{\partial^2 s_1}{\partial x^k \partial x^i}(x_0)\frac{\partial s}{\partial x^j}(x_0) + H_{p_j}\left(x_0, \frac{\partial s}{\partial x}(x_0)\right)\frac{\partial^2 s}{\partial x^i \partial x^j}(x_0)\right\}. \tag{4}$$

Due to homogeneity of H(x, p) we have

$$H_{p_j x^i}\left(x_0, \frac{\partial s}{\partial x}(x_0)\right)\frac{\partial s}{\partial x^j}(x_0) = m\, H_{x^i}\left(x_0, \frac{\partial s}{\partial x}(x_0)\right);$$

$$H_{p_j p_k}\left(x_0, \frac{\partial s}{\partial x}(x_0)\right)\frac{\partial s}{\partial x^j}(x_0) = (m-1)\, H_{p_k}\left(x_0, \frac{\partial s}{\partial x}(x_0)\right).$$

Taking into account the equality

$$H_{x^i}\left(x_0, \frac{\partial s}{\partial x}(x_0)\right) + H_{p_k}\left(x_0, \frac{\partial s}{\partial x}(x_0)\right)\frac{\partial^2 s_1}{\partial x^i \partial x^k}(x_0) = 0,$$

which can be proved by differentiation of the relation $H\left(x, \frac{\partial s_1(x)}{\partial x}\right) = 0$ with respect to x^i, we reduce (4) to the form

$$\ddot{s}(x_0) = H_{p_i}\left(x_0, \frac{\partial s}{\partial x}(x_0)\right)\left\{H_{x^i}\left(x_0, \frac{\partial s}{\partial x}(x_0)\right) + H_{p_j}\left(x_0, \frac{\partial s}{\partial x}(x_0)\right)\frac{\partial^2 s}{\partial x^i \partial x^j}(x_0)\right\}.$$

The latter expression does not vanish due to condition (3).

Let us consider the function s(x) on Y as a function of variables (α, t), where α are coordinates on char X and t is a parameter along the ray. The point t = 0 is a zero of second order in $\alpha = \alpha_0$ (α_0 corresponding to the point x_0). Therefore there exists such a neighbourhood of zero in the plane [t] that the function s(x) has exactly two zeros with regard to their multiplicity, α being close to α_0. If we take into account that t = 0 is a zero of the function s(x) of the second order, we can see that in the described neighbourhood s(x) has no zeroes except t = 0, i.e. X and Y intersect near char X only in the points of char X. It completes the proof of the Lemma.

Remark 1. In the proof of the Lemma we used the fact that manifolds are tangent in some point iff this point lays in the projection of intersection of corresponding Lagrangean manifolds.

Let N^*(char X) be a conormal bundle of char X in the symplectic space $T_0^*([x]) = [x,p] \setminus \{p = 0\}$, i.e. $N^*(\text{char } X) = \{(x,p) \mid s(x) = 0,\ H(x, \frac{\partial s}{\partial x}) = 0,\ p = \lambda\frac{\partial s}{\partial x} + \mu \frac{\partial}{\partial x} H(x, \frac{\partial s}{\partial x}),\ |\lambda|^2+|\mu|^2 \neq 0,\ \lambda\in\mathbb{C}, \mu\in\mathbb{C}\}$. We define the set of characteristic co-vectors char (char X) in the bundle N^*(char X) as the set of co-vectors (x,p) such that $H(x,p) = 0$. This set is defined by the equation

$$P(x,\lambda,\mu) \overset{def}{=} H\left(x, \lambda \frac{\partial s(x)}{\partial x} + \mu \frac{\partial}{\partial x} H\left(x, \frac{\partial s(x)}{\partial x}\right)\right) = 0. \tag{5}$$

The left-hand side of the latter relation is a homogeneous polinomial of the order m with respect to (λ, μ). Thus, the set char (char X) is homogeneous with respect to (λ, μ). If the point $x_1 \in$ char X is fixed, the set char(char X) $\cap \{x = x_1\} / \mathbb{C}_*$ contains no more than m points. Indeed, the equation (5) has the root $\mu = 0$ (due to characteristicity of X) and this root is single due to Condition 1:

$$\frac{\partial}{\partial\mu} P(x,\lambda,\mu)\Big|_{\substack{\mu=0\\ \lambda=1}} = H_p\left(x, \frac{\partial s(x)}{\partial x}\right) \frac{\partial}{\partial x} H\left(x, \frac{\partial s(x)}{\partial x}\right) \neq 0.$$

So the polinomial $P(x, \lambda, \mu)$ has the form

$$P(x,\lambda,\mu) = \sum_{k=1}^{m} a_k(x) \lambda^{m-k} \mu^k,$$

where $a_1(x) \neq 0$. The roots of the equation (5) except $\mu = 0$ lie in the chart $\mu = 1$ of the space $\mathbb{CP}^1 = [\lambda,\mu]\setminus\{(0,0)\}/\mathbb{C}_*$. These roots are determined thereby with the help of the equation of the power m-1 with non-vanishing highest-order coefficient.

Let us consider now the system of trajectories of the vector field V(H), emerging from char (char X). The projection of a trajectory on the space X (i.e. the corresponding ray) evidently depends only on the class of equivalence of the origin (x, λ, μ) modulo the action of the group $\mathbb{C}_*$. The union of the rays, described above, we denote by Z. We suppose the following condition to be valid.

Condition 2. The projection of the trajectory emerging from the point of char (char X) which projects on x_0 is transversal to char X.

It means analytically that

$$H_p\left(x_0, \lambda_1 \partial s/\partial x(x_0) + \mu_1 \partial/\partial x\, H(x, \partial s(x)/\partial x)\big|_{x=x_0}\right) \partial/\partial x\, H(x, \partial s(x)/\partial x)\big|_{x=x_0} \neq 0,$$

where (λ_1, μ_1) is a solution of the equation (5) for $x = x_0$.

Lemma 2. If the Condition 2 is valid then:

1) All the roots of equation (5) are single.

2) The set Z decomposes into m irreducible components; each component is generated by system of trajectories emerging from the points of char (char X) corresponding to one of the roots of equation (5). Each of these components is an analytical manifold in a neighbourhood of x_o.

Proof. The first assertion follows due to Condition 2 from the equality

$$\frac{\partial P}{\partial \mu}(x_0,\lambda_1,\mu_1)=H_p\left(x_0,\lambda_1\frac{\partial s}{\partial x}(x_0)+\mu_1\frac{\partial}{\partial x}H\left(x,\frac{\partial s(x)}{\partial x}\right)\Big|_{x=x_0}\right)\frac{\partial}{\partial x}H\left(x,\frac{\partial s}{\partial x}\right)\Big|_{x=x_0}\neq 0$$

for each root (λ_1, μ_1) of the equation (5) for $x = x_o$. The second assertion follows from the first one and from Condition 2 as in the proof of Lemma 1. This completes the proof.

Let $\tilde{Y}$ be an analytical set which goes through the point x_o and does not contain the whole set X. We define the space $A_q(X, \tilde{Y})$ as the space of multiple-valued analytical functions such that for any point $x_1 \in X \setminus \tilde{Y}$ there exists a neighbourhood $W(x_1)$ and a constant C for which the inequality

$$|f(x)| \leq C|s(x)|^q, \quad x \in W(x_1)$$

holds. Note that a function $f \in Aq(X,\tilde{Y})$ can have singularities on $\tilde{Y}$ of arbitrary type.

The paper aims at constructing the asymptotics of the solution u(x) of the equation (1) in the space Aq(X,Y) modulo the space $A_{q+N}(X,\tilde{Y})$ for any N.

2. Some geometrical aspects. Let us consider the function S(y,p,t) defined as the solution of the Cauchy problem

$$\begin{cases}\dfrac{\partial S(y,p,t)}{\partial t}=H\left(y,\dfrac{\partial S(y,p,t)}{\partial y}\right)\\ S(y,p,0)=p\cdot y\end{cases}\tag{6}$$

Here $y = (y^1,\dots, y^n)$. If $|p| < 1$ the problem (6) has the solution for $|t| < \varepsilon$. But in this case one can verify by means of direct calculation that the function $\lambda^{-1}S(y, \lambda p, \lambda^{1-m} t)$ satisfies the problem

$$\begin{cases}\dfrac{\partial}{\partial t}\{\lambda^{-1}S(y,\lambda p,\lambda^{1-m}t)\}=H\left(y,\dfrac{\partial}{\partial y}\{\lambda^{-1}S(y,\lambda p,\lambda^{1-m}t)\}\right)\\ \lambda^{-1}S(y,\lambda p,\lambda^{1-m}t)\big|_{t=0}=p\cdot y\end{cases}$$

Thus, there exists a holomorphic solution of the problem (6) in the domain $|t|\cdot|p|^{m-1}<\varepsilon$, which is homogeneous of degree 1 with respect to the action of the group $\mathbb{C}_*$ on $[p,t]$ given by the formula

$$\lambda(p, t) = (\lambda p, \lambda^{1-m} t) \tag{7}$$

We denote the quotient space of $([p] \setminus \{(0, \ldots, 0)\}) \times [t]$ modulo the action (7) of $\mathbb{C}_*$ by Ω. The neighbourhood of zero in the space Ω defined by the inequality

$$|t| \cdot |p|^{m-1} < \varepsilon$$

will be denoted by Ω_ε.

We introduce manifolds

$X = \{(y,p,t) \mid s(y) = 0\}$,

$\operatorname{char} X = \{(y,p,t) \mid s(y) = 0, H(y, \frac{\partial s(y)}{\partial y}) = 0\}$,

$Y = \{(y,p,t) \mid s_1(y) = 0\}$,

$T = \{(y,p,t) \mid t = 0\}$,

$\Sigma_x = \{(y,p,t) \mid S(y,p,t) - p\cdot x = 0\}$

in the space $[y,p,t]$. All these manifolds are invariant with respect to the action (7) of $\mathbb{C}_*$. Their images in $[y] \times \Omega_\varepsilon$ we denote by X^*, $\operatorname{char} X^*$, Y^*, T^*, Σ_x^* respectively.

<u>Lemma 3</u>. The manifold $\Sigma_{x_0}^*$ is tangent to X^* in the point $\{(x_0, \lambda p_0, 0), \lambda \in \mathbb{C}_*\}$, where $p_0 = \partial s/\partial x\,(x_0)$.

<u>Proof</u>. It suffices to verify that Σ_{x_0} is tangent to X in the point $(x_0, p_0, 0)$. The conditions of tangency of Σ_{x_0} and X are

$$\begin{cases} s(y) = 0 \\ s(y, p, t) - px_0 = 0 \\ d\,[S(y,p,t) - px_0] = \lambda\, d\, s(y) \end{cases} \tag{8}$$

for some $\lambda \in \mathbb{C}_*$. Let us substitute $y = x_0$, $p = p_0$, $t = 0$ in (8). As $x_0 \in X$ we have $s(x_0) = 0$; $S(x_0, p_0, 0) - p_0 x_0 = 0$. The equality $\partial S/\partial t\,(x_0, p_0, 0) = 0$ follows from $x_0 \in \operatorname{char} X$ and the equation (6). The equality $\partial/\partial p [S(x_0, p, 0) - px_0] \big|_{p=p_0} = 0$ follows from $\partial S/\partial p (x_0, p, 0) \big|_{p=p_0} = x_0$. The equality $\partial S/\partial y (x_0, p_0, 0) = \lambda \partial s/\partial y (x_0)$ for $\lambda = 1$ follows from relations $p_0 = \partial s/\partial y\,(x_0)$. This completes the proof of the Lemma.

Lemmas 1 and 3 yield that Σ_{x_0} is tangent to the manifolds char X and Y in the point $(x_o, p_o, 0)$.

Now we present the geometrical interpretation of the function $S(y,p,t)$. Let us consider the phase space $[x,p;\ y,q]$ endowed with the structure form $dq \wedge dy - dp \wedge dx$. Let

$$g_t : [y,q] \longrightarrow [x,p]$$

be a canonical transformation defined as a shift by t along the trajectories of the vector field (2). The function g_t evidently satisfies the following homogeneity condition

$$g_{\lambda^{1-m}t}(y, \lambda q) = (x, \lambda p) \text{ if } g_t(y,q) = (x,p).$$

Let $\mathcal{L}$ be a family of Lagrangian manifolds in $[x,p;\ y,\ q;\ t]$ depending on parameter t such that $\mathcal{L} \cap \{t = t_o\} \overset{\text{def}}{=} \mathcal{L}_{t_o} = \text{graph } g_{t_o}$ for any t_o. On $\mathcal{L}$ we let also

$$d\bar{S} = -pdx + qdy - H(y,q)dt, \quad \bar{S}\Big|_{t=0} = 0.$$

In a neighbourhood of $t = 0$ (y,p,t) are the coordinates on $\mathcal{L}$; we set $S(y,p,t) = \bar{S} + px\Big|_L$. Thus,

$$\frac{\partial S}{\partial y} = q(y,p,t), \quad \frac{\partial S}{\partial p} = x(y,p,t), \quad \frac{\partial S}{\partial t} = H(y,q(y,p,t)),$$

$$S(y,\ p,\ 0) = p\cdot y,$$

i.e. $S(y,p,t)$ satisfies the problem (6). We have verified that $S(y,p,t)$ is a generating function of the transformation (9).

Below we shall use the following affirmation.

<u>Proposition 1</u>. Let $\tilde{Y}$ be an analytical manifold in the space $[x]$ which goes through the point x_o. The manifold Σ_x is tangent to $\tilde{Y}$ in the point $(y,\ p,t)$ iff $(x,\ p)$ is the end point of the trajectory with origin in $N^*(\tilde{Y}) \cap \{H(y,\ q) = 0\}$, the natural parameter of the end point being equal to t.

<u>Proof</u>. Let $\tilde{Y} = \{f_1(x) = \ldots = f_k(x) = 0\}$, the differentials $df_1, \ldots, df_k$ being independent on $\tilde{Y}$. The conditions of tangency of Σ_x and $\tilde{Y}$ in the point $(y,\ p,t)$ are

$$f_i(y) = 0,\ i = 1,\ \ldots,\ k$$

$$S(y,\ p,\ t) - p\cdot x = 0 \tag{10}$$

$$\begin{cases} \frac{\partial S}{\partial y}(y,p,t) = \sum_{i=1}^{k} \lambda_i \frac{\partial f_i(y)}{\partial y}, \\ \frac{\partial S}{\partial p}(y,p,t) - x = 0, \\ \frac{\partial S}{\partial t}(y,p,t) = 0. \end{cases} \tag{11}$$

The relation (10) follows from the relations (11) due to homogeneity. The rest of equations one can reduce to the form

$$f_i(y) = 0, \quad i = 1, \ldots, k \tag{12}$$

$$q(y, p, t) = \sum_{i=1}^{k} \lambda_i \frac{\partial f(y)}{\partial y} \tag{13}$$

$$x(y,p,t) = x \tag{14}$$

$$H(y, q(y, p, t)) = 0 \tag{15}$$

due to the geometrical interpretation of the function $S(y, p, t)$ given above. The relation (14) shows that the point (x,p) lies on the trajectory with the origin in the point $(y, q(y, p,t))$, which lies in $N^*(\tilde{Y})$ due to (12), (13) and lies on the zero level surface of the Hamiltonian H due to (15). Therefore the system (12) - (15) yields that (x, p) and (y, q) lie on the same trajectory, that (x, p) corresponds to natural parameter equal to t, and that $(y, q) \in N^*(\tilde{Y}) \cap \{H(y,q) = 0\}$. The inverse affirmation follows also from the relations (12) - (15). This completes the proof.

<u>Proposition 2</u>. Let $\tilde{Y}$ be an analytical submanifold in $[x]$. The manifold $\Sigma_x \cap T$ is tangent to $\tilde{Y}$ in the point (y, p) iff $x = y$, $p = q \in N^*(\tilde{Y})$.

<u>Proof</u>. The equation of $\Sigma_x \cap T$ due to the initial conditions of the problem (6) has the form

$$\Sigma_x \cap T = \{(p, y) \mid p\cdot(x-y) = 0\} .$$

Let $\tilde{Y} = \{ f_1(x) = \ldots = f_k(x) = 0\}$, $df_1, \ldots, df_k$ being independent on $\tilde{Y}$. The conditions of tangency of $\Sigma_x \cap T$ and $\tilde{Y}$ in the point (y, q) are

$$\begin{cases} f_i(y) = 0, \ i = 1, \ldots, k; \\ p\cdot(y-x) = 0 \\ p = \sum_{i=1}^{k} \lambda_i \frac{\partial f_i(y)}{\partial y} \\ y - x = 0 \end{cases}$$

The latter system of equations gives us the proof of the proposition.

Lemma 4. A set of points x such that Σ^*_x is tangent to X^* coincides with the set $Y \subset [x]$. The point of tangency of Σ^*_x and X^* is isolated. In particular, $\Sigma^*_{x_0}$ is transversal to X^* in any point except $(x_0, p_0, 0)$.

Proof. Suppose Σ^*_x is tangent to X^*. Hence Σ_x is tangent to X in some point (y,p,t). Due to Proposition 1 (x,p) is a point on the trajectory of vector field (2) with the origin in $(y,q) \in N^*(X)$, $H(y,q)$ being equal to zero. Therefore $y \in$ char X and $q = \lambda \partial s(y)/\partial y$, $\lambda \in \mathbb{C}_*$, i.e. $x \in Y$. The uniqueness of the point of tangency follows from the fact that the rays which form Y do not intersect. The completes the proof of the Lemma.

Lemma 5. The set of points x such that Σ^*_x is tangent to char X^* coincides with the set $Z \subset [x]$. The manifold $\Sigma^*_{x_0}$ is transversal to char X^* in any point except $(x_0, p_i, 0)$ $i=1,\ldots,\ldots, m$, where $p_i \in N^*(\text{char } X) \cap \{H = 0\}$.

Proof. If (y,p,t) is the point of tangency of Σ_x and char X then due to Condition 1 (x,p) lies on the trajectory with the origin $(y,q) \in$ char $X \cap \{H = 0\}$, (x,p) corresponding to the natural parameter equal to t. Hence $x \in Z$. Each point $x \in Z$ lies no more than on m rays corresponding to different solutions of the equation (5). If $x \in$ char X, these m rays evidently correspond to $t = 0$. This yields the las assertion of the Lemma.

Lemma 6. A set of points x such that Σ^*_x is tangent to Y^* coincides with the set Y. The points of tangency of the manifolds Σ^*_x and Y^* form the projection of the trajectory of vector field $V(H)$ on the space $[y,p,t]$; the trajectory goes through the point $(x, \partial s/\partial x)$ (here (y,q) is the origin of trajectory, $(x,p) = (x, \partial s/\partial x)$ is its end point, and t is the natural parameter).

From now on this projection will be called a ray.

Proof. Suppose (y,p,t) is the point of tangency Σ_x and Y. Due to Proposition 1 (x,p) is the end point of the trajectory with the origin in $(y, \partial s(y)/\partial y)$, (x,p) corresponds to the natural parameter equal to t (the condition $H(y,q) = 0$ is omitted due to the characteristicity of Y). Hence, $p = \partial s(x)/\partial x$ (the manifold L_Y is invariant with respect to $V(H)$) and (y,q) is the point of the trajectory with origin in (x,p), (y,q) corresponding to the natural parameter equal to $-t$. This completes the proof of the Lemma.

We mention the following obvious affirmation.

Lemma 7. The manifolds Σ_x^* and T^* are transversal to each other.

Let us treat the analytical set

$$\Gamma = X^* \cup Y^* \cup T^*$$

in the space $[y] \times \Omega$. Let l^* be the ray mentioned in Lemma 6 for $x = x_0$.

Let us consider the stratification of Γ consisting of strata

$$\begin{aligned}&X^* \setminus (\text{char } X^* \cup T^*),\ \text{char } X^* \setminus T^*,\ Y^* \setminus (\text{char } X^* \cup T^* \cup l^*),\\ &l^* \setminus (x_0, p_0^*, 0),\ \{(x_0, p_0^*, 0)\},\ (X^* \setminus \text{char } X^*) \cap T^*,\\ &(\text{char } X^* \setminus \{(x_0, p_0^*, 0)\}) \cap T^*,\ (Y^* \setminus \text{char } X^*) \cap T^*,\ T^* \setminus (X^* \cup Y^*).\end{aligned} \qquad (16)$$

Lemma 8. There exists an arbitrarily small number $r > 0$ such that the sphere of the radius r with the center $(x_0, 0)$ in $[y] \times \Omega$ transversally intersects all the strata of Γ.

Proof. The function $r^2 = \sum_{i=1}^{n} |y^i - x^i|^2 + |t|^2 |p|^{2m-2}$ is a real-analytical function and therefore it has a finite number of critical values on each stratum. It is sufficient therefore to choose r^2 to be less then the smallest positive critical value. This completes the proof of the Lemma.

Denote by K_r and S_r the sets

$$K_r = \{(y, p^*, t^*) \mid \sum_{i=1}^{n} |y^i - x^i|^2 + |t|^2 |p|^{2m-2} \leq r^2\},$$

$$S_r = \partial K_r$$

r being chosen as in Lemma 8.

The set K_r with a set $K_r \cap \Gamma$ embedded in it admits a stratification. Its strata are the intersections of strata (16) (without taking into account the ray l^*):

$$\begin{aligned}&X^* \setminus (\text{char } X^* \cup T^*),\ \text{char } X^* \setminus T^*,\ Y^* \setminus (\text{char } X^* \cup T^*),\\ &(X^* \setminus \text{char } X^*) \cap T^*,\ \text{char } X^* \cap T^*,\ (Y^* \setminus \text{char } X^*) \cap T^*\\ &T^* \setminus (X^* \cup Y^*)\end{aligned} \qquad (17)$$

with:

a) the interior $\mathring{K}_r$ of the set K_r;

b) the boundary S_r of the set K_r.

We shall call the strata of the type a) analytical strata, the strata of the type b) - boundary strata of the pair $(K_r, \Gamma \cup S_r)$.

Note that if r is sufficiently small, S_r intersects the manifold $\Sigma^*_{x_0}$ transversally. Lemma 8 yields also that S_r is transversal to 1^*.

Let now

$$\Sigma^* = \{(x,y,p^*,t^*) \mid S(y,p,t) - p\cdot x = 0\} \subset [x,y] \times \Omega$$

and let π be a projection

$$[x,y] \times \Omega \longrightarrow [x].$$

Consider the projection of the pair

$$\begin{array}{c}(K_r \times (U(x_0) \times \tilde{Z}),\ ((\Gamma \cup S_r) \times (U(x_0) \setminus \tilde{Z})) \cup \Sigma^*) \\ \downarrow \\ U(x_0) \setminus \tilde{Z}\end{array} \qquad (18)$$

where $\tilde{Z} = Z \cup X$ and $U(x_0)$ is a neighbourhood of the point x_0 in the space $[x]$.

Let us prove the main affirmation of this section.

<u>Proposition 3</u>. There exists a neighbourhood $U(x_0)$ of the point x_0 such that the projection of the pair (18) forms a locally trivial stratified fibration of pairs.

<u>Proof</u>. Due to assertions of Lemmas 3-8 and Proposition 2 the manifold $\Sigma^*_{x_0}$ is transversal to all the boundary strata of the pair $(K_r, \Gamma \cap K_r)$ but $(Y^* \setminus (\mathrm{char}\, X^* \cup T^*)) \cap S_r$. The latter stratum is tangent to $\Sigma^*_{x_0}$ in the points of 1-dimensional submanifold $1^* \cap S_r$ of the manifold S_r. Hence, there exists a sufficiently small neighbourhood $U(x_0)$ of the point x_0 in the space $[x]$ such that the manifold Σ^*_x is transversal to all the boundary strata except the described above for all $x \in U(x_0)$. Due to Lemmas 3-7 and Proposition 1 the manifold Σ^*_x is transversal to all the analytical strata for $x \in U(x_0) \setminus \tilde{Z}$. To complete the proof we must show that Σ^*_x is transversal to $(Y^* \setminus (\mathrm{char}\, X^* \cup T^*)) \cap S_r$ for $x \in U(x_0) \setminus \tilde{Z}$. Suppose Σ^*_x is tangent to $Y^* \cap S_r$ in a point $\alpha \in S_r$. Consider the maximal analytical linear submanifold in the plane tangent to S_r in the point α. This submanifold is a complex plane of codimension 1; we denote it by P_α. It is evident that the intersections of the manifolds Σ^*_x and Y^* with P_α are tangent to each other in the point α. Lemma 8 yields that P_α is transversal to the ray l_α which lays in Σ^*_x and which goes through α. Let us choose the coordinates $(z^1, \ldots, z^{2n+1})$

in the space $[y,p,t]$ in such a way that $z^1 = 0$ is the equation of P_α (the linear change of variables is sufficient). Suppose $\mathcal{H}(z, Q)$ be an expression of the Hamiltonian $E-H_m(y,q)$ with respect to new coordinates (we use the symplectic space x,p,y,q,t,E with the structure form $dy\wedge dq - dx\wedge dp + dt\wedge dE$; the ray l_α mentioned in Lemma 6 evidently corresponds to this Hamiltonian). If $\beta_1(z) = 0$, $\beta_2(z) = 0$ are the equations of Y and Σ_x respectively then the transversality of P_α and l_α can be expressed by an inequality

$$\left.\frac{\partial \mathcal{H}(z,Q)}{\partial Q_1}\right|_{\substack{z = z(\alpha) \\ Q = \partial\beta_2/\partial z(z(\alpha))}} \neq 0. \tag{19}$$

The tangency of $\Sigma_x \cap P_\alpha$ and $Y \cap P_\alpha$ can be expressed by equalities

$$\frac{\partial\beta_1}{\partial z^i}(z(\alpha)) = \frac{\partial\beta_2}{\partial z^i}(z(\alpha)), \; i = 2, \ldots, 2n-1 \tag{20}$$

Since both Σ_x and Y are characteristical with respect to $\mathcal{H}(z,Q)$, we have

$$\mathcal{H}(z, \frac{\partial\beta_1}{\partial z}) = 0, \; \mathcal{H}(a, \frac{\partial\beta_2}{\partial z}) = 0 \tag{21}$$

Now, due to (19), the equation $\mathcal{H}(z, Q) = 0$ has an unique solution with respect to Q_1 (x_1 being sufficiently close to x_0). Hence, due to the relations (20), (21), we have $\partial\beta_1/\partial z^1(z(\alpha)) = \partial\beta_2/\partial z^1(z(\alpha))$, i.e., by virtue of the equalities (20), Σ_x is tangent to Y in the point α. Hence, $x \in \tilde{Z}$, which contradicts our suggestions. Thus, the manifold Σ_x^* is transversal to all the strata (7) for $x \in U(x_0)\setminus\tilde{Z}$. The assertion of the Proposition follows now from the Thom theorem (see [6]).

3. <u>Definition of ramifying classes</u>. In this section we define three ramifying homology classes.

1) $h_1(x) \in H_{2n-2}(\Sigma_x^* \cap T^*, X^*)$ (we always denote the compact homology group of a pair by H with a subscript index). Let x_1 be a point on X which does not belong to Y and let $U(x_1)$ be a neighbourhood of x_1 which does not intersect Y. Suppose that $\partial s/\partial y^1(x_1) \neq 0$. In this case the equation of X in a neighbourhood of x_1 is

$$y^1 = y^1(y'), \; y' = (y^2, \ldots, y^n); \; x_1^1 = y^1(x_1').$$

The equation of the intersection $\Sigma_x \cap T$ is $p\cdot(x-y) = 0$. The conditions of tangency between $\Sigma_x \cap T$ and X in the point x_1 are

$$p \cdot (x_1 - y) = 0,$$
$$d\{p(y - x_1)\} = \lambda \, ds(y).$$

Hence,

$$y = x_1,\ p = \lambda\, \partial s/\partial y\ (x_1)$$

Due to $\partial s/\partial y^1\ (x_1) \neq 0$ we can choose λ in such a way that $p_1 = 1$, $p' = -\partial y^1(y')/\partial y'|_{y'=x_1'}$. We can carry out all calculations in the chart $p_1 = 1$ of the quotient spaces modulo the action of $\mathbb{C}_*$. Since the determinant of the Hessian of the restriction of the determining function of the manifold $\Sigma_x^* \cap T^*$ (i.e. the function $y^1 - y^1(y') + p'(x'-y')$) equals to

$$\begin{vmatrix} -\partial^2 y^1(y')/\partial y'\partial y' & -1 \\ -1 & 0 \end{vmatrix} = \pm 1 \neq 0\ ,$$

the tangency $\Sigma_x^* \cap T^*$ and X^* in the point x_1 has the simple quadratic type. Hence, the intersection

$$\Sigma_x^* \cap T^* \cap X^*$$

is homeomorphic in a neighbourhood of x_1 to a complex quadric for x sufficiently close to x_1 but not lying on X. We denote by $h_1(x)$ a vanishing cycle of this quadric with orientation which continuously depends on a point x near the point x_1.

2) $h_2(x) \in H_{2n-1}(\Sigma_x^*, X^* \cup T^*)$. Let us consider the exact triple

$$\begin{array}{ccc} & H_*(\Sigma_x^*, X^*) & \\ {}^{p}\swarrow & & \nwarrow^{i} \\ H_*(\Sigma_x^*, X^* \cup T^*) & \xrightarrow{\ \partial\ } & H_*(\Sigma_x^* \cap T^*, X^*) \end{array}\ , \qquad (23)$$

all the manifolds being considered in the neighbourhood $U(x_1)$ of the point x_1 which does not intersect Y. Since $x_1 \notin Y$, we have

$$\partial S/\partial t\ (x_1, p_1, 0) \neq 0$$

(here $p_1 = (1, -\partial y^1(y')/\partial y'|_{y'=x_1'})$ as above).

Expressing t by the variables (y,p) with the help of equation of Σ_x: $S(y,p,t) - p \cdot x = 0$ one can show that the pair (Σ_x^*, X^*) is homeomorphic to the pair $(\mathbb{C}^{2n-1}, \mathbb{C}^{2n-2})$ in a neighbourhood of the point x_1. Hence, $H_*(\Sigma_x^*, X^*) = 0$. Due to the exactness of the triple (23) we have that the homomorphism

$$H_*(\Sigma_x^*, X^* \cup T^*) \xrightarrow{\ \partial\ } H_*(\Sigma_x^* \cap T^*, X^*)$$

is an isomorphism. We define $h_2(x)$ with the help of relation

$$\partial h_2(x) = h_1(x). \tag{24}$$

3) $h_3(x) \in H_{2n}([y] \times \Omega_\varepsilon, \Sigma_x^* \cup X^* \cup T^*)$ (in a neighbourhood $J(x_1)$ as above).

We consider the exact triple

$$\begin{array}{ccc} H_*([y] \times \Omega_\varepsilon, X^* \cup T^*) & & \\ p \swarrow & \nwarrow i & \\ H_*([y] \times \Omega_\varepsilon, \Sigma_x^* \cup X^* \cup T^*) & \xrightarrow{\partial} & H_*(\Sigma_x^*, X^* \cup T^*) \end{array} \tag{25}$$

Since in a neighbourhood of the point under consideration X^* and T^* are analytical manifolds transversal to each other, we have $H_*([y] \times \Omega_\varepsilon, X^* \cup T^*) = 0$, and hence, the homomorphism ∂ involved in the triple (25) is an isomorphism. We define the class $h_3(x)$ by the equality

$$\partial h_3(x) = -h_2(x). \tag{26}$$

With the help of embedding $U(x_1) \subset U(x_0)$ the classes $h_i(x)$ can be considered as classes in the following spaces (x still belongs to $U(x_1)$):

$$h_1(x) \in H_{2n-2}((\Sigma_x^* \setminus Y^*) \cap T^*, X^*),$$

$$h_2(x) \in H_{2n-1}(\Sigma_x^* \setminus Y^*, X^* \cup T^*),$$

$$h_3(x) \in H_{2n}(([y] \times \Omega_\varepsilon) \setminus Y^*, \Sigma_x^* \cup X^* \cup T^*)$$

because $U(x_1)$ does not intersect Y^*. With the help of Proposition 3 the classes $h_i(x)$ can be extended as ramigying classes from $x \in U(x_1)$ to $x \in U(x_0) \setminus \tilde{Z}$. Evidently the extension preserves relations (24), (26).

4. <u>Construction of regulizer</u>. We consider the form

$$\omega = (m-1)t\, dp_1 \wedge \ldots \wedge dp_n + \sum_{k=1}^{n} (-1)^{k-1} p_k\, dt \wedge dp_1 \wedge \ldots \wedge \widehat{dp_k} \wedge \ldots \wedge dp_n$$

in the space $[p,t]$ (the sign $\wedge$ over a differential shows that it must be omitted).

The following Lemma is valid (see [7]).

<u>Lemma 9</u>. If $F(p,t)$ is homogeneous of degree $m-n-1$ with respect to the action (7) of the group $\mathbb{C}_*$ then the form $F(p,t)\omega$ is raising of some form $\alpha \in \Lambda^n(\Omega)$, i.e. $\pi^*\alpha = F(p,t)\omega$. Inversely, for each $\alpha \in \Lambda^n(\Omega)$ there exists a homogeneous function $F(p,t)$ of degree $m-n-1$ such that $\pi^*(\alpha) = F(p,t)\omega$.

Let $U(y,p,t)$ be a homogeneous function of degree r with respect to the action (7) of the group $\mathbb{C}_*$. Due to Lemma 9 the form

$$U(y,p,t)\,\frac{f(y)\,dy\wedge\omega}{[S(y,p,t)-px]^{n+1-m+r}} \tag{27}$$

can be regarded as a form on the space $([y]\times\Omega_\varepsilon)\setminus(\Sigma_x^*\cup X^*\cup Y^*)$ for any function f(y) analytical outside $X\cup Y$. We suppose that

$$f(y)\in A_q(X,Y).$$

In the case $n+1-m+r>0$ the form (27) is singular on Σ_x^* having a polar singularity on this manifold. If inversely $n+1-m+r\leqslant 0$ the form (27) is regular on Σ_x^*.

We define the function R f by the formula

$$\hat{R}[f]=(-1)^{n-m+r}(n-m+r)!\int_{h_2(x)}\operatorname{Res}\frac{U(y,p,t)f(y)\,dy\wedge\omega}{[S(y,p,t)-px]^{n+1-m+r}} \tag{28}$$

for any $f(y)\in A_q(X,Y)$, $q>-1$ in case $n+1-m+r>0$, and

$$\hat{R}[f]=\frac{1}{(m-r-n-1)!}\int_{h_3(x)}U(y,p,t)[S(y,p,t)-px]^{m-r-n-1}f(y)\,dy\wedge\omega \tag{29}$$

in case $n+1-m+r\leqslant 0$.

Note that the function f(y) is differentiated as a result of calculation of the residue, and its singularity on X increases. In case this singularity becomes non-integrable in a neighbourhood of X the integral (28) must be understood as the right-hand side of the equality

$$\int_{h_2(x)}\operatorname{Res}\frac{U(y,p,t)f(y)\,dy\wedge\omega}{[S(y,p,t)-px]^{n+1-m+r}}=\int_{\delta h_2(x)}\frac{U(y,p,t)f(y)\,dy\wedge\omega}{[S(y,p,t)-px]^{n+1-m+r}}\cdot\frac{1}{2\pi i}$$

(δ is the Leray coboundary [8]). If the mentioned singularity is integrable, the right-hand and the left-hand sides of the latter equality coincide.

<u>Proposition 4</u>. The operator $\hat{R}[f]$ is the operator of order r-m with respect to the scale A_q.

<u>Proof</u>. Let us consider an arbitrary point $x^*\in X\setminus Y$ and the sheet of Riemannian manifold of the function $\hat{R}[f]$ on which $h_1(x)$ is a vanishing cycle of the quadric described in section 3 in the con-

struction of $h_1(x)$. As follows from the construction of the classes $h_i(x)$ (Sec. 3) representatives of classes $h_i(x)$ can be chosen inside an arbitrarily small neighbourhood of the point $(y,p,t) = (x^*, 1, \partial y^1/\partial y' \ (x^*), 0)$ for x sufficiently close to x^*, $x \notin \widetilde{Z}$ (we suppose the equation of X in a neighbourhood of x^* to have the form $y^1 = y^1(y')$). Hence, integrals (28) and (29) can be rewritten in the chart $p_1 = 1$ in the space Ω , i.e.

$$\hat{R}[f] = (-1)^{n-m+r}(n-m+r)! \int_{h_1(x)} \operatorname{Res} \frac{U(y,1,p',t)f(y)dy\wedge dt\wedge dp'}{[S(y,1,p',t)-x^1-p'x']^{n+1-m+r}} =$$

$$= (-1)^{n-m+r} \frac{(n-m+r)!}{2\pi i} \int_{\delta h_1(x)} \frac{U(y,1,p',t)f(y)dy\wedge dt\wedge dp'}{[S(y,1,p',t)-x^1-p'x']^{n+1-m+r}} \qquad (30)$$

in case $n+1-m+r > 0$ and

$$\hat{R}[f] = \frac{1}{(m-r-n-1)!} \int_{h_3(x)} U(y,1,p',t)[S(y,1,p',t)-x^1-p'x']^{m-n-r-1} f(y)dy\wedge dt\wedge dp'$$

in case $n+1-m+r \leqslant 0$.

We note first of all that it is sufficient to prove the assertion for $f(y) \in A_q(X,Y)$, q being sufficiently large.

Indeed, using the notation

$$If(y) = \int_{y^1(y')}^{y^1} f(\alpha,y')d\alpha$$

we have $If(y) \in A_{q+1}(X,Y)$, $\partial/\partial y^1 \, If(y) = f(y)$ (Lemma 1.1 of paper [3]). Replacing $f(y)$ by $\partial/\partial y^1 \, If(y)$ in the integrals (30), (31) and integrating by parts we arrive at following results:

A. $n+1-m+r < 0$ $\quad (g(y) \overset{\text{def}}{=} If(y))$.

$$\hat{R}[f] = \frac{1}{(m-r-n-1)!} \int_{h_3(x)} U(y,1,p',t)[S(y,1,p',t)-x^1-p'x']^{m-r-n-1} \times$$

$$\times \frac{\partial g(y)}{\partial y^1} dy\wedge dt\wedge dp' = \frac{1}{(m-r-n-1)!} \int_{h_3(x)} d\{U(y,1,p',t)[S(y,1,p',t)-$$

$$-x^1-p'x']^{m-r-n-1} g(y)dy'\wedge dt\wedge dp'\} - \frac{1}{(m-r-n-1)!} \int_{h_3(x)} \frac{\partial U}{\partial y^1}(y,1,p',t) \times$$

$$\times [S(y,1,p',t)-x^1-p'x']^{m-r-n-1} g(y)\, dy'\wedge dt\wedge dp' - \frac{1}{(m-r-n-2)!} \times$$

$$\times \int_{h_3(x)} U(y,1,p',t)\frac{\partial S}{\partial y^1}(y,1,p',t)\left[S(y,1,p',t)-x^1-p'x'\right]^{m-r-n-2}\times$$

$$\times g(y)\,dy\wedge dt\wedge dp' = \hat{R}_1[g]+\hat{R}_2[g]$$

where R_1 and R_2 are operators of type (29). For $\hat{R}_1$ U is replaced by $-\partial U/\partial y^1$ and for $\hat{R}_2$ U is replaced by $-U\partial S/\partial y^1$ and the number r increased by a unit.

B. n+1-m+r = 0.

$$\hat{R}[f] = \int_{h_3(x)} U(y,1,p',t)\frac{\partial g}{\partial y^1}dy\wedge dt\wedge dp' = \int_{h_3(x)} d\{U(y,1,p',t)\times$$

$$\times g(y)dy'\wedge dt\wedge dp'\} - \int_{h_3(x)}\frac{\partial U}{\partial y^1}(y,1,p',t)g(y)dy\wedge dt\wedge dp' =$$

$$= \int_{h_2(x)} U(y,1,p',t)g(y)|_{\Sigma_x^*}dy'\wedge dt\wedge dp' - \int_{h_3(x)}\frac{\partial U}{\partial y^1}(y,1,p',t)g(y)\times$$

$$\times dy\wedge dt\wedge dp' = -\int_{h_3(x)}\frac{\partial U}{\partial y^1}(y,1,p',t)g(y)dy\wedge dt\wedge dp' +$$

$$+\int_{h_2(x)} \mathrm{Res}\,\frac{U(y,1,p',t)\,\partial S/\partial y^1(y,1,p',t)g(y)dy\wedge dt\wedge dp'}{S(y,1,p',t)-x^1-p'x'} =$$

$$= \hat{R}_1[g]+\hat{R}_2[g].$$

All the assertions of A are valid except for the fact that R_2 becomes an operator of type (28) but not (29).

C. n+1-m+r > 0.

$$\hat{R}[f] = (-1)^{n-m+r}(n-m+r)!\int_{\delta h_2(x)}\frac{U(y,1,p',t)\,\partial g/\partial y^1(y)dy\wedge dt\wedge dp'}{[S(y,1,p',t)-x^1-p'x']^{n+1-m+r}} =$$

$$=(-1)^{n-m+r}\Big\{(n-m+r)!\int_{\delta h_2(x)} d\left[\frac{U(y,1,p',t)g(y)dy'\wedge dt\wedge dp'}{[S(y,1,p',t)-x^1-p'x']^{n+1-m+r}}\right] -$$

$$-(n-m+r)!\int_{\delta h_2(x)}\frac{\partial U/\partial y^1(y,1,p',t)g(y)dy\wedge dt\wedge dp'}{[S(y,1,p',t)-x^1-p'x']^{n+1-m+r}} +$$

$$+(n-m+r+1)!\int_{\delta h_2(x)}\frac{U(y,1,p',t)\,\partial S/\partial y^1(y,1,p',t)g(y)dy\wedge dt\wedge dp'}{[S(y,1,p',t)-x^1-p'x']^{n+2-m+r}}\Big\} =$$

$$= -(-1)^{n-m+r}(n-m+r)!\int_{\delta h_2(x)} \frac{\partial U/\partial y^1(y,1,p',t)\,g(y)\,dy\wedge dt\wedge dp'}{[S(y,1,p',t)-x^1-p'x']^{n+1-m+r}} +$$

$$+\frac{\partial}{\partial x^1}\left\{(-1)^{n-m+r}(n-m+r)!\int_{\delta h_2(x)} \frac{U(y,1,p',t)\,\partial S/\partial y^1(y,1,p',t)}{[S(y,1,p',t)-x^1-p'x']^{n+1-m+r}}\times\right.$$

$$\times\, g(y)\,dy\wedge dt\wedge dp' = \hat{R}_1[g] + \partial/\partial x^1\,\hat{R}_2[g].$$

In contrast with cases A and B for both R_1 and R_2 the number r converges.

With the help of assertions A, B, C and Lemma 1.1 of paper [3] (lemma 1.1 of paper [3] states that the operator $\partial/\partial x^1$ is a generating operator of order 1 in the scale A_q) we reduce the inequalities to the case when q is arbitrarily large (and r is bounded).

Note also that it is sufficient to consider only the case of the operators of type (28). Indeed, the function $\hat{R}[f]$ given by (31) vanishes on X. Therefore (due to differentiation formulae (10.5), (10.6) of paper [8]), there exists a derivative of $\hat{R}[f]$ with respect to x^1 given by an integral of the form (30); we have

$$\hat{R}[f] = (I)^k\,(\partial/\partial x^1)^k\,\hat{R}[f],$$

and we must use once more Lemma 1.1 of paper [3] for reducing the inequalities of integrals (31) to the inequalities of integrals (30).

Hence, it is sufficient to evaluate the integral

$$\hat{R}[f] = (-1)^{n-m+r}(n-m+r)!\int_{h_2(x)} \operatorname{Res}\frac{U(y,1,p',t)\,f(y)\,dy\wedge dt\wedge dp'}{[S(y,1,p',t)-x^1-x'p']^{n+1-m+r}} \qquad (32)$$

with $f(y) \in Aq\ (X, Y)$, q being sufficiently large.

Under the assumptions above we calculate the residue of the form under the integral in the right-hand side of (32). The form under residue sign (including (n-m+r)!) is cohomological to the form

$$\left[\frac{\partial}{\partial y^1}\,\frac{1}{\frac{\partial S}{\partial y^1}(y,1,p',t)}\right]^{n-m+r}\left\{U(y,1,p',t)\,f(y)\right\}\frac{dy\wedge dt\wedge dp'}{S(y,1,p',t)-x^1-x'p'}$$

This yields (as $\partial S/\partial y^1 \neq 0$ in the neighbourhood under consideration)

$$(n-m+r)!\ \operatorname{Res} \frac{U(y,1,p',t)f(y)\,dy\wedge dt\wedge dp'}{[S(y,1,p',t)-x^1-p'x']^{n+1-m+r}} =$$

$$= \left\{ \sum_{j=0}^{n-m+r} U_j(y,1,p',t) \frac{\partial^j f(y)}{(\partial y^1)^j} \right\} \Bigg|_{\Sigma_x} dy'\wedge dt\wedge dp'$$

and we reduce the integral (32) to a sum of integrals

$$\sum_{j=0}^{n-m+r} \hat{R}_j[f] = \sum_{j=0}^{n-m+r} \int_{h_2(x)} \left[U_j(y,1,p',t) \frac{\partial^j f(y)}{(\partial y^1)^j} \right] \Bigg|_{\Sigma_x} dy'\wedge dt\wedge dp'. \tag{33}$$

Let us evaluate now the volume of the representative $\gamma(x)$ of the class $h_2(x)$ in a neighbourhood of the point x^*. We have

$$S(y,p,t) = p\cdot y + H(y,p)t + O(t^2)$$

in a neighbourhood of the point $t = 0$. Note that we have

$$H(y,p) \neq 0 \tag{34}$$

in a sufficiently small neighbourhood of the point under considera-tion. The equation of Σ_x^* is

$$\Sigma_x^* = \left\{ (y,p',t) \,\middle|\, (x^1-y^1)+p'(x'-y') + H(y,1,p')t + O(t^2) = 0 \right\}.$$

The equation of the intersection $\Sigma_x^* \cap X^*$ is

$$y^1(y') = x^1 + p'(x'-y') + H(y,1,p')t + O(t^2). \tag{35}$$

Using the Morse lemma due to relations (22), (34) one can choose coordinates $(z^1, \ldots, z^{2n-2},\ \ )$ in a neighbourhood of the point $(x^*, 1, \partial y^1/\partial y'\,(x^*), 0)$ such that equation (35) transforms to

$$\sum_{j=1}^{2n-2} (z^j)^2 + \tau = x^1 - y^1(x'),$$

the equation of T^* in the new coordinates being $\tau = 0$. The representative $\gamma(x)$ of the class $h_2(x)$ in the new coordinates can be described with the help of relations

$$\begin{aligned} \tau &= t(x^1 - y^1(x')), \\ z^j &= u^j \sqrt{x^1 - y^1(x')},\quad j = 1, \ldots, 2n-2, \\ u^j &\in \mathbb{R},\ \sum_{j=1}^{2n-2} (u^j)^2 \leq 1-t\ . \end{aligned} \tag{36}$$

Due to relations (36) the volume of the representative $\gamma(x)$ of the class $h_2(x)$ in the coordinats (Z,τ) allows an evaluation

$$\mathrm{Vol}\ \gamma(x) \leq c\,|x^1 - y^1(x')|^{\,n} \tag{37}$$

The inequality (37) is valid also in the coordinates (y,p,t) due to the smoothness of the change of variables.

On the contour $\gamma(x)$ we have

$$y^1 - y^1(y') = x^1 - y^1(x) + p'(x'-y') + H(y,1,p')t + O(t^2) = x^1 - y^1(x) -$$
$$- \sum_{j=1}^{2n-2} (z^j)^2 - \tau = (x^1 - y^1(x))\Big[(1-t) + \sum_{j=1}^{2n-2} (u^j)^2\Big]$$

and hence, on $\gamma(x)$ the inequality

$$|y^1 - y^1(y')| \leq C\,|x^1 - y^1(x')|$$

holds. Since $f(y) \in A_q(X, Y)$ and q is sufficiently large, we have

$$\left|\frac{\partial^j f(y)}{(\partial y^1)^j}\right| \leq C\,|y^1 - y^1(y')|^{q-j} \leq C_1\,|x^1 - y^1(x')|^{q-j}$$

on the contour $\gamma(x)$. For integral (33) we get the inequality

$$|\hat{R}[f]| = \Big|\sum_{j=0}^{n-m+r} \int_{h_2(x)} \Big[U_j(y,1,p',t)\frac{\partial^j f}{\partial y^j}\Big]\Big|_{\Sigma_x} dy' \wedge dt \wedge dp'\Big| \leq$$

$$\leq C|x^1 - y^1(x')|^{q-n-r+m}\,\mathrm{Vol}\,\gamma(x) \leq C_1\,|x^1 - y^1(x')|^{q-(r-m)}$$

i.e.

$$\hat{R}\,[f] \in A_{q-(r-m)}(X,Y)$$

This completes the proof of the Proposition.

The calculations analogous to calculations A, B, C in the proof of Proposition 4 show the validity of following relations (we specify in round brackets amplitudes and phase functions of the operators, i.e. the functions U and S involved in formulae (28), (29)):

$$\hat{R}[U,S]\left[\frac{\partial f}{\partial y^i}\right] = -\hat{R}\left(\frac{\partial U}{\partial y^i}, S\right)[f] - \hat{R}\left(U\frac{\partial S}{\partial y^i}, S\right)[f]$$

for $f \in A_q(X,Y)$, $q > 0$.

Furthermore, it is evident that

$$\hat{R}(U,\ S)\ \big[a(y)f(y)\big] = \hat{R}(U\cdot a,\ S)\ [f]\ .$$

By induction we obtain in usual way the formula

$$\hat{R}(U,S)\circ\hat{H} = (-1)^m\Big\{\hat{R}\Big(U\cdot H\Big(y,\frac{\partial S}{\partial y}\Big),S\Big) + \hat{R}(\hat{\mathcal{L}}_1U, S) +$$

$$+ (-1)^{m-1}\hat{R}\Big(U\cdot H_{m-1}\Big(y,\frac{\partial S}{\partial y}\Big),S\Big) + \dots\Big\}$$

for $f \in A_q(X, Y)$, $q > m-1$ where by dots we denote the sum of operators of the form $\hat{R}(U_j, S)$ with the order $\leqslant r-2$ in the scale A_q. Here

$$\hat{\mathcal{L}}_1U = \Big[H_{p_i}\Big(y,\frac{\partial S}{\partial y}\Big)\frac{\partial}{\partial y^i} + \frac{1}{2}\frac{\partial^2 S}{\partial y^i\partial y^j}H_{p_ip_j}\Big(y,\frac{\partial S}{\partial y}\Big)\Big]U(y,p,t).$$

Due to the relation (28) for r=m we have

$$\hat{R}\Big(U\cdot H\Big(y,\frac{\partial S}{\partial y}\Big),S\Big) = \hat{R}\Big(U\cdot\frac{\partial S}{\partial t},S\Big) = (-1)^n n!\times$$

$$\times\int_{h_2(x)}\operatorname{Res}\frac{U(y,p,t)\,\partial S/\partial t(y,p,t)\,f(y)\,dy\wedge\omega}{[S(y,p,t)-p\cdot x]^{n+1}} = (n-1)!\times$$

$$\times\int_{h_2(x)}\Big\{\partial^*\operatorname{Res}\frac{U(y,p,t)f(y)\,dy\wedge\tilde{\omega}}{[S(y,p,t)-p\cdot x]^n} - (-1)^n\operatorname{Res}\frac{\partial U/\partial t\,(y,p,t)}{[S(y,p,t)-p\cdot x]^{n+1}}\times$$

$$\times f(y)\,dy\wedge\omega\Big\}$$

where

$$\tilde{\omega} = \sum_{j=1}^{n}(-1)^{j-1}p_j\,dp_1\wedge\dots\wedge\widehat{dp_j}\wedge\dots\wedge dp_n$$

and ∂^* is a homomorphism of a triple (see [8]):

$$\begin{array}{ccc} & H^*(\Sigma^*_x, X^*) & \\ {\scriptstyle i^*}\swarrow & & \nwarrow{\scriptstyle p^*} \\ H^*(\Sigma^*_x\cap T^*, X^*) & \xrightarrow{\ \partial^*\ } & H^*(\Sigma^*_x, X^*\cup T^*) \end{array}$$

Due to duality and the relation (24) we have

$$\hat{R}\Big(U\cdot H\Big(y,\frac{\partial S}{\partial y}\Big),S\Big)[f(y)] = \hat{R}\Big(\frac{\partial U}{\partial t},S\Big)[f(y)] + (n-1)!\times$$

$$\times\int_{h_1(x)}\operatorname{Res}U(y,p,0)\frac{f(y)\,dy\wedge\tilde{\omega}}{[p\cdot(y-x)]^n}.$$

Hence

$$\hat{R}(U,S)\circ\hat{H}\,[f(y)] = (n-1)!\int\limits_{h_1(x)} \operatorname{Res} \frac{U(y,p,0)f(y)\,dy\wedge\tilde{\omega}}{[p\cdot(y-x)]^n} +$$

$$(-1)^m\,\hat{R}\Big(\frac{\partial U}{\partial t}+\hat{\mathcal{L}}_1 U - H_{m-1}\big(y,\frac{\partial S}{\partial y}\big)U,\ S\Big)[f(y)] +$$

$$\sum_{j=2}^{m}\hat{R}(U_j,S)\,[f(y)], \tag{38}$$

where the orders of operators $\hat{R}(U_j, S)$ do not exceed -2.

Suppose now that the function $U(y,p,t)$ satisfies the problem

$$\begin{cases} \dfrac{\partial U}{\partial t}+\hat{\mathcal{L}}_1 U - H_{m-1}\big(y,\dfrac{\partial S}{\partial y}\big)U = 0, \\ U\big|_{t=0} = \dfrac{1}{(2\pi i)^n}. \end{cases} \tag{39}$$

In this case the second term in the formula (38) vanishes and its first term equals $f(y)$ due to the results of the paper [4]. We have proved the following affirmation.

Proposition 5. Let the function S satisfy the problem (6) and the function U satisfy the problem (39). In this case the equality

$$\hat{R}(U,S)\circ\hat{H} = \hat{1} + \sum_{j=2}^{m}\hat{R}(U_j,S)$$

holds, where $\hat{1}$ is the identity operator, U_j are holomorphic functions and the orders of the operators $\hat{R}(U_j, S)$ do not exceed -2.

5. The main theorem. Let us formulate the main theorem of the paper.

Theorem 1. Suppose the function $u(x)\in A_{q+m}(X,Y)$ (where $q > -1$) satisfies the equation (1), the Condition 2 being valid.

Then the relation

$$u(x) = \hat{R}(U,S)\,[f] \tag{40}$$

is valid modulo the space $A_{q+m+1}(X,Z)$, where U and S satisfy problems (39) and (6) respectively.

Proof. Let us apply the operator $\hat{R}(U,S)$ to both sides of the equality (1). Due to Proposition 5 we have

$$u(x) = \hat{R}(U,S)\,[f] - \sum_{j=2}^{m}\hat{R}(U_j,S)\,[u]. \tag{41}$$

The last sum belongs to $A_{q+m-1}(X,Z)$ due to Proposition 4. This completes the proof of the theorem.

Formula (41) permits us to evaluate the solution u(x) in neighbourhoods of points of the manifold Y lying in some neighbourhood of the set char X. Namely, let us define the space Fq(Y,X) as a set of functions g(x) with singularities no more than on $X \cup Y$ for which for any point $x_1 \in Y \setminus \text{char } X$ there exist a neighbourhood U and a constant C subject to the conditions

$$|D^{\alpha} g(x)| \leq C_{\alpha}, \quad |\alpha| \leq [q],$$
$$|D^{\alpha} g(x)| \leq C_{\alpha} |s_1(x)|^{-1+q-[q]}; \quad |\alpha| = [q] + 1$$

(here $[q]$ is the gratest whole number $\leq q$). The affirmation is valid.

<u>Theorem 2</u>. The inclusion $u(x) \in F_{q+m-\frac{1}{2}}(Y,X)$ is valid and $u(x) = R(U,S)[f] (\text{mod } F_{q+m+3/2}(Y,X))$.

The proof of this theorem can be carried out by direct evaluation of the operators involved in the right-hand side of the formula (41).

References

1. Maslov V.P. Teoriya vozmushchenii i asymptoticheskie metody (Theory of Perturbations and Asymptotic Methods) Moscow State University, Moscow, 1965 (in Russian) (French translation: Théorie des perturbations et methodes asymptotiques, Dunod, Ganthier Villars, Paris, 1972).

2. Sternin B.Yu. Shatalov V.E. On an integral transformation of complex analytic functions, Izvestija Akad. Nauk SSSR, v. 5 (1986).

3. Sternin B.Yu., Shatalov V.E., Laplace-Radon integral operators and singularities of solutions of differential equations on complex manifolds (See this volume).

4. Sternin B.Yu., Shatalov V.E. On an integral transformation of complex analytic functions, Doklady Akad. Nauk SSSR, 280 (1985) N 3 = Soviet Math. Dokl. Vol. 31 (1985) No. 1.

5. Mistchenko A.S., Sternin B.Yu., Shatalov V.E. Lagranjevy mnogoobraziya i metod kanonitcheskogo operatora (Lagrangian Manifolds and Canonical Operator Method), Nauka, Moscow, 1978 (in Russian).

6. Pham F., Introduction a l'étude topologique des singularités de Landau, Paris, 1967.

7. Leray, J. Un prolongement de·la transformation de Laplace qui transforme la solution unitaire d'un opérateur hyperbolique en sa solution élémentaire (Problème de Cauchy, IV). Bull. Soc. Math. de France, 1962, vol. 90, fasc I.

8. Leray J. Le calcul différentiel et integral sur une variété analitique complexe (Problème de Cauchy, III). Bull. Soc. Math. de France, 1959, Vol. 87, fasc. II.

Translated from Russian by M.G. Shatalova

IMAGE OF PERIOD MAPPING FOR SIMPLE SINGULARITIES

A.N.Varchenko
Department of Mathematics
I.M.Gubkin Oil and Gas Institute
Leninski prospekt 65
117917, Moscow, USSR

1. Formulations.

1.1. Hyperelliptic integrals. Consider a family of complex algebraic curves which are defined by the equation

$$y^2 = x^{\mu+1} + \lambda_\mu x^{\mu-1} + \ldots + \lambda_1 \tag{1}$$

and depend on parameters $\lambda = (\lambda_1, \ldots, \lambda_\mu)$. A genus of a nonsingular curve of the family is equal to $[\mu/2]$. A curve has 2 punctured points if μ is odd and 1 punctured point if μ is even. On a nonsingular curve consider basic one-dimensional cycles $\gamma_1(\lambda), \ldots, \gamma_\mu(\lambda)$ which depend continuously on λ (a cycle $\gamma_j(\lambda)$ is a many-valued function of λ). In this article it is proved that the cycles can have arbitrary areas and moreover that there are infinitely many hyperelliptic curves with basic cycles having the given areas.

Theorem 1. Consider the period mapping

$$\lambda \longmapsto \left(\int_{\gamma_1(\lambda)} \omega, \ldots \int_{\gamma_\mu(\lambda)} \omega \right) \tag{2}$$

$w = ydx$. Then the image of the period mapping is $C^\mu \setminus 0$ (for example the image contains vectors with real coordinates). Moreover, every non-zero vector is realized as a vector of integrals on an infinite set of values of parameters, if $\mu > 1$.

There is a generalization of the theorem on the case of a simple singularity of a function with even number of variables. An analog of the first statement of the theorem is valid for a simple singularity of a function with odd number of variables.

The theorem is proved in § 1.5.

1.2. Period mapping. Consider a locally trivial bundle. Associated to the bundle are vector bundles of homologies and cohomologies with complex coefficients of fibres of initial bundle (a base is the same). These bundles possess canonically associated local trivialization (a deformation of an integral cycle in a neighbouring fibre uniquely defined on a cohomological level). This trivialization is called the Gauss-Manin connection.

An isotopy of a base of a fibre bundle with the help of the Gauss-Manin connection uniquely defines an isotopy of homological and cohomological bundles. Therefore an isotopy of a base defines an isotopy of sections of these bundles. Two sections are called equivalent if there is an isotopy of a base which moves the first section to the second section.

Consider a smooth locally trivial bundle and a differential form defined on a space of the bundle and closed on any fibre. The form defines a section of a cohomological bundle: a value of a section in a point of a base is a cohomological class of a restriction of a form on the fibre over a point. This section is called the period mapping of a form. The period mappings of two forms are called equivalent if they are equivalent as sections. The image of the period mapping is the set of all values of the corresponding section, transported in a distinguished fibre along all curves in a base. The image of the period mapping is a subset of a distinguished fibre which is invariant under the monodromy of the Gauss-Manin connection.

Example. Consider the family of all nonsingular curves defined by equation (1) and the form ydx. The family forms a bundle over the space of its parameters. An integral basis of one-dimensional homologies of a fibre continuously depending on a point of a base is covariantly constant in the Gauss-Manin connection and defines many-valued coordinates in cohomologies of fibres. In these coordinates the period mapping is defined by formula (2). In these coordinates the image of the period mapping coincides with the set of all vectors of integrals which stand in the right side.

We shall consider the period mapping in the bundle of zero levels of functions which form a miniversal deformation of a germ of a holomorphic function.

1.3. The period mapping in Milnor fibration (see [2, § 15], [4,

5,6]). Let $f : C^n,0 \longrightarrow C,0$ be a germ of a holomorphic function in a critical point with Milnor number μ. Let $F : C^n \times C$, $0 \times 0 \longrightarrow C_1 0$ be a representative of miniversal deformation of the germ, $F(\cdot ,0) = f$. Let us fix a sufficiently small ball B centred in the origin of C^n and then a sufficiently small (dependind on B) ball Λ, centred in the origin of C^{μ}. A point λ of Λ is called a discriminant point if a level manifold $X_{\lambda} = \{x \in B \mid F(x,\lambda) = 0\}$ is singular. Discriminant points form a set Σ called discriminant. Over a compliment $\Lambda \setminus \Sigma$ of the discriminant manifolds $\{X_{\lambda}\}$ generate a locally trivial bundle which is called central Milnor fibration. We are interested in the fibrations of reduced (n-1)-dimensional homologies and cohomologies with complex coefficients of fibers of the Milnor fibration. Note that these (co)homologies have a dimension μ [3].

Consider on $B \times \Lambda$ a holomorphic differential (n-1)-form. This form is closed on any fibre and defines the period mapping in the cohomological central Milnor fibration. In the considered special case we shall define isotopies and equivalences with the help of local holomorphic diffeomorphisms of Λ defined in the neighbourhood of the origin and preserving the origin and the discriminant.

Suppose that a germ f is quasihomogeneous. Then according to [4] the period mappings of almost all holomorphic differential (n-1)-forms are equivalent. Consequently the period mapping of a form in a general position is a characteristic of a germ.

The period mapping has interesting properties. In some cases it is used to uniformize a base of Milnor fibration and to describe a topology of a base [7]. In other cases the period mapping is used to transfer structures of cohomologies of a fibre to a base of the fibration [2,4,18]. For example an intersection form in cohomologies induces on a base a symplectic structure or a metric with zero curvature (if the number n is even or odd accordingly). In the case of function of two variables this symplectic structure can be analytically extended onto the discriminant. The strata of the discriminant have special lagrangian properties. If the number of variables is even and greater than 2 then the symplectic structure on a whole base can be constructed with the help of so-called adjoined period mappins (see[2,4]).

In this article the image of the period mapping for simple germs is decribed. The problem of a description of the image of the period mapping is mentioned in [5].

<u>Example</u>. Let $f = x_1^2 + \ldots + x_n^2$ be a germ in the origin of type

A_1. The versal deformation is $F = f + \lambda$. (n-1)-dimensional homologies of the Milnor fibre are generated by the vanishing cycle. The period mapping of a (n-1)-form is defined by an integral along the vanishing cycle. An integral has an expansion $C_0 \lambda^{n/2} + C_1 \lambda^{n/2+1} + \ldots$ A form is called a form in a general position if $C_0 \neq 0$. The period mappings of forms in general position are equivalent because the expansion is $\lambda^{n/2}$ after a suitable holomorphic change of a coordinate. The image of the period mapping contains a small punctured neighbourhood of the origin and a degree of the period mapping is equal to n/2.

1.4. Formulation of the results.

Theorem 2. The image of the period mapping is a punctured neighbourhood of the origin. Moreover, for any natural N there exists a punctured neighbourhood of the origin such that any vector in it is the image of more than N points. These statements are valid for any simple germ of the types A_μ, D_μ, E_6, E_7, E_8 of a function with an even number of variables, for any sufficiently small balls B, Λ participating in the definition of the Milnor fibration, for any differential (n-1)-form in a general position. An exception is a germ of type A_1 for which the second statement is not valid, see the example above.

Theorem 3. The closure of the image of the period mapping contains a small neighbourhood of the origin. This statement is valid for any simple germ of types A_μ, D_μ, E_6, E_7, E_8 of a function with an odd number of variables, for any small balls B, Λ, for any differential (n-1)-form in a general position. Moreover, if a germ, the small ball B, a differential form in general position are fixed and the ball Λ is taken sufficiently small, then any vector of the image of the period mapping is the image of no more than $((n+1)/2)^\mu$ points.

It is probable that a small vector in a general position has exactly $((n+1)/2)^\mu$ preimages.

Theorem 2 is proved in §§ 2.1-2.9, theorem 3 is proved in § 2.10.

1.5. Remarks (1).

Probably the closure of the image of the period mapping contains a neighbourhood of the origin for any isolated critical point of a function.

(2) With any period mapping there are associated adjoined period mappings which also have interesting properties, see [2,4-9]. The adjoined period mappings are numerated by natural numbers. The adjoined

period mapping with number k is obtained from the initial period mapping by differentiations k times in the Gauss-Manin connection along a general direction. Problem: to describe the image of the adjoined period mappings.

Example. Consider a versal deformation $-y^2 + x^3 + \lambda_1 x + \lambda_2$ of the germ $-y^2 + x^3$ of the type A_2 and the period mapping of the form ydx. The first adjoined period mapping is defined by the form dx/y. An elementary theory of elliptic curves states that the image of the first adjoined period mapping is the set of all vectors (Z_1, Z_2) with the property $\operatorname{Im}(Z_1/Z_2) > 0$ (in a suitable integral basis of one-dimensional cohomologies of the Milnor fibre).

I would like to point out that this example did not allow to guess the formulations of theorems 1,2.

(3) The main result of the article is theorem 2. Theorem 1 is a consequence of theorem 2. Theorem 3 is an easy consequence of the Brieskorn description of a discriminant of simple singularities, see [10]

(4) A holomorphic differential (n-1)-form on $B \times \Lambda$ is called a form in general position if the differential of the form, restricted on the ball $B \times 0$ is not equal to zero in the origin, see [2,p.314] To prove theorems 2 and 3 it is enough to prove the statements for one form in general position because the period mappings of forms in general position are equivalent in the case of a simple germ. Let us fix a form. Let $x_1, \ldots, x_n$ be coordinates in which f is quasihomogeneous. Let us fix the miniversal deformation

$$F(x, \lambda) = f(x) + \lambda_1 g_1(x) + \ldots + \lambda_\mu g_\mu(x) \qquad (3)$$

where $g_1, \ldots g_\mu$ are monomials and the image of these monomials in the local algebra $C[[x]]/(\partial f/\partial x)$ generates a basis over C. Let us fix the potential $x_1 dx_2 \wedge \ldots \wedge dx_n$ of a volume form as a form in a general position. The simple germs are quasihomogeneous, see [1].

(5) Theorem 1 is a consequence of theorem 2 for the case A_μ. In fact the family of nonsingular curves defined by (1) is isomorphic with the central Milnor fibration of the germ $-y^2 + x^{\mu+1}$ of the type A_μ. The form ydx is in general position. Consequently theorem 2 is valid for this form. Theorem 1 follows from quasihomogeneity of the form and its integrals along cycles.

(6) The zero vector of C^μ does not belong to the image of the period mapping of the potential of the volume form(and so to the image of the period mapping of any form in general position). Actually if the zero vector is in the image then a whole curve in a base is map-

ped in the origin (because of the quasihomogeneity of integrals of the potential). This contradicts the nonsingularity of the period mapping, see [11, § 10].

(7) Theorem 2 is proved by an induction on the Milnor number. There are the following relations of the theorems: $A_2 \Rightarrow A_4 \Rightarrow \dots \Rightarrow A_{2k}$, $A_1 \Rightarrow A_3 \Rightarrow \dots \Rightarrow A_{2k+1}$, $A_{\mu-3} = D_\mu$, $A_2 \Rightarrow E_6$, $A_3 \Rightarrow E_7$, $A_4 \Rightarrow E_8$.

2. Proofs.

2.1. Restriction, projectivization, their images. Consider a bundle, an associated cohomological bundle and a period mapping in it. Restrict a period mapping on a linear connected subset of a base. The big image of a restriction is the set of all values of a restriction transported in a distinguished fibre of a cohomological bundle along all curves in a base. The proper image of a restriction is the set of all values of restriction transported in a distinguished fibre over the subset along all curves inside the subset. Consider a projectivization of a cohomological bundle. Distinguish the subset of all points in a base in which a value of the period mapping is not zero. Over this subset a projectivization of the period mapping is defined, namely a section of a projectivization of a cohomological bundle generated by the initial mapping. The image of the projectivization, big and proper images of a restriction of a projectivization on a subset in a base are defined in a similar way.

2.2. Reduction to projectivization. For any simple germ with the Milnor number greater than 1 distinguish a one-dimensional stratum with the following property: the sum of Milnor numbers of all singular points of the zero level of functions which are parametrized by the stratum is smaller by 1 than the Milnor number of the initial singularity and moreover at least one of these singularities has the type A_1. For A_μ it is astratum $A_{\mu-2} + A_1$, for D_μ it is $A_{\mu-3} + A_1 + A_1$, for E_6 it is $A_2 + A_2 + A_1$, for E_7 it is $A_3 + A_2 + A_1$, for E_8 it is $A_4 + A_2 + A_1$. Distinguish a point of the distinguished stratum and a connected neighbourhood of this point in a compliment of a discriminant. Restrict on this neighbourhood the projectivization of the period mapping of a form in general position.

Theorem 4. If a point of the stratum is chosen sufficiently closely to the origin of Λ, then the big image of the restriction of the projectivization of the period mapping coincides with a distinguished fibre. Moreover, the preimage of any point of a distinguished fibre regarding the projectivization of the period mapping has infinitely many connected components in the compliment of the discriminant.

The theorem is proved in § 2.7.

Lemma 1. Theorem 2 is a corollary of theorem 4.

Proof. Take the potential of the volume as a form in general position. The potential has a positive degree of the quasihomogeneity in the natural quasihomogeneous structure of the (x, λ)-space. Consequently its integrals have a positive degree of the quasihomogeneity considered as functions on λ -space. So according to theorem 4 the image of the period mapping as a subset of a distinguished fibre of a cohomological fibration has the following property: an intersection of the image with any line passing through the origin contains a punctured neighbourhood of the origin. The image of the period mapping of a form in general position is open [11, § 10] The first statement of theorem 2 follows from these remarks. The second statement follows in the same obvious way from theorem 4, a quasihomogeneity of the potential of the volume form and a non-degeneration of the period mapping [11, § 10].

2.3. Remarks. (1). To prove theorem 4 it is sufficient to prove it for the potential of the volume form, because the period mappings of forms in general position are equivalent and because an isotopy of Λ preserving a discriminant preserves all strata of a discriminant [12]
(2). Any one-dimensional stratum distinguished in theorem 4 is a quasihomogeneous orbit, so it is sufficient to prove theorem 4 for one point of a stratum. We shall fix it. Fix the standard coordinates for simple germs:

$$\begin{aligned}
A_\mu &: x_1^{\mu+1} - x_2^2 - Q\\
D_\mu &: -x_1^{\mu-1} + x_1x_2^2 - Q\\
E_6 &: x_1^4 - x_2^3 - Q\\
E_7 &: x_2x_1^3 - x_2^3 - Q\\
E_8 &: x_1^5 - x_2^3 - Q
\end{aligned}$$

where $Q = x_3^2 + \ldots + x_4^2$. Fix miniversal deformations in form (3) where monomials are the same if the germs have the same type but a different number of variables. For $n = 2$ fix real point of the distinguished stratum such that a real part of a corresponding zero level has a picture described in fig.1. If $n > 2$ fix the same point of the distinguished stratum. Denote by N the distinguished stratum, by ν the distinguished point.

Lemma 2. For a point λ of stratum N the space $H_{n-1}(X_\lambda, C)$ is one-dimensional and is generated by the cycle $\delta(\lambda)$ described below. An integral of the potential of the volume form along this cycle is a quasihomogeneous function on N. This function is positive in the point ν.

The cycle $\delta(\nu)$ will be described explicitly. For any other λ the cycle $\delta(\lambda)$ is defined by a quasihomogeneous action on the cycle $\delta(\nu)$. If $n = 2$ then the cycle $\delta(\nu)$ is the oriented boundary of the figure shaded in fig.1. If $n > 1$ then $\delta(\nu)$ is defined by the following condition:

$\delta(\nu) = \{(x_1,\ldots,x_n, \nu) \mid F(x, \nu) = 0, (x_1 x_2)$ belongs to the figure shaded in fig.1$\}$.

Proof of lemma 2. $\dim H_{n-1}(X_\lambda) = 1$ because the sum of the Milnor numbers of singular points of X_λ equals $\mu - 1$. The cycle $\delta(\nu)$ generates $H_{n-1}(X_\nu)$ because an integral of the potential along $\delta(\nu)$ is positive.

2.4. Distinguished basis compatible with a stratum. A subdivision of a graph is a removal of several of its vertices and all edges connected with these vertices.

Let $F(x, \lambda)$ be a versal deformation of a germ of a holomorphic function in an isolated critical point, Π be a stratum of a base of the deformation corresponding to k singular points of the given types on the zero level. Consider a point λ of the compliment of the discriminant. Suppose that λ lies near Π. In $H_{n-1}(X_\lambda)$ there are distinguished subspaces $V_1,\ldots,V_k$ of homologies vanishing in corresponding critical points as $\lambda \to \Pi$. A basis of $H_{n-1}(X_\lambda)$ is compatible with the stratum if any of $V_1,\ldots,V_k$ is generated by a suitable subset of basis vectors. Consider a basis compatible with the stratum and its Dynkin diagram (the definition of Dynkin diagram see in [2]). Compatibility defines a subdivision of Dynkin diagram namely a removal from the diagram of all vertices corresponding to basis vectors which do not belong to $V_1 + V_2 + \ldots + V_k$ (which do not vanish as $\lambda \to \Pi$). The described subdivision will be called a correct subdivision.

Let us return to the versal deformation of a simple germ. In this case according to Grothendieck theorem (see [13-15]) there is a distinguished basis compatible with a stratum whose Dynkin diagram is standard, see fig.2 (see the definition of the distinguished basis in [2]).

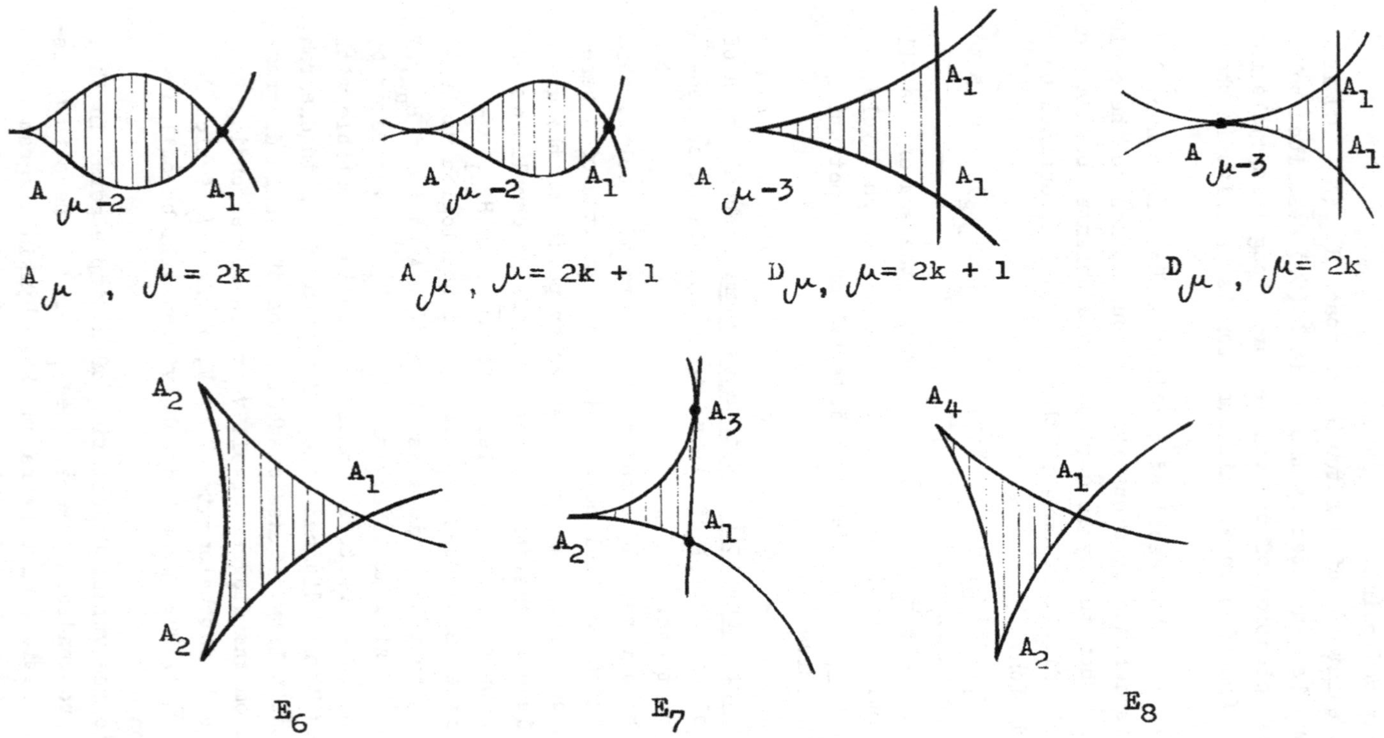
A μ-2
A1
A μ, μ = 2k
A μ-2
A1
A μ, μ= 2k + 1
A μ-3
A1
A1
D μ, μ= 2k + 1
A μ-3
A1
A1
D μ, μ= 2k
A2
A2
A1
E6
A3
A1
A2
E7
A4
A1
A2
E8

Fig.1

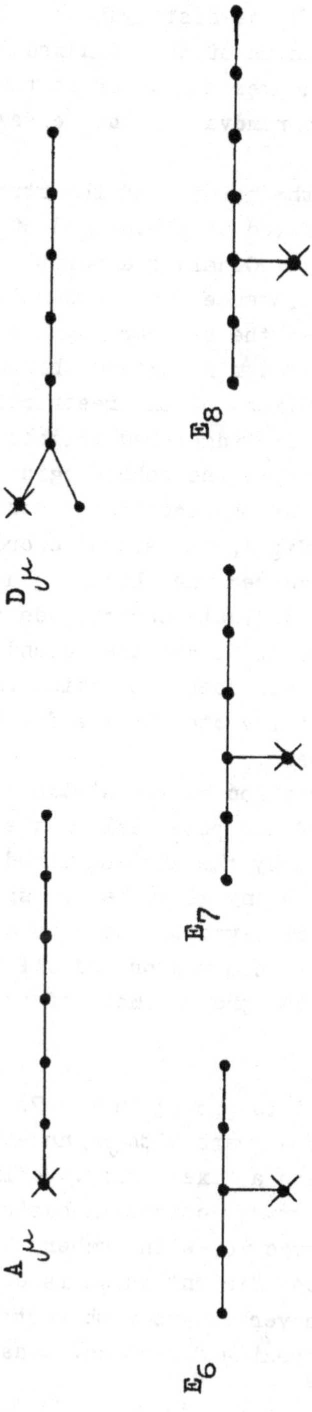

Fig.2

In § 2.3. the one-dimensional stratum N was distinguished. The stratum corresponds to the following subdivision of the standard Dynkin diagram: for A_μ it is the removal of the vertex which is the second from the end, for D_μ, E_μ it is the removal of the vertex which has three edges.

2.5. Refinement of theorem 4. Consider the point ν of the stratum N distinguished in § 2.3. Consider a connected neighbourhood of the point ν in a compliment of a discriminant. Consider a point λ of this neighbourhood which is at a short distance from ν. Consider in $H_{n-1}(X_\lambda)$ a distinguished basis which has the standard Dynkin diagram and wich is compatible with the stratum N. Extend this basis to the basis of covariantly constant sections of the restriction of the homological Milnor fibration on the distinguished neighbourhood. The values of sections define coordinates in the cohomologies of the Milnor fibres. Let the first coordinate be the coordinate corresponding to the vertex marked by a star in fig.2, the second coordinate be the coordinate corresponding to the vertex neighbouring upon the star in fig.2. Specifically the first coordinate corresponds to the singularity A_1 which appears on the stratum N, and the second coordinate corresponds to the basis cycle which does not vanish on the stratum. A vector will be called a special vector if its first coordinate is positive and others are non-zero.

Theorem 5. A proper image of a restriction on the distinguished neighbourhood U of the period mapping of the potential of the volume form contains in coordinates, defined by the distinguished basis, a special vector which is proportional to any given before special vector. Moreover, for any M and any special vector there is a proportional special vector with the property: this vector and all vectors which are sufficiently close to it are the proper image of at least M points of the neighbourhood U.

Lemma 3. Theorem 5 implies theorem 4.

Theorem 5 id proved in § 2.9. Lemma 3 is proved in § 2.7.

2.6. Auxiliary linear algebra. Consider a graph with μ numerated and oriented edges. Consider a space C^μ with a fixed basis. Define an action on C^μ of a group generated by transvections in basis vectors. Namely to the j-th basis vector a transvection with number i adds the i-th basis vector multiplied by a coefficient which is equal to the number of edges going from the j-th vertex to i-th vertex minus the number of edges going in the opposite direction. Consider a dual action in C^{n^*}.

Lemma4. Suppose that a graph is a tree. Then an orbit in C^{n^*} of the set of all vectors with all non-zero coordinates coincides with $^{n^*} \setminus 0$.

Proof is obvious.

.7. Proof of lemma 3. According to lemma 4 and theorem 5 the big image of the restriction on U of the projectivization coincides with the distinguished fibre. This gives the first part of theorem 4. To prove the second part note that the period mapping is non-degenerated. So the preimage of any point of projective space with respect to the projectivization of the period mapping is the union of no more than a countable set of quasihomogeneous orbits. The second part of theorem 5 implies that this set of orbits is infinite if all homogeneous coordinates of an initial point are non-zero. Now theorem 4 follows from lemma 4.

.8. Structure of the period mapping near a stratum of a discriminant. Consider a function $F(\cdot\,,\nu)$, $\nu \in \Lambda$. Suppose that the function $F(\cdot\,,\nu)$ has critical points with the zero critical value. Denote them by $a_1,\dots,a_k$. In the neighbourhood of the point ν there is a (non-canonical) decomposition $\Lambda = \Lambda_o \times \Lambda_1 \times\dots\times \Lambda_k$ of a base of a versal deformation F which has the following property. There are local coordinates λ_j on Λ_j such that in the neighbourhood of any point $(a_j,\nu) \in B \times \Lambda$, $j = 1,\dots,k$, a function $F(x\,,\nu)$ has in suitable coordinates $x = x(Z,\lambda)$ a form $F_j(Z,\lambda_j)$ where F_j is a universal deformation of a critical point a_j of $F(\cdot,\nu)$, see [16].

There is an analogous statement for the period mapping. Namely, consider a small neighbourhood of the point ν in the compliment of the discriminant. Over this neighbourhood there is a naturally defined fibration of cohomologies vanishing in a critical point a_j of the function $F(\cdot\,,\nu)$, $j = 1,\dots,k$. Consider a holomorphic differential (n-1)-form on $B \times \Lambda$. The form defines a period mapping P_j in a fibration of cohomologies vanishing in a_j. Suppose that the initial germ f is quasihomogeneous, differential form is in general position, point ν is sufficiently close to the origin of Λ, germs in points $a_1,\dots,a_k$ of $F(\cdot\,,\nu)$ are quasihomogeneous. Then there is the following lemma.

Lemma 5. A decomposition $\Lambda = \Lambda_o \times \Lambda_1 \times\dots \Lambda_k$ can be chosen in such a way that any P_j, $j = 1,\dots,k$, depends only on λ_j.

The lemma is an easy consequence of the previous statement and a theorem on equivalence of period mappings for quasihomogeneous germs, see [2,4].

In the case of the point $\nu \in N$ distinguished in § 2.3 there are 2 (case A_μ) or 3 (case D_μ, E_μ) critical points on the zero level. One of these points is distinguished and has the type A_1. Λ_0 has dimension 1. A curve $\Lambda_0 \times 0 \times \ldots \times 0$ which goes through ν coincides with quasihomogeneous orbit of ν.

2.9. Proof of theorem 5. Theorem 2 for A_1 is obvious. Consider the case A_μ, $\mu > 1$. Other cases are analogous.

Consider a restriction P of the period mapping of the volume potential on a neighbourhood of the point ν. Let $P = (p_1, \ldots, p_\mu)$ be written in the basis defined above (before theorem 5). p_1 is an integral along the cycle vanishing in A_1, $p_3, \ldots, p_\mu$ are integrals along the cycles vanishing in $A_{\mu-2}$, p_2 is an integral along a non-vanishing cycle. In the neighbourhood of ν consider a decomposition $\Lambda = \Lambda_0 \times \Lambda_1 \times \Lambda_2$ described in lemma 5. In this decomposition Λ_0, Λ_1 have the dimension 1, Λ_2 has the dimension $\mu - 2$. p_1 is a function of λ_1, $p_3, \ldots, p_\mu$ are functions of λ_2, p_2 is a function of λ_0, λ_1, λ_2.

Fix any special vector $(v_1, \ldots, v_\mu)$. Prove theorem 5 for this vector. According to theorem 2 for A_1 and $A_{\mu-2}$ there are germs of holomorphic curves $\lambda_1 = \lambda_1(t)$, $\lambda_2 = \lambda_2(t)$ with the property: $p_1(\lambda_1(t)) = v_1(t^m)$, $(p_3(\lambda_2(t)), \ldots, p_\mu(\lambda_2(t))) = t^m \cdot (v_3, \ldots, v_\mu)$ for a suitable natural m. Map in Λ a product $D \times [0,1]$ of a complex disk and a segment by formula $\gamma : (Z, t) \mapsto (\lambda_0 = Z_1, \lambda_1 = \lambda_1(t), \lambda_2 = \lambda_2(t))$.

Lemma 6 (the main lemma). For any M if $t \in [0,1]$ is sufficiently small then there are such points $Z_1, \ldots, Z_M \in D$ that any vector $P(\gamma(Z_j, t))$, $j = 1, \ldots, M$, will be equal to $t^m(v_1, \ldots, v_\mu)$ after a suitable number of transvections in the first basis vector.

Theorem 5 obviously follows from lemma and from the non-degeneration of the period mapping.

Proof of lemma. Consider the function $p_2(\gamma(\cdot))$. It is continuous by Malgrange theorem [16]. Its value a in the point $0 \times 0 \in D \times [0,1]$ is positive by lemma 2. Values of $p_2(\gamma(\cdot))$ on the disk $D \times 0$ cover a small circle H around a by lemma 2. So the values of $p_2(\gamma(\cdot))$ on the disk $D \times t$ also cover H if it is sufficiently small.

On a complex line consider the arithmetic progression $t^m v_2$, $t^m v_2 + t^m v_1$, $t^m v_2 + 2t^m v_1, \ldots$. If t is sufficiently small then at least M members of this progression are in H. Take such $(Z_1, t), \ldots, (Z_M, t) \in D \times t$ in which p_2 is equal to these members of the progression. These points are suitable for the lemma, because the transvection in the first basis vector subtracts the first coordinate from the second.

2.10. Proof of theorem 3. Consider a singularity of theorem 3. Let μ be its Milnor number. In C^{μ} consider a complexification of the standard action in R^{μ} of a corresponding group generated by reflections. According to [7,9,10] the orbit manifold is holomorphically equivalent to a base of a miniversal deformation of the singularity. This isomorphism maps a submanifold of irregular orbits onto the discriminant of the singularity, the homogeneous strucure of C^{μ} goes to the quasihomogeneous structure of the base of the deformation if the deformation has a form (3). Denote this isomorphism by . Consider the period mapping P and the mapping $P \circ \pi$. $P \circ \pi$ is univalued. $P \circ \pi$ has an analytic continuation on the complexification of the hyperplanes of the reflections. $P \circ \pi$ is homogeneous. The degree of a homogeneity is (n+1)/2. So the number of preimages of a point is no more than $((n + 1)/2)^{\mu}$.

Remark. It is probable that at least one integral of the potential is not zero as a point of the base tends to a point of the discriminant different from the origin (this is almost obvious for A_{μ}). This implies that the number of preimages of a general point is equal to $((n + 1)/2)^{\mu}$.

REFERENCES

1. Arnold V.I., Varchenko A.N., Gusein-Zade S.M. Singularities of differentiable mappings.I. - Moscow, 1982 (in Russian).

2. Arnold V.I., Varchenko A.N., Gusein-Zade S.M. Singularities of differentiable mappings.II. - Moscow, 1984 (in Russian).

3. Milnor J. Singular points of complex hypersurface. - Princeton, 1968.

4. Varchenko A.N., Givental' A.B. Period mapping and form of intersections. - Funct.Anal.Appl., 1982, v.16, N.1 (in Russian).

5. Arnold V.I. Singularities of Ray Systems. - Uspekhi matem.nauk, 1983, v.38, N.2 (in Russian).

6. Arnold V.I. Singularities of Ray Systems. - Proceedings on the International Congress of Mathematicians. Warszawa, 1983, v.I.

7. Looijenga E. A period mapping of certain semiuniversal deformations. - Comp. math., 1975, v.30, N.3.

8. Looijenga E. Homogeneous spaces associated to certain semi-universal deformations. - Proc. Intern. Congress Math., 1978, Helsinki.

9. Looijenga E. Isolated Singular Points on Complete Intersections.- Cambridge, 1984.

10.Brieskorn E. Singular elements of semi-simple algebraic groups.- Actes du Congres Intern. de Math., 1970, t.2, Nice.

11. Varchenko A.N. Hodis asymptotic structure in vanishing cohomologies. - Izvestiya AN SSSR, seriya matem., 1981, v.45, N.3 (in Russian).

12. Varchenko A.N., Gusein-Zade S.M. Topology of caustics, wave fronts and degeneration of critical points. - Uspekhi matem. nauk, 1984, v.39, N.2 (in Russian).

13. Demasure M. Classification des germs a point critique isole et a nombre de modules 0 ou 1. - Lect. Notes Math., 1975, v.431.

14. Lyashko O.V. Disintegration of simple singularities of functions.- Funct.Anal.Appl., 1976, v.10, N.2 (in Russian).

15. Lyashko O.V. Geometry of bifurcation diagrams. - Itogi nauki, Sovremennye problemy matematiki, 1983, v.22 (in Russian).

16. Teissier B. Cycles evanescent, sections planes et conditions de Whitney. - Asterisque, 1978, t.7.

17. Malgrange B. Integrales asymptotiques et monodromie. - Ann. Sci. Ecole Norm. Super., 1974, 7.

18. Saito K. On the Periods of Primitive Integrals. - RIMS, 1982, Tokyo.

THE GEOMETRY OF THE NONHOLONOMIC SPHERE FOR THREE-DIMENSIONAL LIE GROUP

A.M.Vershik, V.Ya.Gershkovich
Department of Mathematics and Mechanics
Leningrad State University
Petrodvorets, 198904, Leningrad, USSR

Introduction.

Let M be a smooth manifold and V be a smooth distribution on M i.e. a subbundle of the tangent bundle TM. In other words a distribution V is a smooth family of linear subspaces V(x) of tangent space T_xM. The vector field ξ on M is called admissible with respect to distribution V iff $\xi(x) \in V(x)$ for all $x \in M$. The distribution V is called absolutely nonholonomic at the point $x \in M$ iff the linear span of Lie brackets $[\xi_{i_1}, \ldots, [\xi_{i_{l-1}}, \xi_{i_l}]$ of some families of admissible vector fields $\{\xi_{i_1}, \ldots, \xi_{i_l}\}$ coincides with the tangent space T_xM.

In papers [1] and [6] it was shown that if the distribution V is absolutely nonholonomic, then any pair of points $x,y \in M$ can be connected by an admissible curve. If M is a Riemannian manifold with metric ρ, then any absolutely nonholonomic distribution V defines the following nonholonomic metric ρ_V on M: for any $x,y \in M$

$$\rho_V(x,y) \overset{\text{def}}{=} \inf_{\gamma \in S} \int_0^1 \langle \dot\gamma, \dot\gamma \rangle^{1/2},$$

where $S = \{ \gamma \mid \gamma(0)=x,\ \gamma(1)=y,\ \dot\gamma \in V \}$.

Nonholonomic metrics arise naturally in the variation calculus, in the theory of differential operators (see for example [19]), in control theory [14] and in algebraic problems related with the growth of words number for discrete and for Lie groups etc. Therefore, it is necessary to investigate the following objects associated with nonholonomic metrics: nonholonomic geodesic flow, nonholonomic exponential map, nonholonomic balls, spheres and wave fronts and some others. It will be proved, that in contrast to Riemannian geometry, in nonholonomic geometry even the local problems, such as the description of the structure of ε-spheres and wave fronts, are not trivial. These descriptions constitute the main purpose of the present paper.

As in the classical geometry, in nonholonomic geometry the case of Lie groups and their homogeneous spaces is especially important and interesting. If the manifold is a homogeneous space, it is natural to assume that both the metric and the distribution are left-invariant. The simplest of non-trivial cases is the case of a three-dimensional Lie group with a two-dimensional left-invariant distribution and a left-invariant metric. Below, we describe the topological and the metric structure of nonholonomic ε- spheres and nonholonomic wave fronts for all three-dimensional Lie groups.

Our results are based on investigations of nonholonomic exponential map; here we include only those notions and results concerning these maps which are necessary for our main purpose. For more general and complete exposition of results concerning exponential map see [18], nonholonomic geodesic flows are investigate in [9].

The present paper has the following structure. In section I the basic notations and definitions are introduced; in s.2 we describe two-dimensional nonholonomic left-invariant distributions for all three-dimensional Lie groups; in the same sections we classify left-invariant metric tensors for all such distributions. In section 3, we obtain the normal forms of equations of nonholonomic geodesics. In s.4, we define nonholonomic exponential maps and wave fronts and describe the relation between nonholonomic wave fronts and spheres. In the next section we study the projection of the nonholonomic geodesic flow on the base of skew-product. Further (in s.6) we describe topological structure of nonholonomic ε-sphere and the nonholonomic wave front and (in s.7) their metric structure. We conclude this paper (s.8) by considering a problem related with nonholonomic sphere, namely the problem of the grouth of powers of Euclidean ball in Heisenberg group (the discrete variant of this problem is the grouth of words number in nilpotent groups).

I. Definitions

Let G be a three-dimensional Lie group. A two-dimensional left-invariant distribution $\mathcal{V}$ on G is uniquely determined by a two-dimensional plane V in Lie algebra $\mathfrak{g}$. The distribution $\mathcal{V}$ is nonholonomic iff $[V,V] = \mathfrak{g}$. By a nonholonomic Lie algebra we understand a pair ($\mathfrak{g}$,V) where $\mathfrak{g}$ is a three-dimensional Lie algebra and V is a nonholonomic plane in $\mathfrak{g}$; a pair $(G, \mathcal{V})$ is called a nonholonomic Lie group. Two nonholonomic Lie algebras ($\mathfrak{g}_I, V_I$) and ($\mathfrak{g}_2, V_2$) are called isomorphic iff there exists a Lie algebras isomorphism $\varphi : \mathfrak{g}_I \longrightarrow \mathfrak{g}_2$ for which $\varphi(V_I)=V_2$.

Let $(G, \mathcal{V})$ be a nonholonomic Lie group. The curve $\gamma : R^I \longrightarrow G$ will be called admissible if $\dot{\gamma} \in \mathcal{V}$ i.e. $\gamma^{-I}\dot{\gamma} \in V$. A left-invariant metric ρ_v on nonholonomic Lie group $(G, \mathcal{V})$ is determined by quadratic form (i.e. a scalar product) on V: for x,y $\subset$ G

$$\rho_v(x,y) = \inf_{\gamma \in S} \int_0^1 \langle \dot{\gamma}, \dot{\gamma} \rangle^{I/2} dt$$

where $S = \{ \gamma : R^I \longrightarrow G \mid \gamma(o) = x, \gamma(I)=y, \dot{\gamma} \in \mathcal{V} \}$.
Denote by $S^v_\varepsilon(x)$ (resp. $D^v_\varepsilon(x)$) the sphere (resp.the ball) with radius ε and centre x in the metric ρ_v. We begin our investigations of the structure of $S^v_\varepsilon(x)$ and $D^v_\varepsilon(x)$ with classification of nonholonomic three-dimensional Lie groups.

2. Nonholonomic Lie groups.

All two-dimensional planes in Lie algebra $\mathfrak{g}$ form the Grassmann manifold $Gr^2_3(\mathfrak{g})$; all nonholonomic planes form the open subset $\overset{o}{Gr}{}^2_3(\mathfrak{g})$ of $Gr^2_3(\mathfrak{g})$. Let us fix a three-dimensional Lie-algebra $\mathfrak{g}_o$. The set of all classes of isomorphic nonholonomic Lie algebras ($\mathfrak{g}$,V) with $\mathfrak{g} = \mathfrak{g}_o$ coincides with the set $Cl\, \mathfrak{g}_o$ of orbits $\overset{o}{Gr}{}^2_3(\mathfrak{g}_o)/Aut\, \mathfrak{g}_o$, where $Aut\, \mathfrak{g}_o$ is the automorphism group of Lie algebra $\mathfrak{g}_o$.

In the persent section we compute $Cl\, \mathfrak{g}_o$ for all three-dimensional Lie algebras. The classification of three-dimentional Lie algebras is well-known (see for example [3,4,7], the full list of such algebras is placed below:

A) The three-dimentional commutative algebra, i.e. three-dimensional torus.

N) The nilpotent Lie algebra (-the Heisenberg algebra) N. Its basis ξ_1, ξ_2, ξ_3 can be chosen in such a way that ξ_3 belongs to the centre of N and $[\xi_1, \xi_2] = \xi_3$.

S) The family of solvable Lie algebras R_α, where parameter α is an arbitrary (2 x 2)-matrix. The basis of R can be chosen in the form ξ_1, ξ_2, ξ_3, where $[\xi_1, \xi_3] = 0$ and $ad\, \xi_2|_{Lin(\xi_1, \xi_3)} = \alpha$.

The set of solvable algebras can be subdivided into four parts in accordance with the type of matrix :

S1) Matrix α has distinct real eigenvalues, i.e. $\alpha = B\begin{pmatrix} 1 & 0 \\ 0 & 2 \end{pmatrix} B^{-1}$, where $\lambda_1 \neq \lambda_2 \in R^1$ and $B \in GL_2\mathbb{R}$.

S2) Matrix α has complex eigenvalues, i.e. $\alpha = B\begin{pmatrix} \cos\alpha & \sin\alpha \\ -\sin\alpha & \cos\alpha \end{pmatrix} \cdot B^{-1}$, where $\alpha \in R^1/2\pi$.

S3) $\alpha = B\begin{pmatrix} \lambda & 0 \\ 0 & \lambda \end{pmatrix} B^{-1}$, where R^1.

S4) $\alpha = B\begin{pmatrix} \lambda & 1 \\ 0 & \lambda \end{pmatrix} B^{-1}$, where $\lambda \in R^1$ and $B \in GL_2\mathbb{R}$.

SS) Semi-simple Lie algebras.

Modulo isomorphism, there exist two semi-simple algebras.

1) The Lie algebra so(3) of all three-variable skew-symmetric matrices, corresponding to Lie group of all ortogonal matrices. The basis of so(3) can be chosen in the form ξ_1, ξ_2, ξ_3 where $[\xi_1, \xi_2] = \xi_3$, $[\xi_2, \xi_3] = \xi_1$, $[\xi_3, \xi_1] = \xi_2$.

2) The Lie algebra $sl_2\mathbb{R}$ of all (2 x 2)-matrices with zero trace corresponding to Lie group $SL_2\mathbb{R}$ of all (2 x 2)-matrices with unit determinant. The algebra $sl_2\mathbb{R}$ has the basis ξ_1, ξ_2, ξ_3, where $[\xi_1, \xi_2] = \xi_3$, $[\xi_1, \xi_3] = 2\xi_1$, $[\xi_2, \xi_3] = -2\xi_2$.

The following proposition describes the set $Cl\,\mathfrak{g}$ for all three-dimensional Lie algebras.

<u>Proposition 1</u>. 1. For commutative algebra and in case c) for solvable algebras, the set $Cl\,\mathfrak{g}$ is empty.

2. For nilpotent algebra; in cases a),b) and d) for solvable algebras and for semi-simple algebra SO(3), the set $Cl\,\mathfrak{g}$ is the one-point set.

3. For semi-simple algebra $sl_2\mathbb{R}$, the set $Cl_{\mathfrak{g}}$ consists of two points.

Remark. Let us present the nonholonomic planes mentioned in proposition I. In the basis of Lie algebras chosen above, these planes are described as follows.

I) For Heisenberg group $V=Lin(\xi_1, \xi_2)$.

2) For solvable Lie algebras

$V=Lin(\xi_2, \xi_1+\xi_3)$ in case SI;

$V=Lin(\xi_1, \xi_2)$ in other cases.

3a) For semi-simple Lie algebra so(3) we have $V=Lin(\xi_1, \xi_2)$

3b) For semi-simple Lie algebra $sl_2\mathbb{R}$ there exist two non-equivalent nonholonomic planes:

$V_1=Lin(\xi_1, \xi_2)$ and

$V_2=Lin(\xi_3, \xi_1+\xi_2)$.

Now we pass on to the discription of the scalar product on nonholonomic planes. For our purpose, i.e. the investigation of the structure of nonholonomic spheres, it is sufficient to classify only classes of equivalence of these scalar product with respect to the action of the two following groups. The first group is the automorphism group of a nonholonomic Lie algebra $(\mathfrak{g}, V)$ and the second group is the multiplication group of all positive real numbers. The following proposition describes the set of equivalent classes $Cl_{(\mathfrak{g},V)}$ for all nonholonomic three-dimensional Lie algebras $(\mathfrak{g}, V)$.

Proposition 2. I. For Heisenberg group and for solvable groups in cases a), b) and d) the set $Cl_{(\mathfrak{g},V)}$ has only one point.

2) For any semi-simple nonholonomic Lie group, the set $Cl_{(\mathfrak{g},V)}$ coincides with the set of all positive real numbers R^1_+.

Remark. In the basis of the nonholonomic plane V, mentioned in remark I, the scalar product g_{ij} has the following form:

I) For Heisenberg algebra and for all solvable Lie algebras $g=\begin{pmatrix} 1 & 0 \\ 0 & 1 \end{pmatrix}$.

2) For all semi-simple nonholonomic Lie algebras, i.e. for (so(3),V), (sl_2R, V_1) and $(sl_2\mathbb{R}, V_2)$, $g_m=\begin{pmatrix} m & 0 \\ 0 & 1 \end{pmatrix}$, where m is an arbitrary positive real number.

In the next section we use these classifications to obtain the normal forms of the equations of nonholonomic geodesics.

3. The normal forms of nonholonomic geodesic equations.

By a nonholonomic geodesics we understand an admissible curve on nonholonomic Lie group (G,) that is locally the shortest. Thus a nonholonomic geodesic is the solution of the conditional variational problem (I):

$$\inf_{\gamma\in S} \int_0^1 \langle \dot\gamma , \dot\gamma \rangle^{I/2}, \qquad \text{(I)}$$

$$S = \{\gamma : R^I \longrightarrow G, \ \gamma(0) = x, \ \gamma(I)=y, \ \dot\gamma \in \mathcal{V}\}.$$

It is well known (see for example [6]), that this solution is described by the system of differential equations given by the following proposition. (We assume that the metric tensor on V is extended arbitrary to $\mathfrak{g}$).

<u>Proposition 3</u>. The following system

$$\begin{cases} \nabla_{\dot\gamma} \dot\gamma = \dot\lambda \omega + \lambda \dot\gamma . \rfloor dw \\ \langle \dot\gamma , w \rangle = 0 \end{cases} \qquad (2)$$

determines the nonholonomic geodesics. Here ∇ is the covariant derivative, corresponding to left-invariant Riemannian metric on G; w is a left-invariant I-form on G annihilating the distribution V and λ is a Lagrange factor.(We denote by ($\dot\gamma \rfloor dw$) the I-form, defined as follows;for vector field ξ , we have ($\dot\gamma \rfloor dw$) (ξ)= $\stackrel{def}{=} dw(\dot\gamma , \xi)$ (cf [6])).

The solutions of system (2) form the nonholonomic geodesic flow on G x ($V \oplus R^I$), where R^I is the set of Lagrange factors. Note that ($V \oplus R^I$) can be identified with $V \oplus V^{\perp}$, where $V^{\perp}$ is the annihilator of V in the coalgebra $\mathfrak{g}^*$. System (2) shows that the nonholonomic geodesic flow is a skew product of the flow on the base (i.e. ($V \oplus V^{\perp}$)) defined by the last three equations of system (2) and flow on the fibre that is the Lie group G.

We transform equations of system (2),using the following considerations.Let us fix the basis ξ_I, ξ_2 of nonholonomic plane V,mentioned in remark I. Since $\xi_3=[\xi_I, \xi_2] \notin V$, we may choose such extension of the scalar product on $\mathfrak{g}$ that $\xi_3 \perp V$. Since γ is admissible curve, one has $\dot\gamma = v_I \xi_I + v_2 \xi_2$. Let's denote the structural constant of Lie algebra $\mathfrak{g}$ by c_{ij}^k i.e.

$$c_{ij}^{\ k} = \langle [\xi_i, \xi_j], \xi_k \rangle , \qquad i,j,k=I,2,3.$$

Christoffel symbols $\Gamma_{ij}^{\ k} = \langle \nabla_{\xi_j} \xi_i, \xi_k \rangle$ of Riemannian connection on G can be expressed in terms of structural constants: $\Gamma_{ij}^{\ k} = \frac{I}{2}(C_{ji}^{\ k}+C_{ki}^{\ j}+C_{kj}^{\ i})$ (see for example [3]). Beginning with this point, we shall consider only the nonholonomic geodesics satisfying the condition $|\dot{\gamma}| = I$, i.e. $V_I^{\ 2}+V_2^{\ 2}=I$. Therefore, we can assume $v_I(t)=\sin \varphi(t)$, $v_2(t)=\cos \varphi(t)$. We obtain the following system of differential equations:

$$\dot{\gamma} = (\sin \varphi) \cdot \xi_I + (\cos \varphi) \cdot \xi_2$$

$$\dot{\varphi} = \lambda \tag{3}$$

$$\dot{\lambda} = \frac{I}{2}(\sin 2\varphi)(C_{3I}^{\ 2}+C_{32}^{\ I}) + (\sin^2 \varphi) \cdot C_{3I}^{\ I} + (\cos^2 \varphi) \cdot C_{32}^{\ 2} +$$
$$+ (C_{I3}^{\ 3} \sin \varphi + C_{23}^{\ 3} \cos \varphi).$$

In order to simplify the last equation of system (3), we need the description of algebraic variety of structural constant of nonholonomic three-dimensional Lie algebras. The structural constant form the algebraic set $S \subset R^{27}$. It is determined by the following equations:

$$\begin{cases} c_{ij}^{\ k} = - c_{ji}^{\ k}; \\ \sum_{r=I}^{3} c_{ij}^{\ r} c_{rk}^{\ l} + c_{jk}^{\ r} c_{ri}^{\ l} + c_{ki}^{\ r} c_{rj}^{\ l} = o, \ (i,j,k) \text{ is an any} \qquad (4) \\ \text{transposition of } (I,2,3); \\ C_{I2}^{\ 3} = I, \ C_{I2}^{\ 2} = C_{I2}^{\ 3} = 0. \end{cases}$$

The first group of equations corresponds to anticommutative law, the second group of equations expresses the Jacobi identity and the last group is the condition of nonholonomity.

Equalities $C_{23}^{\ 2} = - C_{I3}^{\ I}$ and $C_{23}^{\ 3} = 0$ follow from system (4).

Thus one has

$$\begin{cases} \dot{\gamma} = \sin\varphi \cdot \xi_I + \cos\varphi \cdot \xi_2 \\ \dot{\varphi} = \lambda \\ \dot{\lambda} = \frac{I}{2}\sin 2\varphi\,(C_{3I}{}^2 + C_{32}{}^I) + C_{I3}{}^I\cos 2\varphi + \lambda \cdot C_{I3}{}^3 \sin\varphi \end{cases} \quad (5)$$

The following proposition describes the structure of the algebraic variety S in all details.

Proposition 4.I. The variety S has two non-reductible components: $S=P_3 \cup P_2$.

2. P_3 is a three-dimensional affine plane determined by the condition $C_{I3}{}^3 = 0$; P_2 is the two-dimensional affine plane determined by the condition $C_{23}{}^I = C_{23}{}^2 = 0$, and $P_I = P_3 \cap P_2$ is the affine strai-straight line.

3. Points of $(P_3 \setminus P_I)$ correspond to semi-simple Lie algebras; the points of $(P_2 \setminus P_I)$ correspond to solvable Lie algebras and the points of P_I correspond to nilpotent Lie algebra.

System (5) together with proposition 4 gives us the normal form of equations of nonholonomic geodesics for all three-dimensional nonholonomic Lie groups. Since the first and the second equations of system (5) are the same for all Lie groups, we shall omit them.

Theorem I. (On normal form of equations of nonholonomic geodesics) The list of normal forms of equations of nonholonomic geodesics for all nonholonomic three-dimensional Lie groups is as follows:

I) $\dot{\lambda} = 0$ for nilpotent Lie group;

2) $\dot{\lambda} = \frac{I}{2}(\det\alpha)\cdot\sin 2\varphi - \lambda(\mathrm{Sp}\,\alpha)\cdot\sin\varphi$ for any solvable Lie algebra $R\alpha$;

3) $\dot{\lambda} = f(m)\cdot\sin 2\varphi$ for semi-simple Lie algebras (i.e. $(SO(3),V)$, $(SL_2\mathbb{R}, V_I)$ and (SL_2R, V_2)), where

$$f(m) = \begin{cases} (m^{-I}-I)/2 & \text{for } (SO(3),v), \\ 2(I-m^{-I}) & \text{for } (SL_2\mathbb{R}, V_I), \\ 2(m+I) & \text{for } (SL_2 R, V_2). \end{cases}$$

(Here $m \in R_+^I$ is the parameter of metric tensor on V, (see proposition 2).

4. The nonholonomic exponential map and the wave fronts.

The system of differential equations (5) defines on M the nonholonomic geodesic flow. The initial data, uniquely determining a nonholonomic geodesic, include initial point $x \in M$, initial velocity $v \in \vartheta$ and initial value of Lagrange factor $\lambda \in R^I$ (see also [I8]). We assumed that $|\dot\gamma| = I$, thus $\dot\gamma(t) \in S^I \subset V$, where S^I is the unit circle in the plane V. Combining all these data for all geodesics, starting from $x \in M$, we can define the nonholonomic exponential map:

$$\mathrm{Nexp} : R^I \times (S^I \times R^I) \longrightarrow G$$

where $\mathrm{Nexp}\{t, (\varphi, \lambda)\} = \gamma_{\varphi,\lambda}(t)$, here $\gamma_{\varphi,\lambda}$ is the nonholonomic geodesic beginning at x with initial velocity $v(o) = \xi_I \cdot (\sin\varphi) + \xi_2 \cdot (\cos\varphi)$ and initial value of Lagrange factor $\lambda(o) = \lambda$. By $\mathrm{Nexp}_\varepsilon$ we understand the restriction of the map Nexp on the set $\varepsilon \times (S^I \times R^I)$.

By the wave front $A^v_\varepsilon(x)$ on nonholonomic Lie group (G, ϑ) we understand the set of all end-points of nonholonomic geodesics of length ε, beginning at the point x, i.e. the wave front $A^v_\varepsilon(x)$ is the image of cylinder $S^I \times R^I$:

$$A^v_\varepsilon(x) = \mathrm{Nexp}_\varepsilon (S^I \times R^I).$$

For Riemannian geometry case the wave front A^V_ε coincides with the ε-sphere S_ε for all sufficiently small ε; for nonholonomic case $A^v_\varepsilon(x)$ and $S^v_\varepsilon(x)$ do not coincide but the second set is always included in the first one.

<u>Proposition 5</u>. $S^v_\varepsilon(x) \subsetneq A^v_\varepsilon(x)$ for all sufficiently small ε.

<u>Proof</u>. If $y \in S^v_\varepsilon(x)$ then from the theorem of Rashevski and Chow [I,6] follows the existence of an admissible curve, connecting x and y, then the existence of the shortest nonholonomic geodesic γ, connecting x and y follows from Fillippov's lemma [20]. To prove the inclusion, we have to show that the length of γ equals ε. It is valid because the nonholonomic metric and the Riemannian metric generate the same topologies (of [I7]). Finally,

the equality $\lim_{\lambda\to\infty} \mathrm{Nexp}(v_o,\lambda)=x$ shows that sets $A^v_\varepsilon(x)$ and $S^v_\varepsilon(x)$ are not equal.

To describe the wave front, we have to study the projection of nonholonomic geodesic flow on $V \oplus V^\perp$, determined by the two last equations of system (5).

5. The flow on the base of skew-product.

The flow on the base of skew-product is determined by the following equations:

$$\begin{cases} \dot\varphi = \lambda \\ \dot\lambda = \mu_I \sin 2\varphi + \mu_2 \lambda \sin\varphi . \end{cases} \tag{6}$$

We describe below the phase portrait of system (6).

Proposition 6. I) The topological structures of the phase portraits of system (6) are the same for all $\mu_I \neq 0$ and all μ_2.

2) In the case $\mu_I \neq o$ there exist four fix points $\{z_i\}_{i=I}^4$ of system (6) on the cylinder $\{(\varphi,\lambda) (\varphi\in S^I, \lambda\in R^I$, where $z_i=(\varphi_i,o)$, $\varphi_i=\frac{\pi i}{2}$. There exist four separatrices connecting points z_o and z_2 if $\mu_I > o$ and points z_I and z_3 if $\mu_I < o$. All other solutions of system (6) are periodic.

3). If $\mu_I=o$ then topological structures of the phase portraits of system (6) are the same for all μ_2. The set of fixed points form the unit circle determined by the condition $\lambda=o$. All other solutions of system (6) are periodic.

The phase portrait of system (6) for $\mu_I \neq o$ is represented in figure I. Every solution intersects either the straight line $\varphi=o$ or $\varphi=\pi$ or both. Denote by α_λ the solution of system (6) intersecting the line $\varphi=o$ (or $\varphi=\pi$) at the point $(o,\lambda)\in S^I \times R^I$ (resp. $(\pi,\lambda)\in S^I \times R^I$). Let τ_λ be the period of α_λ. Separatrix is a limit of periodic trajectories. Therefore, it is natural to consider it as a periodic trajectory with the period equal to the limit of those periods, i.e. ∞ . Let us denote intersections of separatrix and the lines $\varphi=o$ and $\varphi=\pi$ by λ_∞. The dependence of τ_λ on λ is described by the following proposition.

Proposition 7. Let $\mu_I \neq o$. Then

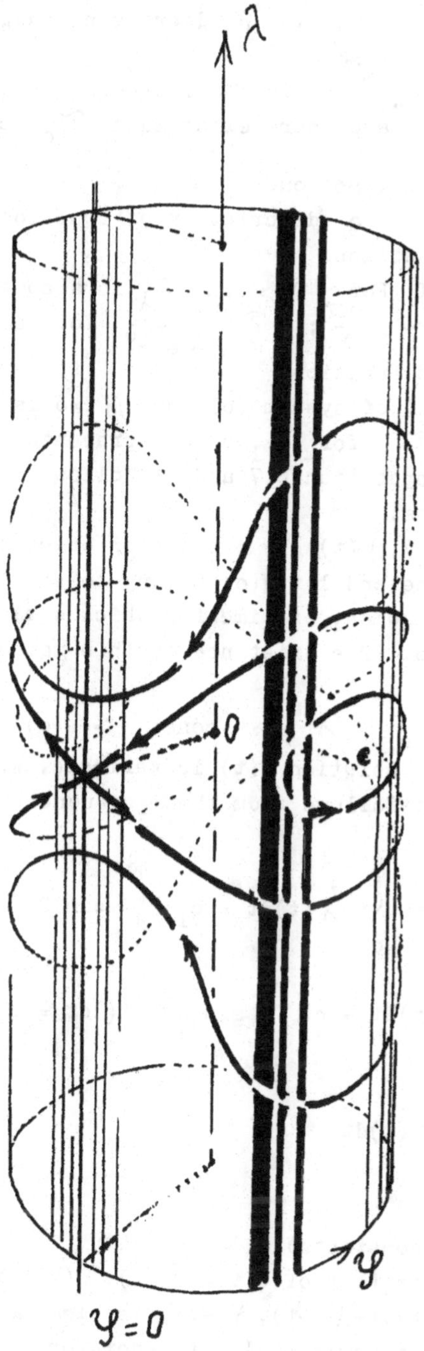

Fig.I

I. For $\lambda > \lambda_\infty$, τ_λ is a decreasing continuous function and $\tau_\lambda \sim \frac{2\pi}{\lambda}$, $\tau_\lambda \xrightarrow[\lambda\to\lambda_\infty]{} \infty$

2. For $0 < \lambda < \lambda_\infty$, τ_λ is an increasing continuous function; $\tau_\lambda \xrightarrow[\lambda\to\lambda_\infty]{} \infty$, and there exist such $\tau_0 > 0$ that $\tau_\lambda \xrightarrow[\lambda\to 0]{} \tau_0$.

3. τ_λ is an even function.

In the case $\mu_I=0$, a description of function τ_λ is given by the following proposition.

Proposition 7'. I. For $\lambda > 0$, τ_λ is a continuous decreasing function; $\tau_\lambda \sim \frac{2\pi}{\lambda}$; $\tau_\lambda \xrightarrow[\lambda\to 0]{} \tau_0$ for some $\tau_0 > 0$.

2. τ_λ is an even function.

The phase portrait of system (6) for $\mu_I=0$ is represented in fig 2. The graph of τ_λ for $\mu_I \neq 0$ is represented in fig.3.

The proofs of proposition 6,7 and 7' follow from the following sequences of statements.

I) Let $q(t)=(\varphi(t),\lambda(t))$ be a solution of system (6) situated in the upper half-plane and let $q(0)$ belong to straight line $\varphi=0$. Denote by T the first moment of time for which $q(t)$ intersects the line $\varphi=0$ for $t > 0$. (The existence of this moment follows from the inequalities:

$$\dot\varphi = \lambda \geqslant \text{const} > 0)$$

To prove that the solution $q(t)$ is periodic, we have to show that $\lambda(0)=\lambda(T)$. It follows from the sequence of equalities:

$$\lambda(T) - \lambda(0) = \int_0^T \dot\lambda(t)dt = c_I \int_0^T \sin 2\varphi(t)dt +$$

$$+ c_2 \int_0^T \dot\varphi \sin\varphi(t)dt = c_I \int_0^T \sin 2\varphi(t)dt = c_2 \Big|_0^T \sin (t) =$$

$$= c_I \int_0^T \sin 2\dot\varphi(t)dt \quad .$$

(Here c_I, c_2 are some constants).

It remains to note that of $q(t)=(\varphi(t),\lambda(t))$ is a solution of system (6) then $q(t)=(-\varphi(-t),\lambda(-t))$ is also a solution. These solutions have the same initial data, therefore, they coincide. Thus $\varphi(-t)=-\varphi(t)$, i.e. $\varphi(t)$ is an odd function on the circle S^I and its integral over the period equal zero.

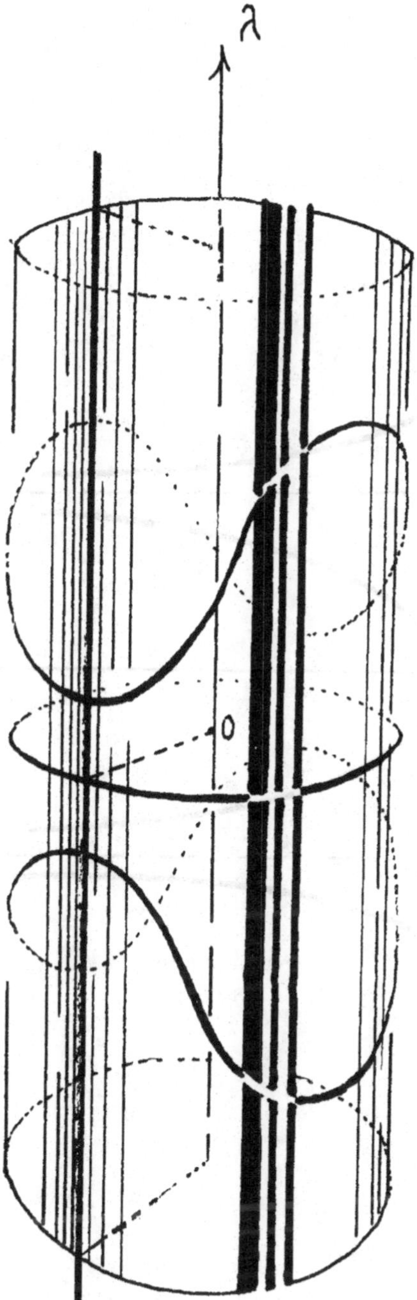

Fig.2

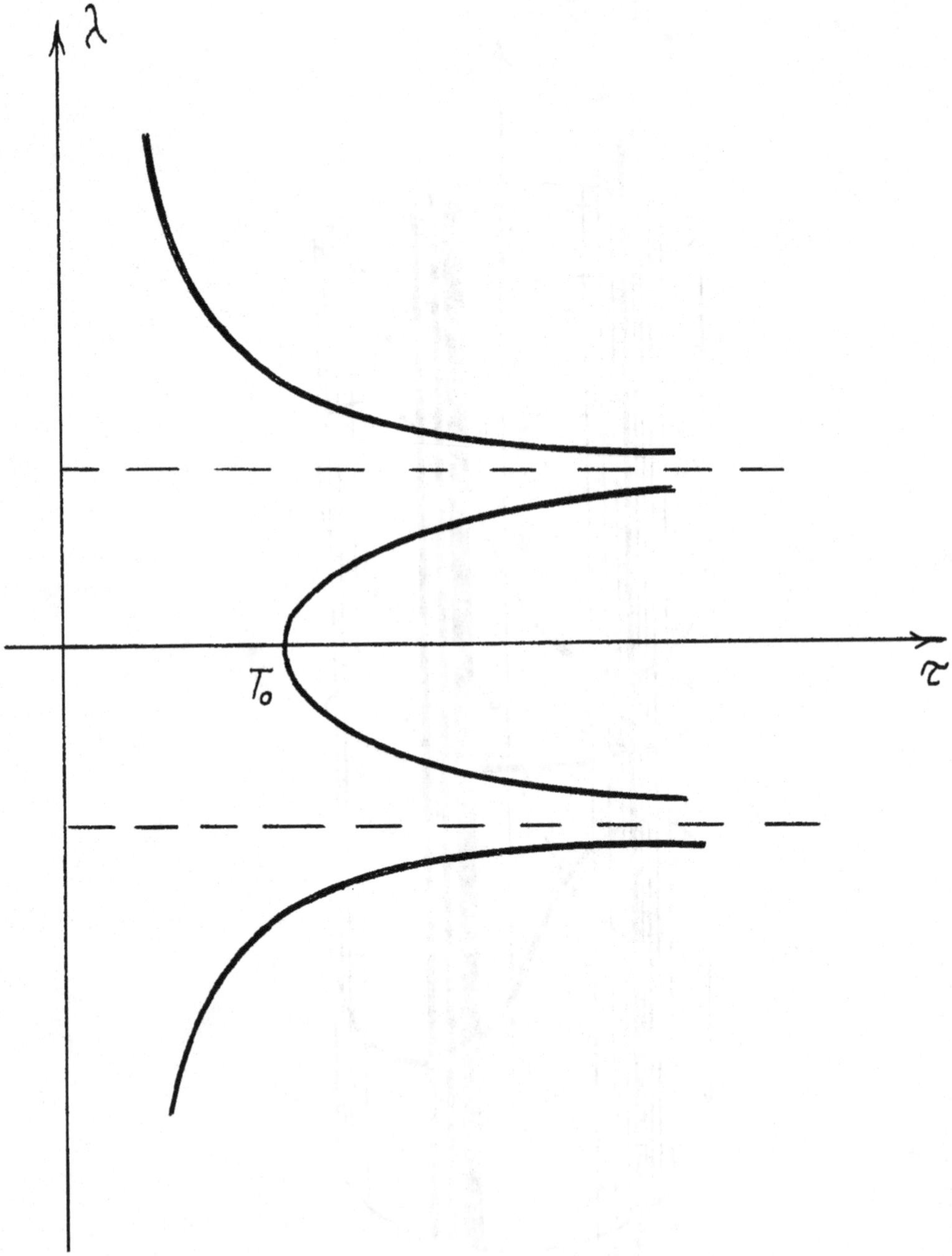

Fig.3

2. It can be shown (by analogy with the previous statement) that all solutions of system (6), except fixed points and separatrices, are periodic.

3. Now we show that τ_λ is strictly decreasing for $\lambda > \lambda_\infty$. Let $q_1(t)=(\varphi_1(t), \lambda_1(t))$ and $q_2(t)=(\varphi_2(t), \lambda_2(t))$ be two solutions of system (6) with $\varphi_1(o)=\varphi_2(o)$ and $\lambda_2(o) > \lambda_1(o)$. If $\varphi_1(t_1)=\varphi_2(t_2)$, then the uniqueness theorem for system of differential equations shows that $\lambda_1(t_1) < \lambda_2(t_2)$ i.e. the second solution lies upper than the first one. Therefore the angular velocity $\dot\varphi_2(t)=\lambda_2(t)$ of the second solution is greater than the angular velocity $\dot\varphi_1(t)=\lambda_1(t)$ of the first solution. Thus $\tau_{\lambda_1} > \tau_{\lambda_2}$.

6. The topological structure of the wave front and the ε-sphere.

The wave front is the image of a cylinder under the nonholonomic exponential map: $NExp_\varepsilon$. To describe the differential and topological structure of wave front $A^V_\varepsilon(x)$, we have to investigate the singularities of the map Nexp . Since all solutions $(\varphi(t), \lambda(t))$ of system (6) are periodic on the base, the first equation of system (5) transforms into the following equation:

$$\gamma^{-1}\dot\gamma = r(t),$$

where $r(t)=\xi_1 \sin\varphi(t) + \xi_2 \cos\varphi(t)$ is a periodic curve on Lie algebra $\mathfrak{g}$. The key idea of the following is the use of the Floquet theory; from this theory it follows (see [8]) that $Nexp_\varepsilon$ maps any curve $q_{n,\varepsilon}$ with $\tau = n\varepsilon$ (where n is an integer number different from zero) into one point $z_{n,\varepsilon}$. Denote by Λ the set $(\bigcup_{n\in\mathbb{Z}\setminus o} q_{n,\varepsilon})$. Proposition 7 shows that the curves $q_{n,\varepsilon}$ divide the cylinder $S^1 \times R^1$ into the domains C_i, $i\in\mathbb{Z}$. The domain C_i is bounded by curves $q_{i,\varepsilon}$ and $q_{i+1,\varepsilon}$ for $i > o$; the domain C_o is bounded by $q_{-1,\varepsilon}$ and $q_{+1,\varepsilon}$; for $i < 0$, C_i is bounded by the curves $q_{-i-1,\varepsilon}$ and $q_{-i,\varepsilon}$.

<u>Theorem 2</u>. Let $(G, \mathcal{V})$ be any three-dimensional nonholonomic Lie group. Then for all sufficiently small positive ε the following statements hold:

I) The map $Nexp_\varepsilon$ is regular at any point $x \in (S^1 \times R^1)\setminus\Lambda$

2) For any $n \in \mathbb{Z} \setminus \{0\}$, $Nexp_\varepsilon$ maps the curve $q_{n,\varepsilon} \subset S^I x R^I$ into one point $Z_{n,\varepsilon} \in G$. The germ of the wave front $A^v_\varepsilon(x)$ at the point $Z_{n,\varepsilon}$ is diffeomorphic with the germ of standart cone K in R^3 (i.e. $K = \{(x,y,z,)\ (z^2 = x^2+y^2)\}$).

3) For any $i \in \mathbb{Z}$, the image of the closune of $\bar{C}_i$ (i.e. $Nexp_\varepsilon(\bar{C}_i)$) is homeomorphic with two-dimensional Euclidean sphere.

4) The nonholonomic ε - sphere $S^v_\varepsilon(x) \subset A^v_\varepsilon(z)$ is the image of C_0.

5) $Nexp_\varepsilon$ maps the set $(S^I x R^I)$ C_0 into the interior of the nonholonomic ε-ball $D^v_\varepsilon(x)$.

7. The metric structure of nonholonomic sphere.

We have described the topological structure of nonholonomic wave fronts and nonholonomic ε-spheres. It turns out that ε - spheres are similar (in some natural sense) for all nonholonomic three-dimensional Lie groups (in the case if ε is sufficiently small). The present section contains the proof of this similarity. First, we prove that any nonholonomic sphere is close to the simplest of these spheres - the nonholonomic sphere in Heisenberg group and then investigate this sphere.

Let (G, ϑ) be the nonholonomic three-dimensional Lie group and $\xi_I, \xi_2,$ be the basis of vector fields of distribution ϑ, mentioned in Remark I. Choose such coordinate system $\{x_I, x_2, x_3\}$ in some neighborhood of the unit of G, in which vector fields ξ_I and ξ_2 are described by the following formulas:

$$\xi_I = \frac{\partial}{\partial x_I}, \quad \xi_2 = \frac{\partial}{\partial x_2} + x_I \frac{\partial}{\partial x_3} + \dots .$$

Thus $\xi_3 = [\xi_I, \xi_2]$ is given by the formula $\xi_3 = \frac{\partial}{\partial x_3} + \dots$. Let us define the vector field ξ_2 by the formula

$\xi_2 = \frac{\partial}{\partial x_2} + x_I \frac{\partial}{\partial x_3}$. Then the Lie algebra generated by the vector fields ξ_I and ξ_2 is isomorphic to Heisenberg algebra. To compare ε - spheres of the Lie group G and of the Heisenberg group, we consider the following two systems of differential equations.

I) The system of differential equations

$$\begin{aligned} \dot{\gamma}_I &= \sin \varphi_I \cdot \xi_I + \cos \varphi_I\ \xi_2 \\ \dot{\varphi}_I &= \lambda_I \\ \dot{\lambda}_I &= \mu_I \sin 2\varphi_I + \mu_2 \lambda_I \sin \varphi_I \end{aligned} \tag{7}$$

determines the nonholonomic geodesics on Lie group G.

2) The following system determines the nonholonomic geodesics on Heisenberg group

$$\begin{cases} \dot{\gamma}_2 = \mathrm{Sin}\, \varphi_2 \, \xi_I + \mathrm{Cos}\, \varphi_2 \, \xi_2 \\ \dot{\varphi}_2 = \lambda_2 \\ \dot{\lambda}_2 = 0 \end{cases} \qquad (7^I)$$

Proposition 8. Let γ_I be a solution of system (7) and γ_2 be a solution of system (7^I) and γ_I and γ_2 have the same initial data. Then $\rho(\gamma_I(\varepsilon), \gamma_2(\varepsilon)) = \mathcal{O}(\varepsilon^3)$, where ρ is the usual metric on R^3. Thus, ε - spheres and wave fronts coincide to within ε^3 for all threedimensional nonholonomic Lie groups. It remains to describe - spheres and wave fronts for the Heisenberg group. The Heisenberg group N is, as a manifold, diffeomorphic to R^3. We'll obtain now from (7^I) the formulas for ε - sphere in R^3. Let be the nonholonomic geodesic with initial data (φ, λ) and let $(x_{\varphi,\lambda}, Y_{\varphi,\lambda}, Z_{\varphi,\lambda})$ be the coordinates of $\gamma_{\varphi,\lambda}(\varepsilon)$. Then the wave front is determined by the following equations:

$$x_{\varphi,\lambda} = \frac{\sin(\theta + \varphi) - \sin \varphi}{0}$$

$$y_{\varphi,\lambda} = \frac{\mathrm{Cos}\, \varphi - \mathrm{Cos}\,(\theta + \varphi)}{\theta} \qquad (8)$$

$$z_{\varphi,\lambda} = \varepsilon^2 \, \frac{2 - (\sin(2\theta + 2\varphi) - \sin 2\varphi) + 4 \sin \varphi (\cos(\theta + \varphi) - \cos \varphi)}{4\theta^2}$$

The equations of nonholonomic wave front for Heisenberg group show thet it has the following properties:

I) The wave front $A_\varepsilon^V(x)$ and the ε - sphere $S_\varepsilon^V(x)$ are quasihomogeneous surfaces, i.e. $S_{r\varepsilon}^V(x) = \delta_r S_\varepsilon^V(x)$ where δ_r is the following transformation of R^3

$$\delta_r(x,y,z) = (rx, ry, r^2 z).$$

2) The symmetry group of $A_{\varepsilon}^{v}(x)$ and $S_{\varepsilon}^{v}(x)$ consists of four elements, this group is homeomorphic with $Z_2 + Z_2$. Any non-unit symmetry changes signs of some pair of coordinates in R^3. In fig.4, the wave front, ε - sphere and the nonholonomic exponential map for Heisenberg group are schematically represented. It is necessary to give some comments. For Heisenberg group the curves $q_{n,\varepsilon}$ coincide with circles on the cylinder $S^I x R^I$ determined by the equations $\lambda = \frac{2n\pi}{\varepsilon}$. The wave front is the union of "beads" B_i. Every "bead" B_i is an image of the close domain $\overline{C}_i$ (see section 6); beads B_i for all i with $i \neq o$, lie inside the main "bead" B that coincides with the nonholonomic sphere S_{ε}^{v}. The singular points $Z_{n,\varepsilon}$ that are the images of circles $q_{n,\varepsilon}$ have the following coordinates:

$$Z_{k,\varepsilon} = (0,\ 0,\ \frac{\varepsilon^2}{4_k \pi}).$$

The metric structure of the wave front and the ε - sphere is represented in fig.5 (for the case $\varepsilon = I$). Since the vector of quasi-homogeneous degrees is (I,I,2), the proposition 8 shows that - spheres for all nonholonomic Lie groups are similar.

8. The nonholonomic ball and the sequence of powers of Enclidean ball in Heisenberg group.

Nonholonomic balls naturally arise in problems related with the growth of powers of set in Lie group. This problem is the continuous version of the problem of words number growth in Lie group [IO,I2]. We begin with the following proposition.

Proposition 9. Let $D_{\varepsilon}^{v}(e)$ be the nonholonomic ball in the Heisenberg group N. Then its powers are also the nonholonomic balls and we have:

$$(D_{\varepsilon}^{v}(e))^n = D_{n\varepsilon}^{v}(e).$$

Using the quasi-homogeneous contraction δ_r (see section 7) the previous formula can be rewritten as follows:

$$\delta_{\frac{I}{n}}\left(D_{\delta}^{v}(e)\right) = D_{\delta}^{v}(e) \qquad (9).$$

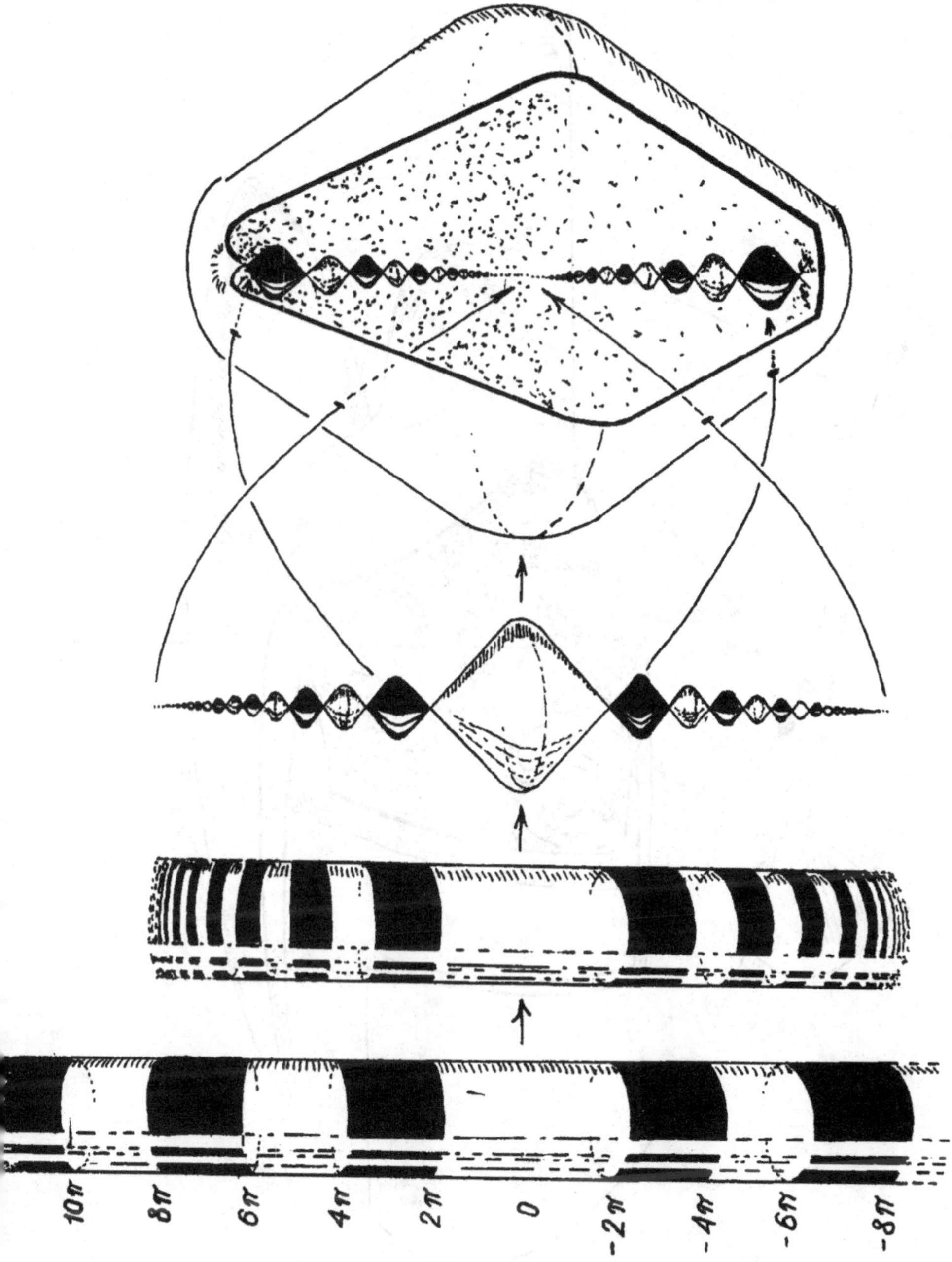

Fig.4

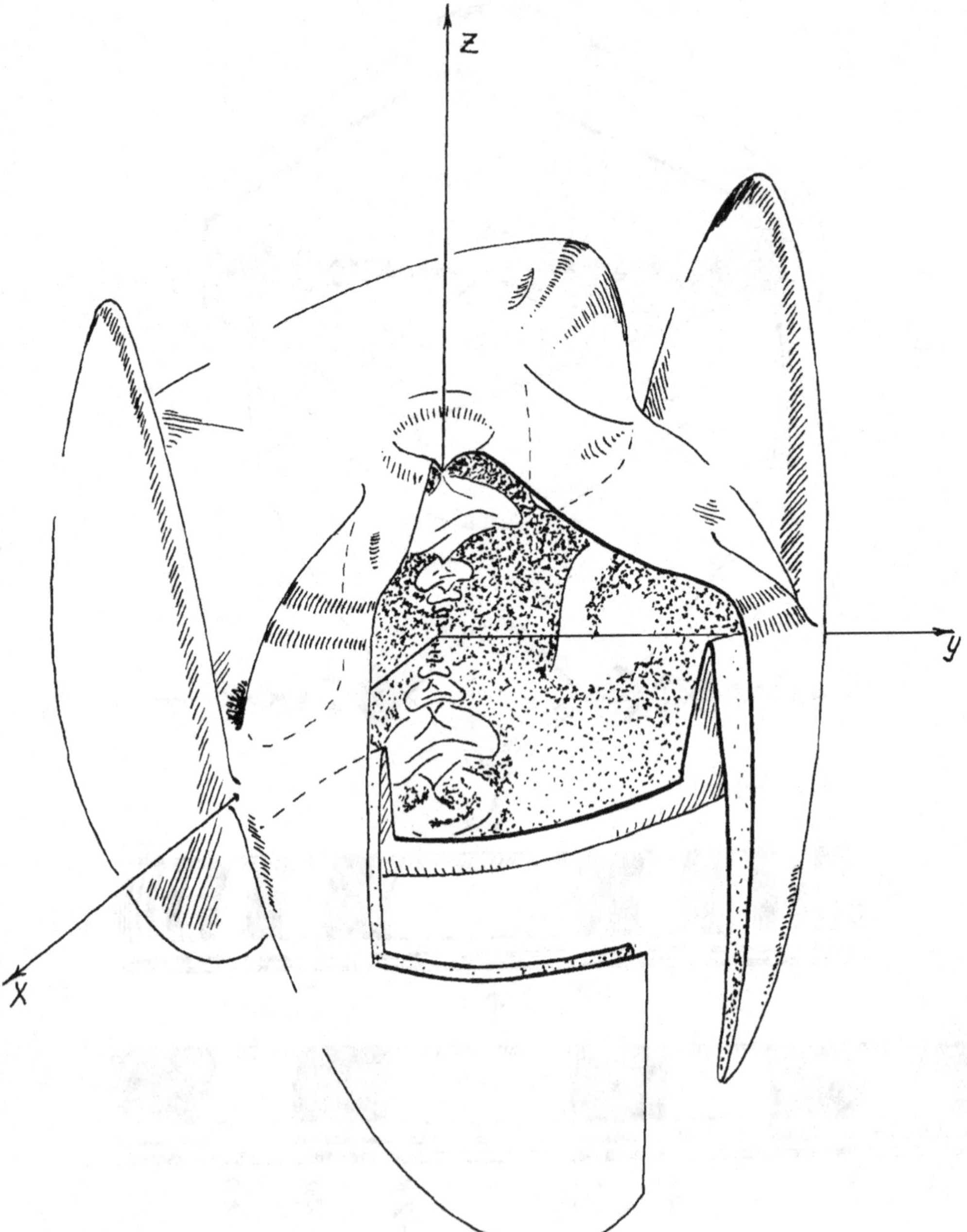

Fig.5

It appears, that formula (9) remains valid if we replace the Euclidean ball by any bounded subset $B \subset N$ with the same projection on the plane V. Denote by F_V the orthogonal projector of the Lie algebra $\mathfrak{g}$ on the plane V and by k_δ the disk with radius δ in the plane V.

<u>Theorem 3</u>. Let b be a bounded subset in the Heisenberg algebra and the following conditions hold:

I. $b \supset k_\delta$,

2. $F_V b = k_\delta$,

If B=expb, then $\lim \rho(\Psi_{n-I} B^n, D^V(e))=o$ (here ρ is a distance between two sets).

<u>Proof</u>. Let b_L be $[-L,L] \cdot \xi_3$, that is the subset of the center of the Lie algebra $\mathfrak{g}$. Denote by B_L the set exp b_L and by K the set exp k_δ. Then there exists such positive number L that $k_\delta \subset b \subset k_\delta \times b_L$ and we obtain the following chains of inclusions:

$$k_\delta \subset B \subset K_\delta \cdot B_L \quad \text{and} \quad K^n_\delta \subset B^n \subset K^n \cdot B_{nL}.$$

Now the proof is completed by two statements.

I. For all sufficient small ε and sufficient large n, the following inclusions take place:

$$D^V_{\delta-\varepsilon}(e) \subset \delta_{n-I} K^n_\delta \subset D^V_\delta(e).$$

(The right inclusion is evident and the left inclusion follows from [I7]).

2. In the assumption of the previous statement the following inclusions take place:

$$D^V_{\delta-\varepsilon}(e) \in \delta_{n-I} B^n \subset D^V_{\delta+\varepsilon}(e).$$

Concluding Remarks

I. Theorem 3 strenghthens the results of [I0]. It can be generalized to the case of Finsler metric. (The definition of Finsler metric see e.g. in [I6]). The most important applications related with Finsler metrics are connected with metrics generated by the finite families of vector fields. These metrics are defined as follows. Let $\{\xi_I, \ldots, \xi_m\}$ be a family of vector fields. The distance between two points x and y is defined as the smallest time necessary to move from x to y along the curve γ satisfying the condition $\gamma(t)=\xi_{i(t)}$ for some integer i with $I \leq i \leq m$.

2. It should be noted that there is a relation between our results about nonholonomic ball as a limit of the sequence of powers of Riemannian ball and the problem usually called "the animal growth". Our results can be interpreted as the determination of asymptotics of increasing configurations of sets in Heisenberg group.

3. We would like to draw the attention of the reader to the relation between our investigations and the problem of random walks on groups. Let μ_I be a characteristic measure of set A and $\mu_n = \mu_{n-I} * \mu_I$ for $n \geq I$. Theorem 3 can be interpreted as investigations of the behaviour of sequences μ_n or more precisely of their supports behaviour. It is interesting to investigate the behaviour of this sequence in more detail.

The authors are indebted to N.V.Ivanov who drew their attention to papers [IO,II] and to N.E.Mnev who prepared the figures.

REFERENCES

I. Rashevskiĭ P.K. On connection of any two points by the admissible curve. - Zapiski Mosc.Inst.K.Libknexta. Ser.phys.-math.,I938,N 2, p.83-94 (in Russian).

2. Vershik A.M.,Gershkovich V.Ya. Nonholonomic problem and distribution geometry. - The addition to Russian translation of [6] .

3. Dubrovin B.A., Novikov S.P.,Fomenko A.T. The modern geometry. M.: Nauka,I979 (in Russian).

4. Auslander L.,Green L.,Hahn F. Flows on homogeneous Spaces. Princeton Univ.Press,I963.

5. Arnol'd V.I., Varchenko A.M.,Gusein-Zade S.M. The smooth map singularities. v.I,2. M.: Nauka,I982,I984 (in Russian).

6. Griffits Ph.A. Exterior differential system and calculus of Variations. Boston: Birkhauser,I983.

7. Bourbaki N. Groups et algebres de Lie. Ch.I-3. Hermann,Paris, I97I,I972.

8. Lefshetz S. Differential equations. Geometric theory.N.Y.: Interscience publ.,I957.

9. Vershik A.M.,Gershkovich V.Ya. Nonholonomic geodesic flow on $SL_2\mathbb{R}$. - Zapiski Sem.LOMI AN USSR, v.I55,I986 (in Russian).

IO. Pansu P. Croissance des boules et des geodesiques fermees dans les nilvarietes. - Ergodic theory and dynamic system I983,3,p.3, 4I5-446.

II. Mitchel J. On Carno-Caratheodori metrics. - Jour.of diff.geom., I985,2I,No I,p.35-42.

I2. Gromov M. Groups of polynomial growth and expanding maps. - Inst. Haytes Sci.Publ.Math., № 53, I98I.

I3. Kaimanovich V.A.,Vershik A.M. Random walks on discrete groups: Boundary and entropy. - The annals of probability,I983,v.II,№3, p.457-490.

I4. Lobry C. Dynamical polysystems and control theory. Geometric methods in system theory. - Proceedings of Nato Adv.Inst.Boston: Reidell Publ.Comp.,I973,p.I-42.

I5. Sussmann H.J. Orbits families of vector fields and integrabilities of distributions. - Trans.Amer.Math.Soc.,I973,I80,I7I-I88.

I6. Rund H. The differential geometry of Finsper spaces. Berlin,I959.

I7. Gershkovich V.Ya. Two-sided estimates of metric generated by absolutely nonholonomic distributions. - Soviet Math.Dokl.,v.30, I984, №2,p.506-5I0.

I8. Vershik A.M.,Gershkovich V.Ya. Nonholonomic dynamic system. Distribution geometry and calculus of variations. - Modern problem in Math.Fundamental branches. Dymanic system -7. VINITI Acad. Sci.USSR,I987 (in Russian).

I9. Nagel A.,Stein E.M.,Wainger S. Balls and metrics defined by vector fields I: Basic properties. - Acta Math.,I55:I-2,p.I03-I48.

20. Filippov A.F. On some problems of control theory. - Vestnik MGU, ser.math.,I959, N 2,p.25-32 (in Russian).

Vol. 1173: H. Delfs, M. Knebusch, Locally Semialgebraic Spaces. XVI, 329 pages. 1985.

Vol. 1174: Categories in Continuum Physics, Buffalo 1982. Seminar. Edited by F.W. Lawvere and S.H. Schanuel. V, 126 pages. 1986.

Vol. 1175: K. Mathiak, Valuations of Skew Fields and Projective Hjelmslev Spaces. VII, 116 pages. 1986.

Vol. 1176: R.R. Bruner, J.P. May, J.E. McClure, M. Steinberger, H_∞ Ring Spectra and their Applications. VII, 388 pages. 1986.

Vol. 1177: Representation Theory I. Finite Dimensional Algebras. Proceedings, 1984. Edited by V. Dlab, P. Gabriel and G. Michler. XV, 340 pages. 1986.

Vol. 1178: Representation Theory II. Groups and Orders. Proceedings, 1984. Edited by V. Dlab, P. Gabriel and G. Michler. XV, 370 pages. 1986.

Vol. 1179: Shi J.-Y. The Kazhdan-Lusztig Cells in Certain Affine Weyl Groups. X, 307 pages. 1986.

Vol. 1180: R. Carmona, H. Kesten, J.B. Walsh, École d'Été de Probabilités de Saint-Flour XIV – 1984. Édité par P.L. Hennequin. X, 438 pages. 1986.

Vol. 1181: Buildings and the Geometry of Diagrams, Como 1984. Seminar. Edited by L. Rosati. VII, 277 pages. 1986.

Vol. 1182: S. Shelah, Around Classification Theory of Models. VII, 279 pages. 1986.

Vol. 1183: Algebra, Algebraic Topology and their Interactions. Proceedings, 1983. Edited by J.-E. Roos. XI, 396 pages. 1986.

Vol. 1184: W. Arendt, A. Grabosch, G. Greiner, U. Groh, H.P. Lotz, U. Moustakas, R. Nagel, F. Neubrander, U. Schlotterbeck, One-parameter Semigroups of Positive Operators. Edited by R. Nagel. X, 460 pages. 1986.

Vol. 1185: Group Theory, Beijing 1984. Proceedings. Edited by Tuan H.F. V, 403 pages. 1986.

Vol. 1186: Lyapunov Exponents. Proceedings, 1984. Edited by L. Arnold and V. Wihstutz. VI, 374 pages. 1986.

Vol. 1187: Y. Diers, Categories of Boolean Sheaves of Simple Algebras. VI, 168 pages. 1986.

Vol. 1188: Fonctions de Plusieurs Variables Complexes V. Séminaire, 1979–85. Edité par François Norguet. VI, 306 pages. 1986.

Vol. 1189: J. Lukeš, J. Malý, L. Zajíček, Fine Topology Methods in Real Analysis and Potential Theory. X, 472 pages. 1986.

Vol. 1190: Optimization and Related Fields. Proceedings, 1984. Edited by R. Conti, E. De Giorgi and F. Giannessi. VIII, 419 pages. 1986.

Vol. 1191: A.R. Its, V.Yu. Novokshenov, The Isomonodromic Deformation Method in the Theory of Painlevé Equations. IV, 313 pages. 1986.

Vol. 1192: Equadiff 6. Proceedings, 1985. Edited by J. Vosmansky and M. Zlámal. XXIII, 404 pages. 1986.

Vol. 1193: Geometrical and Statistical Aspects of Probability in Banach Spaces. Proceedings, 1985. Edited by X. Fernique, B. Heinkel, M.B. Marcus and P.A. Meyer. IV, 128 pages. 1986.

Vol. 1194: Complex Analysis and Algebraic Geometry. Proceedings, 1985. Edited by H. Grauert. VI, 235 pages. 1986.

Vol. 1195: J.M. Barbosa, A.G. Colares, Minimal Surfaces in $\mathbb{R}^3$. X, 124 pages. 1986.

Vol. 1196: E. Casas-Alvero, S. Xambó-Descamps, The Enumerative Theory of Conics after Halphen. IX, 130 pages. 1986.

Vol. 1197: Ring Theory. Proceedings, 1985. Edited by F.M.J. van Oystaeyen. V, 231 pages. 1986.

Vol. 1198: Séminaire d'Analyse, P. Lelong – P. Dolbeault – H. Skoda. Seminar 1983/84. X, 260 pages. 1986.

Vol. 1199: Analytic Theory of Continued Fractions II. Proceedings, 1985. Edited by W.J. Thron. VI, 299 pages. 1986.

Vol. 1200: V.D. Milman, G. Schechtman, Asymptotic Theory of Finite Dimensional Normed Spaces. With an Appendix by M. Gromov. VIII, 156 pages. 1986.

Vol. 1201: Curvature and Topology of Riemannian Manifolds. Proceedings, 1985. Edited by K. Shiohama, T. Sakai and T. Sunada. VII, 336 pages. 1986.

Vol. 1202: A. Dür, Möbius Functions, Incidence Algebras and Power Series Representations. XI, 134 pages. 1986.

Vol. 1203: Stochastic Processes and Their Applications. Proceedings, 1985. Edited by K. Itô and T. Hida. VI, 222 pages. 1986.

Vol. 1204: Séminaire de Probabilités XX, 1984/85. Proceedings. Edité par J. Azéma et M. Yor. V, 639 pages. 1986.

Vol. 1205: B.Z. Moroz, Analytic Arithmetic in Algebraic Number Fields. VII, 177 pages. 1986.

Vol. 1206: Probability and Analysis, Varenna (Como) 1985. Seminar. Edited by G. Letta and M. Pratelli. VIII, 280 pages. 1986.

Vol. 1207: P.H. Bérard, Spectral Geometry: Direct and Inverse Problems. With an Appendix by G. Besson. XIII, 272 pages. 1986.

Vol. 1208: S. Kaijser, J.W. Pelletier, Interpolation Functors and Duality. IV, 167 pages. 1986.

Vol. 1209: Differential Geometry, Peñíscola 1985. Proceedings. Edited by A.M. Naveira, A. Ferrández and F. Mascaró. VIII, 306 pages. 1986.

Vol. 1210: Probability Measures on Groups VIII. Proceedings, 1985. Edited by H. Heyer. X, 386 pages. 1986.

Vol. 1211: M.B. Sevryuk, Reversible Systems. V, 319 pages. 1986.

Vol. 1212: Stochastic Spatial Processes. Proceedings, 1984. Edited by P. Tautu. VIII, 311 pages. 1986.

Vol. 1213: L.G. Lewis, Jr., J.P. May, M. Steinberger, Equivariant Stable Homotopy Theory. IX, 538 pages. 1986.

Vol. 1214: Global Analysis – Studies and Applications II. Edited by Yu.G. Borisovich and Yu.E. Gliklikh. V, 275 pages. 1986.

Vol. 1215: Lectures in Probability and Statistics. Edited by G. del Pino and R. Rebolledo. V, 491 pages. 1986.

Vol. 1216: J. Kogan, Bifurcation of Extremals in Optimal Control. VIII, 106 pages. 1986.

Vol. 1217: Transformation Groups. Proceedings, 1985. Edited by S. Jackowski and K. Pawalowski. X, 396 pages. 1986.

Vol. 1218: Schrödinger Operators, Aarhus 1985. Seminar. Edited by E. Balslev. V, 222 pages. 1986.

Vol. 1219: R. Weissauer, Stabile Modulformen und Eisensteinreihen. III, 147 Seiten. 1986.

Vol. 1220: Séminaire d'Algèbre Paul Dubreil et Marie-Paule Malliavin. Proceedings, 1985. Edité par M.-P. Malliavin. IV, 200 pages. 1986.

Vol. 1221: Probability and Banach Spaces. Proceedings, 1985. Edited by J. Bastero and M. San Miguel. XI, 222 pages. 1986.

Vol. 1222: A. Katok, J.-M. Strelcyn, with the collaboration of F. Ledrappier and F. Przytycki, Invariant Manifolds, Entropy and Billiards; Smooth Maps with Singularities. VIII, 283 pages. 1986.

Vol. 1223: Differential Equations in Banach Spaces. Proceedings, 1985. Edited by A. Favini and E. Obrecht. VIII, 299 pages. 1986.

Vol. 1224: Nonlinear Diffusion Problems, Montecatini Terme 1985. Seminar. Edited by A. Fasano and M. Primicerio. VIII, 188 pages. 1986.

Vol. 1225: Inverse Problems, Montecatini Terme 1986. Seminar. Edited by G. Talenti. VIII, 204 pages. 1986.

Vol. 1226: A. Buium, Differential Function Fields and Moduli of Algebraic Varieties. IX, 146 pages. 1986.

Vol. 1227: H. Helson, The Spectral Theorem. VI, 104 pages. 1986.

Vol. 1228: Multigrid Methods II. Proceedings, 1985. Edited by W. Hackbusch and U. Trottenberg. VI, 336 pages. 1986.

Vol. 1229: O. Bratteli, Derivations, Dissipations and Group Actions on C*-algebras. IV, 277 pages. 1986.

Vol. 1230: Numerical Analysis. Proceedings, 1984. Edited by J.-P. Hennart. X, 234 pages. 1986.

Vol. 1231: E.-U. Gekeler, Drinfeld Modular Curves. XIV, 107 pages. 1986.

Vol. 1232: P.C. Schuur, Asymptotic Analysis of Soliton Problems. VIII, 180 pages. 1986.

Vol. 1233: Stability Problems for Stochastic Models. Proceedings, 1985. Edited by V.V. Kalashnikov, B. Penkov and V.M. Zolotarev. VI, 223 pages. 1986.

Vol. 1234: Combinatoire énumérative. Proceedings, 1985. Edité par G. Labelle et P. Leroux. XIV, 387 pages. 1986.

Vol. 1235: Séminaire de Théorie du Potentiel, Paris, No. 8. Directeurs: M. Brelot, G. Choquet et J. Deny. Rédacteurs: F. Hirsch et G. Mokobodzki. III, 209 pages. 1987.

Vol. 1236: Stochastic Partial Differential Equations and Applications. Proceedings, 1985. Edited by G. Da Prato and L. Tubaro. V, 257 pages. 1987.

Vol. 1237: Rational Approximation and its Applications in Mathematics and Physics. Proceedings, 1985. Edited by J. Gilewicz, M. Pindor and W. Siemaszko. XII, 350 pages. 1987.

Vol. 1238: M. Holz, K.-P. Podewski and K. Steffens, Injective Choice Functions. VI, 183 pages. 1987.

Vol. 1239: P. Vojta, Diophantine Approximations and Value Distribution Theory. X, 132 pages. 1987.

Vol. 1240: Number Theory, New York 1984–85. Seminar. Edited by D.V. Chudnovsky, G.V. Chudnovsky, H. Cohn and M.B. Nathanson. V, 324 pages. 1987.

Vol. 1241: L. Gårding, Singularities in Linear Wave Propagation. III, 125 pages. 1987.

Vol. 1242: Functional Analysis II, with Contributions by J. Hoffmann-Jørgensen et al. Edited by S. Kurepa, H. Kraljević and D. Butković. VII, 432 pages. 1987.

Vol. 1243: Non Commutative Harmonic Analysis and Lie Groups. Proceedings, 1985. Edited by J. Carmona, P. Delorme and M. Vergne. V, 309 pages. 1987.

Vol. 1244: W. Müller, Manifolds with Cusps of Rank One. XI, 158 pages. 1987.

Vol. 1245: S. Rallis, L-Functions and the Oscillator Representation. XVI, 239 pages. 1987.

Vol. 1246: Hodge Theory. Proceedings, 1985. Edited by E. Cattani, F. Guillén, A. Kaplan and F. Puerta. VII, 175 pages. 1987.

Vol. 1247: Séminaire de Probabilités XXI. Proceedings. Edité par J. Azéma, P.A. Meyer et M. Yor. IV, 579 pages. 1987.

Vol. 1248: Nonlinear Semigroups, Partial Differential Equations and Attractors. Proceedings, 1985. Edited by T.L. Gill and W.W. Zachary. IX, 185 pages. 1987.

Vol. 1249: I. van den Berg, Nonstandard Asymptotic Analysis. IX, 187 pages. 1987.

Vol. 1250: Stochastic Processes – Mathematics and Physics II. Proceedings 1985. Edited by S. Albeverio, Ph. Blanchard and L. Streit. VI, 359 pages. 1987.

Vol. 1251: Differential Geometric Methods in Mathematical Physics. Proceedings, 1985. Edited by P.L. García and A. Pérez-Rendón. VII, 300 pages. 1987.

Vol. 1252: T. Kaise, Représentations de Weil et GL_2 Algèbres de division et GL_n. VII, 203 pages. 1987.

Vol. 1253: J. Fischer, An Approach to the Selberg Trace Formula via the Selberg Zeta-Function. III, 184 pages. 1987.

Vol. 1254: S. Gelbart, I. Piatetski-Shapiro, S. Rallis. Explicit Constructions of Automorphic L-Functions. VI, 152 pages. 1987.

Vol. 1255: Differential Geometry and Differential Equations. Proceedings, 1985. Edited by C. Gu, M. Berger and R.L. Bryant. XII, 243 pages. 1987.

Vol. 1256: Pseudo-Differential Operators. Proceedings, 1986. Edited by H.O. Cordes, B. Gramsch and H. Widom. X, 479 pages. 1987.

Vol. 1257: X. Wang, On the C^*-Algebras of Foliations in the Plane. V, 165 pages. 1987.

Vol. 1258: J. Weidmann, Spectral Theory of Ordinary Differential Operators. VI, 303 pages. 1987.

Vol. 1259: F. Cano Torres, Desingularization Strategies for Three-Dimensional Vector Fields. IX, 189 pages. 1987.

Vol. 1260: N.H. Pavel, Nonlinear Evolution Operators and Semigroups. VI, 285 pages. 1987.

Vol. 1261: H. Abels, Finite Presentability of S-Arithmetic Groups. Compact Presentability of Solvable Groups. VI, 178 pages. 1987.

Vol. 1262: E. Hlawka (Hrsg.), Zahlentheoretische Analysis II. Seminar, 1984–86. V, 158 Seiten. 1987.

Vol. 1263: V.L. Hansen (Ed.), Differential Geometry. Proceedings, 1985. XI, 288 pages. 1987.

Vol. 1264: Wu Wen-tsün, Rational Homotopy Type. VIII, 219 pages. 1987.

Vol. 1265: W. Van Assche, Asymptotics for Orthogonal Polynomials. VI, 201 pages. 1987.

Vol. 1266: F. Ghione, C. Peskine, E. Sernesi (Eds.), Space Curves. Proceedings, 1985. VI, 272 pages. 1987.

Vol. 1267: J. Lindenstrauss, V.D. Milman (Eds.), Geometrical Aspects of Functional Analysis. Seminar. VII, 212 pages. 1987.

Vol. 1268: S.G. Krantz (Ed.), Complex Analysis. Seminar, 1986. VII, 195 pages. 1987.

Vol. 1269: M. Shiota, Nash Manifolds. VI, 223 pages. 1987.

Vol. 1270: C. Carasso, P.-A. Raviart, D. Serre (Eds.), Nonlinear Hyperbolic Problems. Proceedings, 1986. XV, 341 pages. 1987.

Vol. 1271: A.M. Cohen, W.H. Hesselink, W.L.J. van der Kallen, J.R. Strooker (Eds.), Algebraic Groups Utrecht 1986. Proceedings. XII, 284 pages. 1987.

Vol. 1272: M.S. Livšic, L.L. Waksman, Commuting Nonselfadjoint Operators in Hilbert Space. III, 115 pages. 1987.

Vol. 1273: G.-M. Greuel, G. Trautmann (Eds.), Singularities, Representation of Algebras, and Vector Bundles. Proceedings, 1985. XIV, 383 pages. 1987.

Vol. 1274: N. C. Phillips, Equivariant K-Theory and Freeness of Group Actions on C^*-Algebras. VIII, 371 pages. 1987.

Vol. 1275: C.A. Berenstein (Ed.), Complex Analysis I. Proceedings, 1985–86. XV, 331 pages. 1987.

Vol. 1276: C.A. Berenstein (Ed.), Complex Analysis II. Proceedings, 1985–86. IX, 320 pages. 1987.

Vol. 1277: C.A. Berenstein (Ed.), Complex Analysis III. Proceedings, 1985–86. X, 350 pages. 1987.

Vol. 1278: S.S. Koh (Ed.), Invariant Theory. Proceedings, 1985. V, 102 pages. 1987.

Vol. 1279: D. Ieşan, Saint-Venant's Problem. VIII, 162 Seiten. 1987.

Vol. 1280: E. Neher, Jordan Triple Systems by the Grid Approach. XII, 193 pages. 1987.

Vol. 1281: O.H. Kegel, F. Menegazzo, G. Zacher (Eds.), Group Theory. Proceedings, 1986. VII, 179 pages. 1987.

Vol. 1282: D.E. Handelman, Positive Polynomials, Convex Integral Polytopes, and a Random Walk Problem. XI, 136 pages. 1987.

Vol. 1283: S. Mardešić, J. Segal (Eds.), Geometric Topology and Shape Theory. Proceedings, 1986. V, 261 pages. 1987.

Vol. 1284: B.H. Matzat, Konstruktive Galoistheorie. X, 286 pages. 1987.

Vol. 1285: I.W. Knowles, Y. Saitō (Eds.), Differential Equations and Mathematical Physics. Proceedings, 1986. XVI, 499 pages. 1987.

Vol. 1286: H.R. Miller, D.C. Ravenel (Eds.), Algebraic Topology. Proceedings, 1986. VII, 341 pages. 1987.

Vol. 1287: E.B. Saff (Ed.), Approximation Theory, Tampa. Proceedings, 1985–1986. V, 228 pages. 1987.

Vol. 1288: Yu. L. Rodin, Generalized Analytic Functions on Riemann Surfaces. V, 128 pages, 1987.

Vol. 1289: Yu. I. Manin (Ed.), K-Theory, Arithmetic and Geometry. Seminar, 1984–1986. V, 399 pages. 1987.